Sustainable Energy Systems Planning, Integration and Management

Sustainable Energy Systems Planning, Integration and Management

Special Issue Editors

Amjad Anvari-Moghaddam
Behnam Mohammadi-ivatloo
Somayeh Asadi
Kim Guldstrand Larsen
Mohammad Shahidehpour

MDPI • Basel • Beijing • Wuhan • Barcelona • Belgrade

MDPI

Special Issue Editors
Amjad Anvari-Moghaddam
Aalborg University
Denmark

Behnam Mohammadi-ivatloo
University of Tabriz
Iran

Somayeh Asadi
University Park
USA

Kim Guldstrand Larsen
Aalborg University
Denmark

Mohammad Shahidehpour
Illinois Institute of Technology
USA

Editorial Office
MDPI
St. Alban-Anlage 66
4052 Basel, Switzerland

This is a reprint of articles from the Special Issue published online in the open access journal *Applied Sciences* (ISSN 2076-3417) from 2018 to 2019 (available at: https://www.mdpi.com/journal/applsci/special_issues/Sustainable_Energy).

For citation purposes, cite each article independently as indicated on the article page online and as indicated below:

LastName, A.A.; LastName, B.B.; LastName, C.C. Article Title. *Journal Name* **Year**, *Article Number*, Page Range.

ISBN 978-3-03928-046-9 (Pbk)
ISBN 978-3-03928-047-6 (PDF)

Cover image courtesy of Shutterstock/FotoIdee.

Contents

About the Special Issue Editors

Amjad Anvari-Moghaddam received his Ph.D. degree (Hons.) from University of Tehran in 2015 in Power Systems Engineering. From 2015 until 2019, he was a Postdoctoral Research Fellow at Aalborg University, Denmark. Currently, he is Associate Professor at the Department of Energy Technology, Aalborg University. His research interests include planning, control, and operation of energy systems, mostly renewable and hybrid power systems with appropriate market mechanisms. Dr. Anvari-Moghaddam serves as the GE/Associate Editor of the IEEE Access, IET Renewable Power Generation, IEEE Transactions on Industry Informatics, Future Generation Computer Systems, Applied Sciences, Electronics, and Sustainability. He is the recipient of 2020 DUO-India Fellowship Award, 2018 IEEE Outstanding Leadership Award (Halifax, Nova Scotia, Canada), the 2017 IEEE Outstanding Service Award (Exeter, UK), and the DANIDA research grant from the Ministry of Foreign Affairs of Denmark in 2018.

Behnam Mohammadi-ivatloo received his B.Sc. degree in Electrical Engineering from University of Tabriz, Tabriz, Iran, in 2006, and M.Sc. and Ph.D. degrees from Sharif University of Technology, Tehran, Iran, in 2008, all with honors. He is currently Associate Professor at the Faculty of Electrical and Computer Engineering, University of Tabriz, Tabriz, Iran. His main area of research is in the economics, operation, and planning of intelligent energy systems in a competitive market environment.

Somayeh Asadi is Assistant Professor in the Department of Architectural Engineering at Pennsylvania State University and holds the Hartz Family Career Development Assistant Professor position. Prior to that, she was Assistant Professor at Texas A&M University. Her research interests cut across the following themes: automated design, critical infrastructure systems, food–water–energy nexus, design of high performance buildings, and environmental sustainability. Dr. Asadi is the recipient of several NSF, DOE, DEP, and Qatar national research foundation awards and has more than 70 journal and conference publications.

Kim Guldstrand Larsen received his Ph.D. degree from Edinburgh University. Since 1993, he has been Full Professor of Computer Science and Director of the ICT-Competence Center of Embedded Software Systems (CISS) which have been conducting several national and international research projects. He has more than 30 years of experience working with modeling, verification, performance analysis, and optimization of distributed, embedded, and intelligent systems. He is member of the Royal Danish Academy of Sciences and Letters, Danish Academy of Technical Sciences, and Academia Europaea. He is Honorary Doctor at ENS Cachan, France, and Uppsala University. He won the CAV Award in 2015 and the Grundfos Prize in 2016. He has won an ERC Advanced Grant that started in 2015, and since January 2008, he has been the Danish National Expert for the European ICT program.

Mohammad Shahidehpour joined Illinois Institute of Technology (IIT) in 1983, where he is presently a University Distinguished Professor. He also serves as the Bodine Chair Professor and Director of the Robert W. Galvin Center for Electricity Innovation at IIT. He is a Fellow of IEEE, Fellow of the American Association for the Advancement of Science (AAAS), Fellow of the National Academy of Inventors (NAI), and an elected member of the US National Academy of Engineering. Dr. Shahidehpour has been the Principal Investigator of $60 million grants and contracts on power system operation and control, smart grid research and development, and large-scale integration of renewable energy. He is the recipient of numerous awards throughout his career, most recently, the IEEE PES Ramakumar Family Renewable Energy Excellence Award, IEEE PES Douglas M. Staszesky Distribution Automation Award, IEEE PES Outstanding Power Engineering Educator Award, IEEE PES Outstanding Engineer Award, and Edison Electric Institute's Power Engineering Educator Award.

applied sciences

MDPI

Editorial

Sustainable Energy Systems Planning, Integration, and Management

Amjad Anvari-Moghaddam [1],*, Behnam Mohammadi-ivatloo [2], Somayeh Asadi [3], Kim Guldstrand Larsen [4] and Mohammad Shahidehpour [5]

[1] Department of Energy Technology, Aalborg University, 9220 Aalborg East, Denmark
[2] Department of Electrical and Computer Engineering, University of Tabriz, Tabriz 5166616471, Iran; bmohammadi@tabrizu.ac.ir
[3] Department of Architectural Engineering, Pennsylvania State University, University Park, PA 16802, USA; asadi@engr.psu.edu
[4] Department of Computer Science, Aalborg University, 9220 Aalborg East, Denmark; kgl@cs.aau.dk
[5] Armour College of Engineering, Illinois Institute of Technology, Chicago, IL 60616, USA; ms@iit.edu
* Correspondence: aam@et.aau.dk

Received: 8 October 2019; Accepted: 11 October 2019; Published: 20 October 2019

1. Introduction

Energy systems worldwide are undergoing a major transformation as a consequence of the transition towards the widespread use of clean and sustainable energy sources. The electric power system in a sustainable future will augment the centralized and large-grid-dependent systems of today with distributed, smaller-scale energy generation systems that increasingly adopt renewable energy sources (e.g., solar and wind) and rely on cyber technologies to ensure resiliency and efficient resource sharing. Basically, this involves massive changes in technical and organizational levels together with tremendous technological upgrades in different sectors ranging from the energy generation and transmission systems down to the distribution systems. These actions constitute a huge science and engineering challenges and demands for expert knowledge in the field to create solutions for a sustainable energy system (both at the energy supply and demand sides) that is economically, environmentally, and socially viable while meeting high security requirements. On energy consumers' side, useful and efficient energy services such as light, heating and cooling, cooking, communication, power, and motion are needed. These services are offered by specific equipment/devices, which use energy blocks either efficiently or inefficiently. Producing energy with high environmental, societal, or health risks is not a cheap way to meet such energy demand, but packages of efficient equipment and energy at least societal costs, which includes external costs, should be the ultimate objective to satisfy the needs of customers. At the supply side, there exists a bunch of opportunities for renewable energy technologies complemented with energy efficiency measures not only to provide local benefits, but also to contribute to sustainable development, which is framed in a three-pillar model: Economy, Ecology, and Society. Thus, the relationship between the use of renewables in energy mix and the sustainable developments goals (SDGs) can be viewed as a set of objectives and constraints that involve both global and local/regional considerations.

2. Sustainable Energy Systems Planning, Integration, and Management

To cover the above-mentioned promising and dynamic areas of research and development, this special issue was launched to allow gathering of contributions in sustainable energy systems planning, integration, and management. In total, 31 papers were submitted to this special issue, out of them 14 were selected for publication which denotes an acceptance rate of 45%. The accepted articles in this special issue cover a variety of topics, ranging from design and planning of small to large-scale energy systems to operation and control of energy networks in different sectors namely electricity 1, heat 9, and transport 13.

Focusing on wave energy harvesting, the paper authored by Eugen Rusu [1] assessed the most relevant patterns of the wave energy propagation in the western side of the Black Sea considering eight different simulating waves nearshore computational domains. Special attention was paid to the high, but not extreme, winter wave energy conditions. The cases considered were focused on the coastal waves generated by distant storms, which means the local wind has not very high values in the targeted areas. From the analysis of the results, it was first noticed that the general patterns regarding the spatial distribution of the wave energy in the basin of the Black Sea do not substantially change. This means that the western side of the sea still remains its most energetic part. Furthermore, a general enhancement of the wave energy is noticed, and it appears that this enhancement is relatively higher in the western side. Moreover, related studies showed that the coastal environment of the Black Sea, and especially its northern and western side, is fully appropriate for the implementation of offshore wind projects.

The work, done by Wang et al. [2] illustrated the role of the multi-criteria decision making in determining the optimal location for solid waste-to-energy plant installation. To this end, the weight of criteria in the multi-criteria decision-making approach was determined using a hybrid fuzzy analysis as well as based on economic factors, technical requirements factors, environment factors, and social factors. In addition, the comprehensive model presented in that research can be used as a guideline to perform the placing mechanism in many countries to determine the optimal location for installing solid waste-to-energy plants.

The works done by Gebresenbet et al. [3] investigated smart logistics system (SLS) for the management of the pruning biomass supply chain. The major components of SLS including smart box, on-board control unit, information platform, and central control unit were studied. The author extended their work to study the role of smart logistics systems in increasing the efficiency of agricultural subscribers to use pruning biomass for generating renewable energy [4]. The smart logistics systems in the presence of renewable sources was shown to have an undeniable role in decreasing logistics cost, increasing pruning marketing opportunity, and decreasing product loss. Four actors of the presented biomass supply chain include producer, traders, transporter, and consumers. In the proposed model, the initial chain, i.e., producers, were the farmers.

In order to consider the uncertainty related to the wind speed as well as electrical power produced through wind farms, a new short-term wind speed forecasting method was introduced to achieve a more satisfying performance in forecasting data by Liu et al. [5]. In this research, to overcome the shortcomings of some traditional models and enhance the forecasting ability, Ensemble Empirical Mode Decomposition model was developed. The authors also used six evaluation indexes, Willmott's Index, Nash–Sutcliffe coefficient, Legates and McCabe Index, mean absolute error, square root of the average of the error square, and the average of absolute percentage error, to assess the effectiveness of the proposed method.

The work done by Peicong Luo [6] elaborated on an efficient approach to dynamically adjust the datacenter load to balance the unstable renewable energy input into the grid. Their experimental results illustrated that the dynamic load management of multiple datacenters could help the smart grid to reduce losses and mitigate stability issues such as bus voltage variations and the overloading of transmission lines and accordingly save operational costs. As a complement to the proposed system, a forecasting method was also designed, based on the concept of neural networks, to predict renewable energy generation in advance so as to adjust the power of the datacenters as soon as possible and reduce the extra losses.

Authors of [7] studied the effects of the different meteorological variables on the power generation via photovoltaic systems. The investigated meteorological data in this paper included solar radiation, outdoor temperature, wind speed, and daylight time. To obtain a statistical representative model of the generated power by photovoltaic systems, the gradient descent optimization (GDO) was used. It was shown that the presented statistical method has the capability to minimize the introduced structure error using a linear Least Square Regression. Using the presented statistical structure, the estimation results' error was about 7% compared to the real data.

In [8] a robust strategy based on particle resampling was proposed to select the load identification features for applications in energy monitoring and measurement systems specially in residential sector. Through incorporation of a 2-D fuzzy membership measurement, it was shown that the feature extracted by the resampling method is closer to that of the actual device, and can be applied to load identification. However, a relatively stable switching period of the devices is needed.

The field measurements and numerical simulation was carried out by Zhu et al. [9] to assess the energy performance of a typical rural residential building in the Ningxia Hui Autonomous Region in Northwest China. It was found that application of solar energy resources is an effective approach to improve the indoor temperature. The other influencing factor such as building layout and proper thermal performance of the building envelope could reduce wind velocities and convective heat loss. Besides, insulation materials and double-glazed windows were found to significantly improve energy performance in new buildings.

Recently, Wang and Zhong [10] developed a three-dimensional simulation model to investigate the building heating load with various irregular heating durations and internal wall layouts. The results of the simulation model were validated with collected experimental data. In comparison of daily heating load with operation hours, results indicated a direct relationship between the daily heating load and the peak value. It was also found that the use of continuous heating when the daily operation hours are more than the threshold values is more economical.

In [11], a novel framework was designed for economic cooling load dispatch in conventional water-cooled chillers and to model the uncertain nature of cooling demand in the day-ahead scheduling. Three decision-making modes including (a) risk-neutral approach, (b) risk-aversion or robustness approach, and (c) risk-taker or opportunistic approach were considered in this study. To determine the optimum operating point of the chiller loading in these three modes, the information gap decision theory was used. The developed model has the capability to enable a system operator to enter the energy cost parameter and reduce the daily energy cost to this critical value and also to identify the increase in maximum cooling demand in the robustness model.

In [12], a new methodology was proposed for internal covering designs to enhance the permeable, semi-permeable, and impermeable internal coverings effect over indoor ambiences. The artificial neural network was used to forecast the indoor ambience. The results of this study had a significant impact on the indoor thermal comfort and energy consumption and suggested the permeable coverings as the suitable approach to minimize energy peak demands in the first hours of occupation and improving thermal comfort condition.

The article, authored by Yuhuan Liu et al. [13], studied the operating organization problem with the multi-type bus, namely pure electric buses and traditional fuel buses, aiming to provide guidance for future application of electric buses. Minimization of the energy consumption of vehicles as well as the waiting and traveling time of passengers were considered as the objectives in the work, while vehicle full load limitation, minimal departure interval, mileage range, and charging time window were taken as constraints. The authors also made a comprehensive analysis on the relationship between the bus driving energy consumption and bus dispatching and bus type matching ratio under the background that pure electric buses gradually replace traditional fuel buses, and many routes are operated with mixing pure electric buses and traditional fuel buses.

Focusing on robust planning of energy and environment systems, a dual robust stochastic fuzzy optimization (DRSFO) model was developed in [14] to analyze the trade-offs between system costs and reliability for planning of energy and environment systems while considering associated risks from the stochastic and fuzzy uncertainties. The model was developed using historical annual electricity data to forecast the future demand. To avoid any deviations from the optimized decision schemes even during the electricity shortage, hourly or seasonal electricity load curves were not considered in the model. It was suggested by the authors to develop a simplified calculation procedure, consider economic aspects of the model, and refine multi-objective models to accurately find the trades-off between energy and environmental systems.

3. Sustainable Energy Systems of the Future

The future energy systems must necessarily match the so-called energy triangle and deliver on all the three dimensions: (1) providing safe, secure, and reliable energy while delivering access to all energy consumers, (2) supplying energy at affordable prices, and (3) assuring sustainable development. Over the past decades, the energy systems that evolved in different sectors have greatly achieved the goal of enabling substantial economic growth. However, due to some reasons, the aforementioned triangle is out of balance in different places around the world. For about 1.1 billion people—14% of the global population—do not have access to electricity according to [15]. Many more suffer from poor quality energy supply. In several places across the globe (such as sub-Saharan Africa and developing Asia), affordability is still an unresolved issue. More importantly, the environmental impact in most countries is beyond what we can sustain, especially as world population heads towards a projected nine billion people. The goal is, however, to create sustainable energy future by installing intelligent, cost-effective, and efficient systems with the lowest ecological footprints.

In the future, integrated models, which take end-use energy efficiency, renewable energy harvesting, and SDGs into account, may be in a favorable position to better link the weak and strong sustainable development paradigms for decision-making processes. By including important and relevant bottom-up indicators in a well-defined structure, integrated models are deemed to explore scenarios for boosting social and economic development and energy access and security as well as mitigating negative environmental and health impacts. In this way, today's energy systems can be expected to transform to sustainable energy systems which are carbon neutral, efficient, accessible, affordable, and secure.

Acknowledgments: We would like to take this opportunity to thank all the authors for their great contributions in this special issue and the esteemed reviewers for their time spent on reviewing manuscripts and their valuable comments helping us improve the articles. We wish also to place on record our appreciation to the dedicated editorial team of *Applied Sciences* for their tremendous support and dedication.

Conflicts of Interest: The authors declare no conflict of interest.

References

1. Rusu, E. Study of the Wave Energy Propagation Patterns in the Western Black Sea. *Appl. Sci.* **2018**, *8*, 993. [CrossRef]
2. Wang, C.; Nguyen, V.; Duong, D.; Thai, H. A Hybrid Fuzzy Analysis Network Process (FANP) and the Technique for Order of Preference by Similarity to Ideal Solution (TOPSIS) Approaches for Solid Waste to Energy Plant Location Selection in Vietnam. *Appl. Sci.* **2018**, *8*, 1100. [CrossRef]
3. Gebresenbet, G.; Bosona, T.; Olsson, S.; Garcia, D. Smart System for the Optimization of Logistics Performance of the Pruning Biomass Value Chain. *Appl. Sci.* **2018**, *8*, 1162. [CrossRef]
4. Bosona, T.; Gebresenbet, G.; Olsson, S.; Garcia, D.; Germer, S. Evaluation of a Smart System for the Optimization of Logistics Performance of a Pruning Biomass Value Chain. *Appl. Sci.* **2018**, *8*, 1987. [CrossRef]
5. Liu, T.; Liu, S.; Heng, J.; Gao, Y. A New Hybrid Approach for Wind Speed Forecasting Applying Support Vector Machine with Ensemble Empirical Mode Decomposition and Cuckoo Search Algorithm. *Appl. Sci.* **2018**, *8*, 1754. [CrossRef]
6. Luo, P.; Wang, X.; Jin, H.; Li, Y.; Yang, X. Smart-Grid-Aware Load Regulation of Multiple Datacenters towards the Variable Generation of Renewable Energy. *Appl. Sci.* **2019**, *9*, 518. [CrossRef]
7. Ruz-Hernandez, J.; Matsumoto, Y.; Arellano-Valmaña, F.; Pitalúa-Díaz, N.; Cabanillas-López, R.; Abril-García, J.; Herrera-López, E.; Velázquez-Contreras, E. Meteorological Variables' Influence on Electric Power Generation for Photovoltaic Systems Located at Different Geographical Zones in Mexico. *Appl. Sci.* **2019**, *9*, 1649. [CrossRef]

8.   He, H.; Lin, X.; Xiao, Y.; Qian, B.; Zhou, H. Optimal Strategy to Select Load Identification Features by Using a Particle Resampling Algorithm. *Appl. Sci.* **2019**, *9*, 2622. [CrossRef]

9.   Zhu, Y.; Fan, X.; Wang, C.; Sang, G. Analysis of Heat Transfer and Thermal Environment in a Rural Residential Building for Addressing Energy Poverty. *Appl. Sci.* **2018**, *8*, 2077. [CrossRef]

10. Wang, S.; Zhong, K. Effects of Configurations of Internal Walls on the Threshold Value of Operation Hours for Intermittent Heating Systems. *Appl. Sci.* **2019**, *9*, 756. [CrossRef]

11. Shi, E.; Jabari, F.; Anvari-Moghaddam, A.; Mohammadpourfard, M.; Mohammadi-ivatloo, B. Risk-Constrained Optimal Chiller Loading Strategy Using Information Gap Decision Theory. *Appl. Sci.* **2019**, *9*, 1925. [CrossRef]

12. Orosa, J.; Vergara, D.; Costa, Á.; Bouzón, R. A Novel Method Based on Neural Networks for Designing Internal Coverings in Buildings: Energy Saving and Thermal Comfort. *Appl. Sci.* **2019**, *9*, 2140. [CrossRef]

13. Liu, Y.; Yao, E.; Liu, S. Energy Consumption Optimization Model of Multi-Type Bus Operating Organization Based on Time-Space Network. *Appl. Sci.* **2019**, *9*, 3352. [CrossRef]

14. Chen, C.; Zeng, X.; Huang, G.; Yu, L.; Li, Y. Robust Planning of Energy and Environment Systems through Introducing Traffic Sector with Cost Minimization and Emissions Abatement under Multiple Uncertainties. *Appl. Sci.* **2019**, *9*, 928. [CrossRef]

15. WEO-2017 Special Report: Energy Access Outlook, International Energy Agency 2017. Available online: https://webstore.iea.org/weo-2017-special-report-energy-access-outlook (accessed on 16 September 2019).

applied sciences

MDPI

Article

Study of the Wave Energy Propagation Patterns in the Western Black Sea

Eugen Rusu

Department of Mechanical Engineering, Faculty of Engineering, "Dunărea de Jos" University of Galati, 47 Domneasca Street, 800008 Galati, Romania; Eugen.Rusu@ugal.ro; Tel.: +40-740-205-534

Received: 26 May 2018; Accepted: 15 June 2018; Published: 17 June 2018

Abstract: The most relevant patterns of the wave energy propagation in the western side of the Black Sea were assessed in the present work. The emphasis was put on the western side because this is also the most energetic part of the Black Sea. The assessments performed relate some recent results provided by a numerical wave modeling system based on the spectrum concept. The SWAN model (acronym for Simulating Waves Nearshore) was considered. This was implemented over the entire sea basin and focused with increasing resolution in the geographical space towards the Romanian nearshore. Furthermore, some data assimilation techniques have also been implemented, such that the results provided are accurate and reliable. Special attention was paid to the high, but not extreme, winter wave energy conditions. The cases considered are focused on the coastal waves generated by distant storms, which means the local wind has not very high values in the targeted areas. This also takes into account the fact that the configuration of the environmental matrix in the Black Sea is currently subjected to significant changes mainly due to the climate change. From this perspective, the present work illustrates some of the most recent patterns of wave energy propagation in the western side of the Black Sea, considering eight different SWAN computational domains. According to most of the recent evaluations, the nearshore of the Black Sea is characterized by an average wave power lower than 6 kW/m. The results of the present work show that there is a real tendency of the wave energy enhancement. This tendency, especially concerns the western side of the basin, where in the high conditions considered, values of the wave power about 10 times greater than the average have been noticed.

Keywords: Black Sea; Romanian coastal environment; wave energy; numerical models; SWAN

---

1. Introduction

Without a doubt, conversion of wave energy into electricity represents one of the greatest challenges of the 21st century. Wave energy is abundant and it has a higher density and predictability than wind or solar energy. Furthermore, there is a wide variety of ways to harness the waves, although the technologies associated with the wave energy conversion are not yet mature enough and none of the existing solutions can be considered now as being the best and the most efficient. Thus, in the struggle against the CO_2 emissions about 100 GW of the ocean energy capacity is expected to be installed by 2050 [1]. From this perspective, ocean energy is expected to play a significant role in the future EU energy system, and by 2050 its potential contribution should cover about 10% of the EU power demand.

An important step forward in extracting renewable energy resources in the marine environment is represented by the recent dynamics of the offshore wind industry. Thus, by receiving momentum from the onshore wind, the development of the offshore wind has had a spectacular advance in the last years. The wind is stronger and steadier offshore, and moreover, while the land is almost saturated, in the marine environment there is practically unlimited space to deploy large wind farms. Furthermore, despite some initial difficulties related to the high cost of installation and operating conditions in the harsh marine environment, very high dynamics in increasing the efficiency of the offshore wind can be also noticed. This is measured, especially through the levelized cost of energy (LCOE), which reached a record value of about 7 c€/kWh in 2017, or less. Thus, 5.5 c€/kWh have been reported at the 700 MW Borssele (The Netherlands) due to government tender and size, and 5c €/kWh (without transmission) at the 600 MW Kriegers Flak (Denmark) [2]. This means the offshore wind becomes now not only the cheapest marine renewable energy resource, but it is also cheaper than some traditional resources. For example, the current average LCOE for atomic energy is still about 11 c€/kWh, or greater [3].

Various studies [4–7] showed that the wind energy resources along the coasts of the Black Sea, and especially in its western side, are comparable with those from many offshore wind farms that are already operational [8–11]. Thus, it is expected that the high dynamics of the offshore wind industry will have as a result also the implementation of some wind farm projects in the nearshore of the Black Sea, in general and in its western side, which is more energetic, in special. Furthermore, most of the studies [12–15] indicate that the climate change will induce significant enhancements of the wind speed in the Mediterranean and Black Sea basins.

On the other hand, although the potential of wave energy in the Black Sea is not comparable with that from the ocean [16–18], the expected advances of the WEC (Wave Energy Converters) technologies may make this coastal environment interesting also for the implementation of the marine energy farms, especially as regards the hybrid wind-wave projects. Many coastal areas in the Black Sea are subjected to high erosion processes and the future nearshore farms can play an important role in the coastal protection, because they extract (or dissipate) part of the wave energy before the waves arriving to the shore [19–24]. Furthermore, this is a general problem for many coastal environments and the marine energy farms, besides providing electricity, can become an effective solution in the struggle against the coastal erosion [25–27].

From this perspective, the present work has as a main objective to assess the most recent wave energy propagation patterns in the western side of the Black Sea. This considers also the fact that there is an obvious dynamics of the wave climate in the Black Sea having as a consequence important changes in the actual patterns of the nearshore wave propagation and inducing significant enhancements of the wave power. For this purpose, a wave modeling system based on the SWAN (Simulating Waves Nearshore) spectral model [28] has been implemented in the Black Sea basin and focused on the Romanian nearshore in a multilevel modeling system with increasing resolution towards the cost.

2. Materials and Methods

2.1. Theory of SWAN Spectral Model

SWAN is a spectral phase averaged wave model that integrates the action balance equation in time, in the geographical space and in the spectral space, which is defined by the relative frequency (σ) and the wave direction (θ) [29,30]:

$$\frac{\partial N}{\partial t} + \nabla(N) + \frac{\partial}{\partial \sigma}\dot{\sigma}N + \frac{\partial}{\partial \theta}\dot{\theta}N = \frac{S}{\sigma}, \tag{1}$$

The wave action (N) is considered in the above equation since in the presence of the currents the action density is conserved while the energy spectrum is not. The wave action is equal to the energy density (E) divided by the relative frequency. For larger scale, the geographical space is represented in spherical coordinates, longitude (λ) and latitude (φ), and the operator (∇) has the expression:

$$\nabla_{\text{Sph}}(N) = \frac{\partial}{\partial \lambda}\dot{\lambda}N + \frac{1}{\cos \varphi}\frac{\partial}{\partial \varphi}\dot{\varphi}N, \tag{2}$$

For coastal applications the Cartesian coordinates (x) and (y) are mostly used and the operator (∇) becomes in this case:

$$\nabla_{Cart}(N) = \frac{\partial}{\partial x}\dot{x}N + \frac{\partial}{\partial y}\dot{y}N, \tag{3}$$

The left side of the governing Equation (1) represents the kinematic part, which indicates the propagation of the wave action in time, geographical and spectral spaces considering also the effect of some relevant phenomena as wave diffraction or refraction. On the right hand side is the source (S) expressed in terms of energy density. In deep water, three components are more relevant, corresponding to the atmospheric input (S_{in}), nonlinear quadruplet interactions (S_{nl}) and whitecapping dissipation (S_{diss}). In intermediate and shallow water some additional terms, corresponding the finite depth effects (S_{fd}) and including phenomena as bottom friction, depth-induced wave-breaking or triad nonlinear wave—wave interactions, may become significant and in this case the total source becomes:

$$S = S_{in} + S_{nl} + S_{diss} + S_{fd} \tag{4}$$

As regards the wave power components (expressed in W/m, i.e., energy transport per unit length of wave front), they are computed in the spectral wave models, with the relationships [31]:

$$P_x = \rho g \iint c_x \, E(\sigma, \theta) d\sigma d\theta \tag{5}$$

$$P_y = \rho g \iint c_y \, E(\sigma, \theta) d\sigma d\theta$$

In the above equation x, y are the problem coordinate system and c_x, c_y are the propagation velocities of the wave energy in the geographical space (absolute group velocity components) defined as:

$$c_x = \frac{dx}{dt}, \; c_y = \frac{dy}{dt} \tag{6}$$

Thus, the absolute value of the energy transport (denoted also as wave power) will be:

$$P = \sqrt{P_{\tilde{x}}^2 + P_{\tilde{y}}^2}, \tag{7}$$

2.2. Computational Levels Defined

A multilevel, SWAN-based wave prediction system has been implemented and focused on the western side of the sea in a downscaling process [32–35]. Various sensitivity tests and validations have been carried out for each computational level [36], taking into account also the computational strategies adopted in other coastal environments where the model results have been intensively validated against in situ measurements and remotely sensed data [37]. A special attention was paid to the coastal areas with more complex coastal dynamics, as for example the mouths of the Danube River where strong interactions occur between the waves and currents generated by the Danube River outflow [38,39].

From this perspective, four SWAN computational levels have been defined in the present work, performing model simulations in eight different areas. The first three levels correspond to the spherical coordinates and their characteristics are provided in Table 1. In this table $\Delta\lambda$ and $\Delta\varphi$ represent the spatial resolution, Δt the time resolution, nf number of frequencies, $n\theta$ number of directions, $n\lambda$ number of grid points in longitude, $n\varphi$ number of grid points in latitude, and np total number of grid points. Thus, the first computation level (L1) corresponds to the generation area, which comprises the entire basin of the Black Sea and the corresponding SWAN domain was denoted as Sph1 and has the resolution in the geographical space $0.08° \times 0.08°$. The second level (L2) reflects the SWAN domain (Sph2) defined to drive the coastal wave transformation in the Romanian nearshore, located on the western side of the basin. The resolution in the geographical space is in this case $0.02° \times 0.02°$. Finally, the third level defined considering the spherical coordinates (L3) comprises three different SWAN domains (denoted as Sph3, Sph4 and Sph5), all of them having the resolution in the geographical space

$0.01° \times 0.01°$. Sph3 is focused on the area at the mouths of the Danube River, while Sph4 and Sph5 is focused on the southern part of the Romanian nearshore—they are denoted as Southern RO1 and Southern RO2. The reason for considering two domains, one larger and another one smaller, focused on the same area in the nearshore of the main Romanian littoral cities is that for the domain Southern RO1 a local multiparameter data assimilation scheme has been implemented [40,41] considering the in situ measurements from the Gloria drilling unit, which is located close to the western boundary of the computational domain. On the other hand, from operational considerations, it was also found useful to define another SWAN domain, computationally more effective, which is focused especially on the Romanian littoral cities. The geographical spaces corresponding to these five computational domains (Sph1–Sph5) are illustrated in Figure 1.

Table 1. Characteristics of the computational domains defined in spherical coordinates for the Simulating Waves Nearshore (SWAN) model simulations focused on the western side of the Black Sea.

Spherical Domains	$\Delta\lambda \times \Delta\varphi$	Δt (min)	nf	$n\theta$	$ng\lambda \times ng\varphi = np$
Sph1—Black Sea (L1)	$0.08° \times 0.08°$	10 non-stat	24	36	$176 \times 76 = 13{,}376$
Sph2—Coastal driver (L2)	$0.02° \times 0.02°$	10 non-stat	24	36	$141 \times 141 = 19{,}881$
Sph3—Danube mouths (L3)	$0.01° \times 0.01°$	10 non-stat	24	36	$71 \times 61 = 4331$
Sph4—Southern RO1 (L3)	$0.01° \times 0.01°$	10 non-stat	24	36	$221 \times 221 = 48{,}821$
Sph5—Southern RO2 (L3)	$0.01° \times 0.01°$	10 non-stat	24	36	$161 \times 141 = 22{,}701$

The fourth computational level (L4) is related to three Cartesian SWAN domains (Cart1, Cart2 and Cart3). The characteristics of these Cartesian areas are given in Table 2. The first two correspond to the high resolution coastal areas in front of the Sulina and Saint George arms of the Danube, while the third to the coastal environment close to Mangalia city, located in the extreme south of the Romanian nearshore.

Table 2. Characteristics of the Cartesian computational domains considered for the SWAN model simulations focused on the western side of the Black Sea.

Cartesian Domains	$\Delta x \times \Delta y$ (m)	Δt (min)	nf	$n\theta$	$ngx \times ngy = np$
Cart1—Sulina (L4)	50×50	60 stat	30	36	$135 \times 216 = 29{,}160$
Cart2—Sacalin (L4)	200×200	60 stat	30	36	$353 \times 251 = 88{,}603$
Cart3—Mangalia (L4)	50×50	60 stat	30	36	$96 \times 107 = 10{,}172$

The physical processes activated in the SWAN simulations, corresponding to the eight computational domains considered are presented in Table 3. In this table: *Wave* indicates the wave forcing, *Tide* the tide forcing, *Wind* the wind forcing, *Curr* the current field input, *Gen* generation by wind, *Wcap* the whitecapping process, *Quad*—the quadruplet nonlinear interactions, *Triad* the triad nonlinear interactions, *Diff* diffraction process, *Bfric* bottom friction, *Set up* wave-induced setup, and *Br* depth-induced wave breaking. For each computational level the most relevant processes have been considered as presented in Table 3.

Table 3. Physical processes activated in the SWAN simulations, corresponding to the eight computational domains defined. X—process activated, 0—process inactivated.

Input/Process Domains	Wave	Wind	Tide	Curr	Gen	Wcap	Quad	Triad	Diffr	Bfric	Set up	Br
Sph1	0	X	0	0	X	X	X	0	0	X	0	X
Sph2	X	X	0	0	X	X	X	X	0	X	0	X
Sph3	X	X	0	X	X	X	X	X	X	X	0	X
Sph4	X	X	0	0	X	X	X	X	0	X	0	X
Sph5	X	X	0	0	X	X	X	X	X	X	0	X
Cart1	X	X	0	X	X	X	X	X	X	X	X	X
Cart2	X	X	0	X	X	X	X	X	X	X	X	X
Cart3	X	X	0	0	X	X	X	X	X	X	X	X

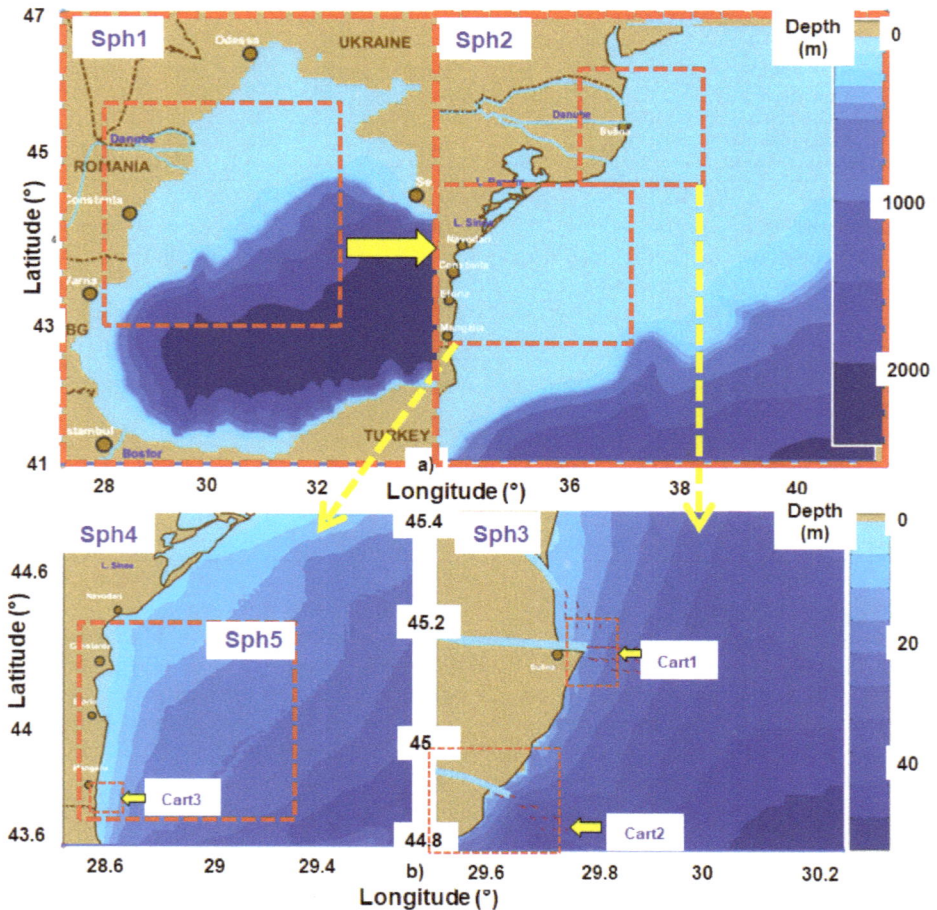

Figure 1. The computational domains defined in spherical coordinates: (**a**) Sph1—Black Sea basin and Sph2 (right side)—western coastal driver; (**b**) Sph3—nearshore area at the mouths of the Danube River (right side); Sph4 and Sph5 (left side)—Southern RO1 and RO2. The positions of the three Cartesian domains are also indicated.

The geographical spaces corresponding to these three Cartesian computational domains (Cart1–Cart3) are illustrated in Figure 2.

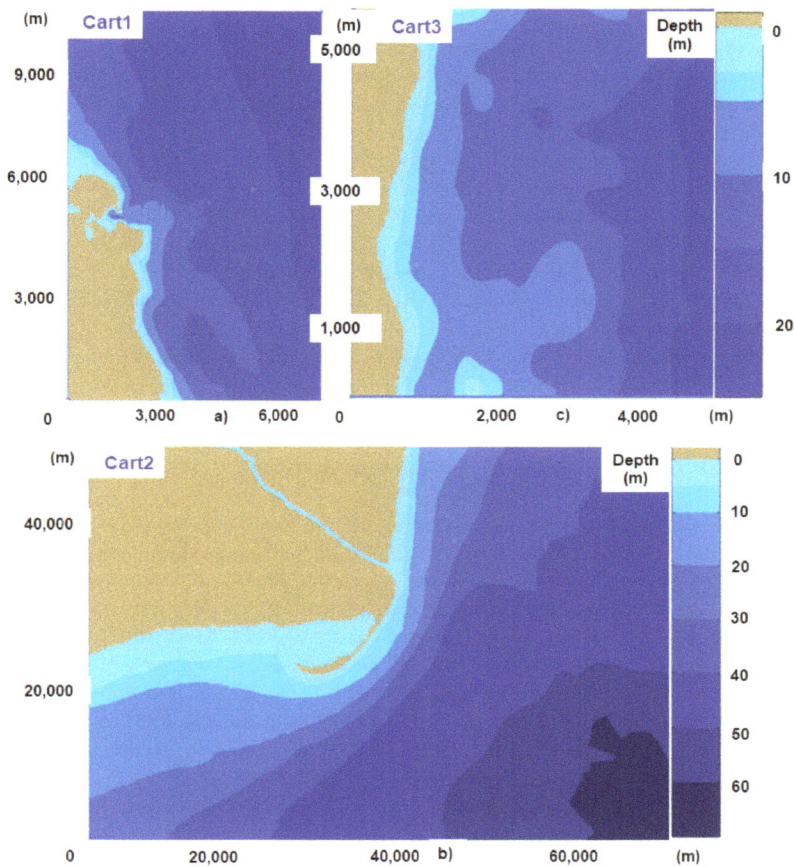

Figure 2. The computational domains defined in Cartesian: (**a**) Cart1—the nearshore in front of Sulina arm of the Danube River; (**b**) Cart2—Sacalin Peninsula and the Saint George arm of the Danube River; (**c**) coastal environment close to Mangalia city, the south of the Romanian nearshore.

2.3. SWAN Model Validations and Implementation of Data Assimilation Techniques

Extensive calibrations and validations have been carried out for the wave modeling system against both 'in situ' and remotely sensed measurements [16,33–36]. While the validations against the satellite data considered only the significant wave height, those carried out against in situ measurements, besides this wave parameter, considered also the mean wave period and the wave direction [15–17,32–35]. As it is known, an increased wave period leads to higher celerity, enhancing as a consequence the wave power. From this perspective a brief discussion will be focused next on the model capability in predicting with an acceptable accuracy this important wave parameter. Since the Black Sea is an enclosed basin, the fetch is significantly smaller than in the ocean environment. This means that the value of the mean period is not so large, an average value for the western coast of the Black Sea being about 4 s. This is almost half the average value of the mean period close to the European oceanic coasts [18]. The results presented in [15–17,32–35], considering validations and calibration studies against data provided by two directional buoys: Gelendzhik (37.98° E, 44.51° N) and Hopa (41.38° E, 41.42° N), show that the wave modeling system implemented predict the mean wave period with a reasonable accuracy, in some statistical parameters, as for example the scatter index the results being systematically better than for the significant wave height. Furthermore, also on the west coast of the Black Sea, a comparison of the SWAN results with the data measured at the Gloria drilling platform, which operates in the western sector (44.31° N, 29.34° E) at a location where the water depth is about 50 m shows also a relatively good accuracy of the model predictions in terms of mean wave periods.

Furthermore, for increasing the accuracy and reliability of the wave predictions, techniques for assimilating the satellite data have been designed and implemented for the first two computational levels corresponding to the SWAN domains Sph1 and Sph2 [42–44]. These assimilation techniques are based on the optimal interpolation method [45]. The main idea behind these assimilation techniques is to combine the information coming from measurements with the results of the numerical models into an optimal estimation of the field of interest. For these two computational domains, the assimilation of the wave data has been performed in terms of the significant wave height (Hs).

It has to be highlighted that the satellite observations provided by the multimission system were considered for data assimilation. These comprise measurements from seven satellites. Thus, the data coming from the satellites: ERS-2 (European Remote Sensing), Poseidon, JASON-1, JASON-2 and GEOSAT (GEOdetic SATellite) Follow-On (GFO) were considered in the assimilation process, while the data from ENVISAT (Environmental Satellite) and TOPEX (TOPography Experiment) were used for validations.

From this perspective, Table 4 presents the statistical results obtained for the Hs values simulated with SWAN and the Hs values obtained after the application of the data assimilation method, against the altimeter measurements used for validation across the Black Sea. The results correspond to the 15-year time interval 1999–2013. The parameters presented in Table 4 are: mean measured and simulated values of the significant wave height, bias, mean absolute error, RMS (root mean square) error, scatter index (SI), correlation coefficient (R), and the regression slope (S), all of them being computed according to their standard definitions. In Table 4, N represents the number of data points considered in the statistical analysis. As it is known the bias, mean absolute error, RMS error and scatter index are better when the values are smaller, while the correlation coefficient and the regression slope are better when they are closer to the unity. From this perspective, it can be seen from Table 4 that the results with data assimilation appeared improved for all parameters. Figure 3 illustrates the corresponding scatter diagrams (the 15-year time interval 1999–2013) for the significant wave height. On the left side of this figure, the results of the SWAN model without data assimilation are presented while on the right side the results after performing the assimilation of the satellite data are given. The results presented in Figure 3 also indicate a clear improvement in the case when the data assimilation technique is used. Detailed information concerning the techniques considered for assimilating satellite data in the Black Sea wave modeling is given in [17,42–44].

Table 4. Statistical results obtained for the Hs values simulated with SWAN and the Hs values obtained after the application of the DA (data assimilation) method, against altimeter measurements used for validation across the Black Sea, results corresponding to the 15-year time interval (1999–2013).

Parameter	Mean$_{Obs}$ (m)	Mean$_{Sim}$ (m)	Bias (m)	MAE (m)	RMSE (m)	SI	R	S	N
SWAN Hs (m)	1.04	0.97	−0.07	0.27	0.35	0.35	0.88	0.98	316,920
SWAN$_{DA}$ Hs (m)		1.00	−0.04	0.21	0.29	0.28	0.91	0.99	

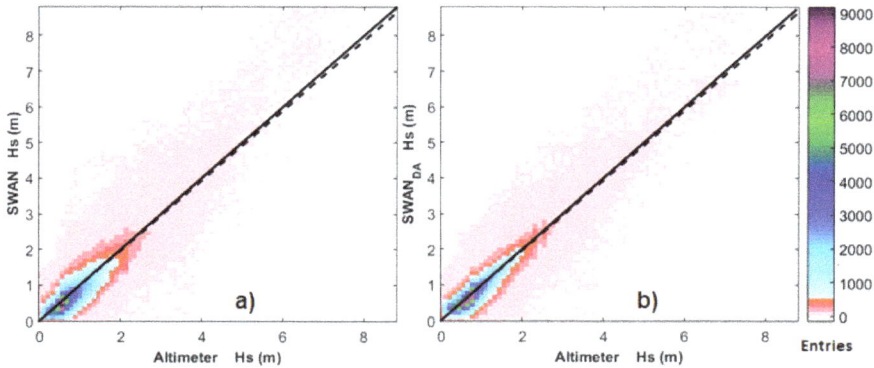

Figure 3. *Hs* scatter diagrams: **a)** SWAN without DA and **b)** SWAN with DA (**right**), results corresponding to the 15-year time interval 1999–2013.

Finally, as previously mentioned, an alternative multi parameter assimilation scheme was also designed, especially for the computational domain Sph4, considering in the assimilation process the in situ measurements carried out at the Gloria drilling unit [40,41]. In this case three different wave parameters have been assimilated. These are significant wave height, mean period and mean wave direction. This approach is based on the successive correction method (SCM) and its main advantage consists in the fact that the corrections are not limited to the significant wave height but operate on other two important wave parameters, which are wave direction and period. As shown in [40,41] this assimilation method improves the reliability of the wave predictions. On the other hand, the main inconvenience in relationship with the assimilation of the satellite data consists in the fact that it is directly related to a permanent source for in situ measurements. That is why from the computational domains considered in the present work, the above assimilation approach is effective only for Sph4 and the subsequent domains Sph5 and Cart3.

3. Results and Discussion

Model system simulations have been performed for the entire year 2017. The emphasis was given to the analysis of the high, but not extreme, wintertime conditions, where wintertime was considered the six-month period from January to March and from October to December. The wind fields considered are those provided by the US National Centres for Environmental Prediction (NCEP). The spatial resolution of the wind data is 0.325 degrees in both latitude and longitude while the temporal resolution is 3 h.

Some of the most relevant results of the simulations performed with the SWAN model for the year 2017 are analyzed and discussed next. These consider the eight computational domains (corresponding to four different levels) defined before. Also, for each domain the physical parameterizations presented in Table 4 have been activated in the model setting. The discussion focuses on the high wave energy conditions.

From this perspective, Figure 4 illustrates high wave conditions in the entire basin of the Black Sea (computational domain Sph1). These model results correspond to the time frame 8 January 2017. Figure 4a presents the significant wave height scalar fields (*Hs*) and the wave vectors and Figure 4b the wave power scalar fields (*Pw*) and the energy transport vectors. The maximum values of the significant wave height and wave power are also indicated, they are: *Hsmax* = 5.2 m and *Pwmax* = 70 kW/m. These correspond to two different locations offshore the Danube Delta, which are very close. At this point, it has to be highlighted that the case study presented in Figure 4 corresponds to one of the most common patterns concerning the spatial distribution of the wave energy in the basin of the Black Sea.

According to this pattern, the tendency of the wave energy is to concentrate in the western part of the basin in the Black Sea. This is also related to the characteristics of the atmospheric circulation, according to which the strongest winds blowing over the Black Sea are usually from the northeast.

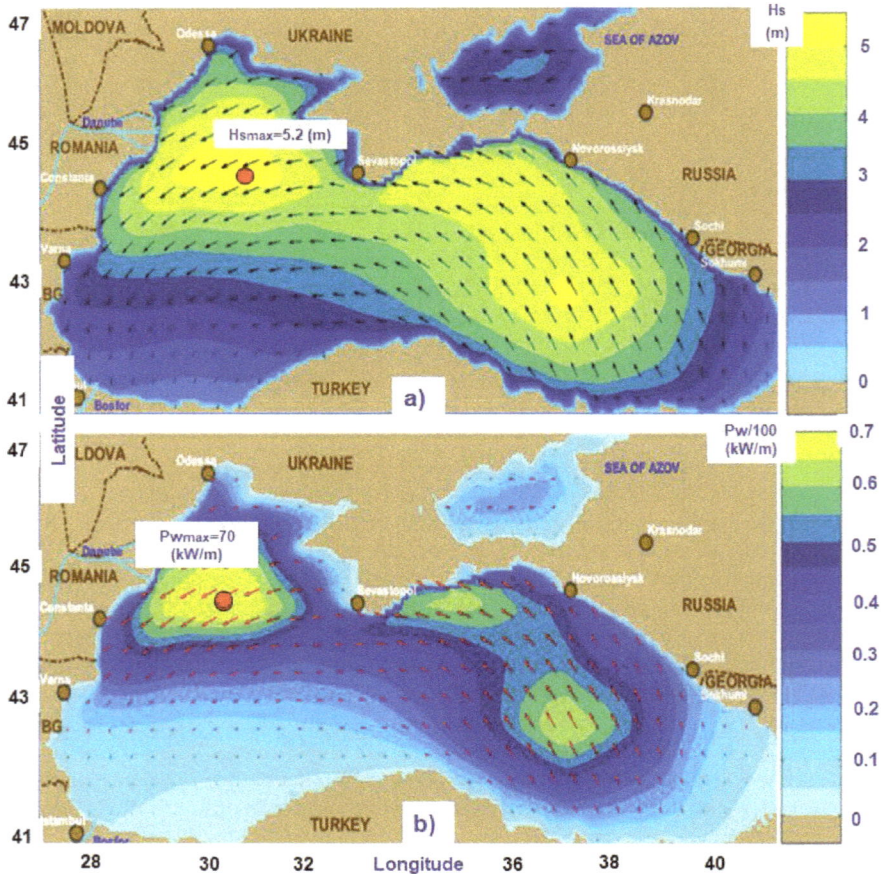

Figure 4. High wave conditions in the Black Sea (computational domain Sph1), model results corresponding to the time frame 8 January 2017. (**a**) Significant wave height scalar fields and wave vectors (represented by black arrows); (**b**) wave power scalar fields and energy transport vectors (represented by red arrows). The maximum values of the significant wave height and wave power are also indicated.

Going to the second level (corresponding to the spherical computational domain Sph2), Figure 5 presents high, but not extreme, wave conditions in the western Black Sea. The model results correspond in this case to the time frame 4 February 2017. Figure 5a illustrates the significant wave height scalar fields and wave vectors while Figure 5b the wave power scalar fields and energy transport vectors. The maximum values of the significant wave height and wave power are $Hsmax = 4.9$ m and $Pwmax = 57$ kW/m. As shown in the figure, these maximums are located in two different points, both close the eastern boundary approximately in the center of the computational domain.

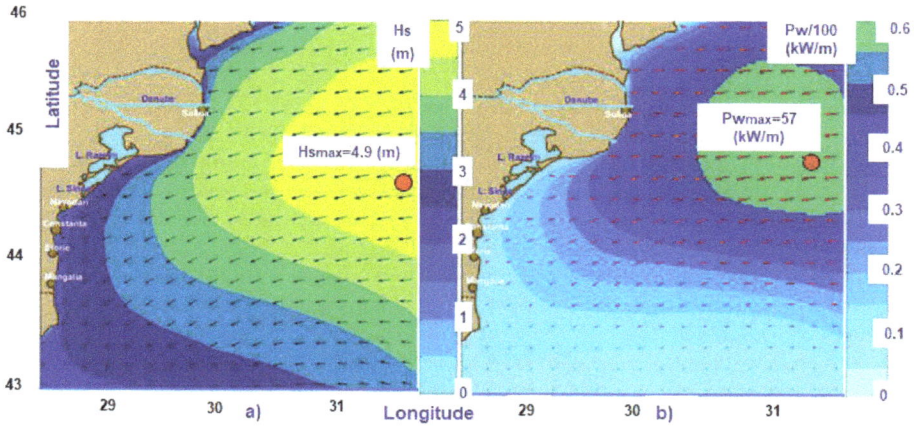

Figure 5. High wave conditions in the western Black Sea (computational domain Sph2), model results corresponding to the time frame 4 February 2017. (**a**) Significant wave height scalar fields and wave vectors (represented by black arrows); (**b**) wave power scalar fields and energy transport vectors (represented by red arrows).

Regarding now level 3, Figure 6 presents high wave conditions at the mouths of the Danube River (computational domain Sph3). The model results correspond to the time frame 22 March 2017. Figure 6a illustrates the significant wave height scalar fields and wave vectors while Figure 6b the wave power scalar fields and energy transport vectors. The maximum values of the significant wave height and wave power are $Hsmax = 5.2$ m and $Pwmax = 58$ kW/m. Although they are both located in the southern side of the domain, their positions are different this time. Thus, the significant wave height has the highest value at the mouth of the Saint George arm while the point with the maximum wave power is located about 0.35 degrees offshore, close to the southern boundary of the SWAN domain. The next two areas defined for this level (L3) are illustrated in Figures 7 and 8 and they correspond to the computational domains Sph4—Southern RO1 and Sph5—Southern RO2, respectively. Thus, Figure 7 presents the model results for the time frame 7 October 2017 and Figure 8 those for the time frame 25 October 2017. The maximum values of the significant wave height and wave power are $Hsmax = 4.0$ m and $Pwmax = 58$ kW/m, for the case presented in Figure 7 and $Hsmax = 4.9$ m and $Pwmax = 76$ kW/m, for the case presented in Figure 8. In both cases the significant wave height and wave power have the maximum values in the same point, which is located in the northeastern side of the SWAN computational domain.

Finally, Figure 9 presents the wave power scalar fields and energy transport vectors in the high resolution Cartesian domains (Cart1–Cart3). Thus, Figure 9a presents an average to high wave energy situation in the computational domain Cart1 corresponding to the time frame 8 November 2017. The maximum values of the significant wave height and wave power are $Hsmax = 4.3$ m and $Pwmax = 54$ kW/m. The points of maximum are located both in front of the Sulina bar, but that corresponding to the maximum wave power is closer to the nearshore. An important characteristic of the SWAN model simulations in this computational domain is related to the higher values of the Benjamin-Feir Index (*BFI*). *BFI*, or the steepness-over-randomness ratio, has been introduced formally by Jansen [43] and is defined as:

$$BFI = \sqrt{2\pi}\, St \cdot Q_p \qquad (8)$$

where *St* represents the integral wave steepness and is computed as the ration between the significant wave height and the wavelength and Q_p represents the peakedness of the wave spectrum and it is defined as:

$$Q_p = 2 \frac{\iint \sigma E^2(\sigma, \theta) d\sigma \, d\theta}{\left(\iint \sigma E(\sigma, \theta) d\sigma \, d\theta\right)^2} \tag{9}$$

This parameter is related to the occurrences of the high waves (the risks for the rogue waves apparition). Hence, *BFI* is a spectral shape parameter that can be related to the kurtosis of the wave height distribution. In particular, for Gaussian-shaped spectra in the narrow band approximation Jansen [46] showed that the kurtosis depends on the square of *BFI*. Furthermore, various experimental results show that for *BFI* = 0.2 the maximum wave heights are very well-described by the Rayleigh distribution while for values of *BFI* greater than 0.9 the ratio *Hmax/Hs* is substantially underestimated. From this perspective, [36] showed that the highest *BFI* values occur in this area between the two points of maximum wave power and significant wave height, respectively, indicated in Figure 9a for significant wave heights of about 3 m and wave directions of 90 degrees (*BFI* = 1.87). This indicates a very high probability of the rogue wave occurrence with very high wave energy. The importance of mentioning the rogue waves in this context is that besides the very high wave power, which they possess, they are also very dangerous for the human activities. These include the operational activities related to the maintenance of the devices, but also devices themselves might be put in danger by such very high and unexpected waves.

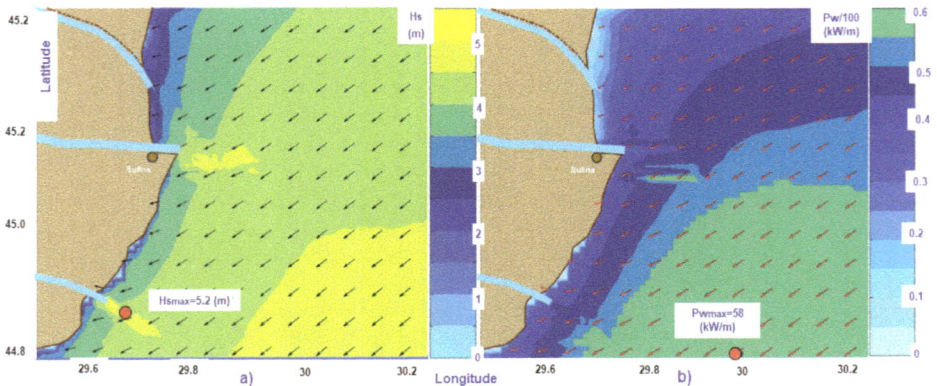

Figure 6. High wave conditions at the mouths of the Danube River (computational domain Sph3), model results corresponding to the time frame 22 March 2017. (**a**) Significant wave height scalar fields and wave vectors (represented by black arrows); (**b**) wave power scalar fields and energy transport vectors (represented by red arrows).

Figure 9b illustrates a high wave energy situation in the computational domain Cart2 corresponding to the time frame 8 November 2017. The maximum values of the significant wave height and wave power are *Hsmax* = 5.1 m and *Pwmax* = 78 kW/m. The points of maximum are located both on the northern boundary, but in different places. Finally, Figure 9c illustrates also a high wave energy situation in the computational domain Cart3 (the coastal environment close to the Romanian city Mangalia) corresponding to the time frame 18 December 2017. The maximum values of the significant wave height and wave power are in this case *Hsmax* = 5.3 m and *Pwmax* = 71 kW/m. These maximums correspond both to the same point, located in the southern right corner of the computational domain.

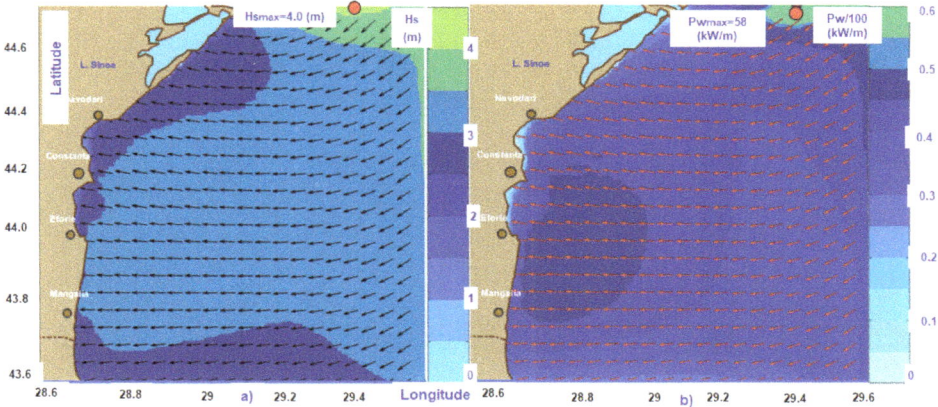

Figure 7. Average to high wave conditions in the southern side of the Romanian nearshore (computational domain Sph4—Southern RO1), model results corresponding to the time frame 7 October 2017. (**a**) Significant wave height scalar fields and wave vectors (represented by black arrows); (**b**) wave power scalar fields and energy transport vectors (represented by red arrows).

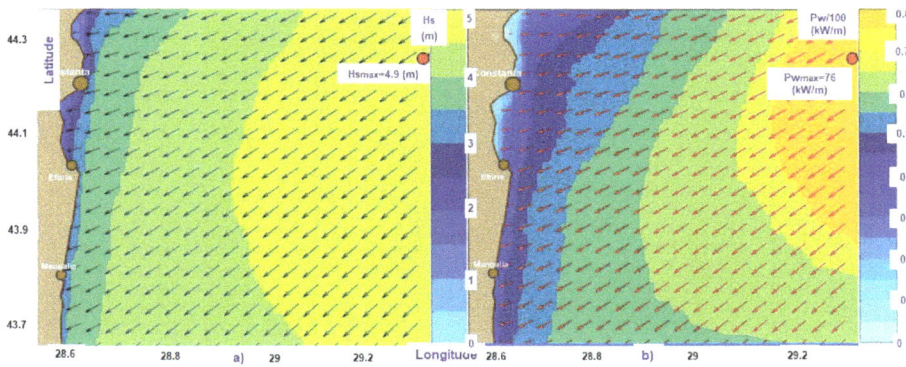

Figure 8. High wave conditions in the southern side of the Romanian nearshore (computational domain Sph5—Southern RO2), model results corresponding to the time frame 25 October 2017. (**a**) Significant wave height scalar fields and wave vectors (represented by black arrows); (**b**) wave power scalar fields and energy transport vectors (represented by red arrows).

Some observations can be made at the end of this section. First of all, as already mentioned, the cases presented are related to high winter time wave conditions. They were selected as representative, but at the same time, although cannot be considered very usual, they do not represent unusual situations since such wave energy conditions are encountered with a certain frequency in the winter period. Another observation is related to the fact that usually the maximum values of the significant wave height and of the wave power do not occur exactly in the same point, but they are very often in closer positions from a geographical perspective. Finally, although there is a direct relationship between the wave power and the significant wave height, a certain significant wave height does not automatically indicates the value of the wave power. Thus, according to the results presented in Figure 4 (for the domain Sph1), a $Hsmax$ = 5.2 m corresponds a $Pwmax$ = 70 kW/m. On the other hand, as Figure 6 shows (for the domain Sph3), a $Hsmax$ = 5.2 m corresponds a $Pwmax$ = 58 kW/m, while according to Figure 8 (related to the domain Sph3), a $Hsmax$ = 4.9 m corresponds to a

$Pwmax$ = 76 kW/m. The explanation of these differences can be found in the definition of the wave power components (Equation (5)). Thus, the components of the wave power depend not only on the energy spectrum, but also on the velocity of the wave group. This means that for the same significant wave height, if the waves travel faster, they will have a higher energy.

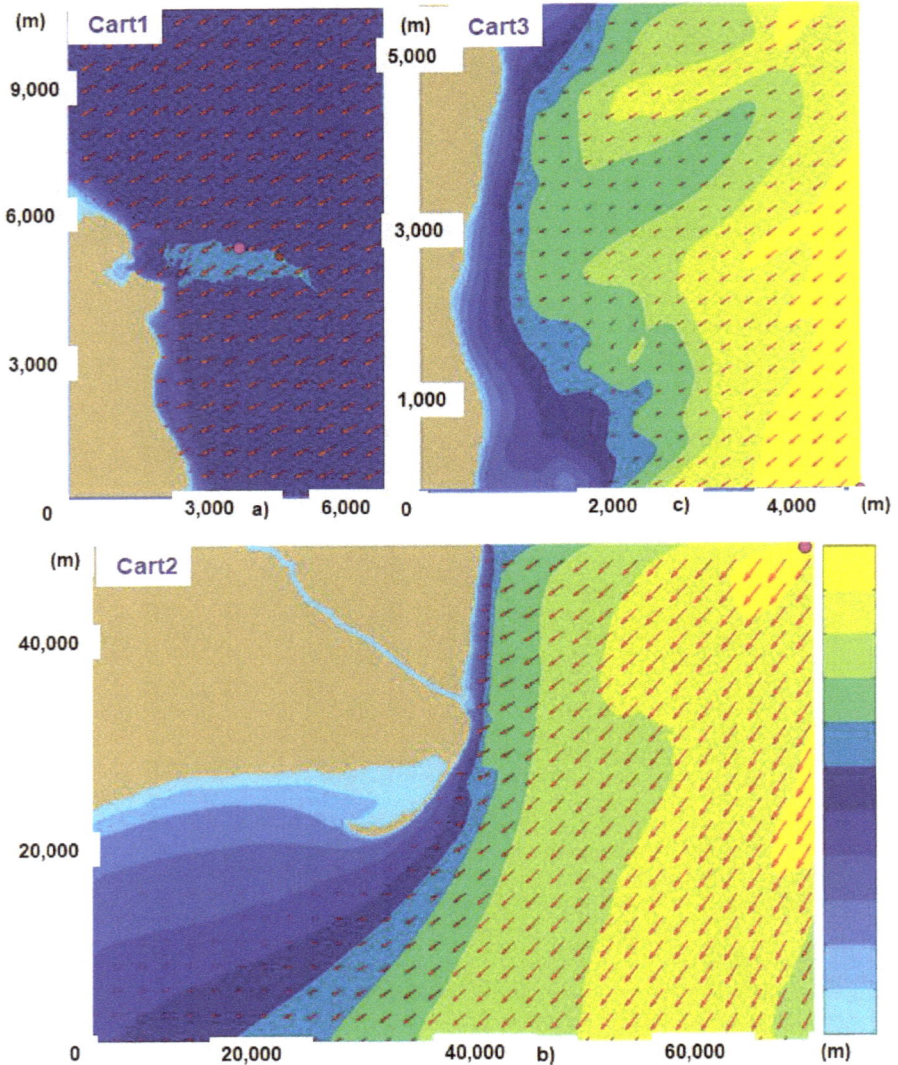

Figure 9. Wave power scalar fields and energy transport vectors (represented by red arrows) in the high resolution Cartesian domains defined: (a) Cart1—Sulina bar, average to high wave energy situation corresponding to the time frame 8 November 2017; (b) Cart2—Sacalin Peninsula, high wave conditions, time frame 28 November 2017. (c) Cart3—Mangalia nearshore, high wave conditions, time frame 18 December 2017.

Finally, at the end of this section it can be highlighted that, although the results of this work are focused on the coastal environment of the Black Sea, most of them reflect the general trends encountered in the coastal environment, indicating the fact that the effects of the climate change is rather similar in many nearshore areas [47,48].

4. Conclusions

This work illustrates some recent results related to the energy propagation patterns in the western side of the Black Sea basin, especially focused on the Romanian nearshore. Thus, simulations with the SWAN spectral phase averaged wave model have been performed for the entire year 2017 and some of the corresponding results have been presented and analyzed. This modeling system was previously validated at various scales, and furthermore, some data assimilation schemes have been also implemented, so that its results can be considered both accurate and reliable. Four computational levels comprising eight different SWAN domains have been considered in the present work. Five of them were defined in spherical coordinates (Sph1–Sph5), while the other three in Cartesians (Cart1–Cart3). The emphasis was given to the high, but not extreme, wave energy conditions generated by distant storms. This means that the local wind has not very high values, the maximum wind speed being usually less than 10 m/s in the nearshore areas considered. The present study also takes into account the fact that, mainly due to the climate change, the configuration of the environmental matrix in the Black Sea is currently subjected to significant changes and such high wave energy conditions may occur now with an increased frequency.

From the analysis of the results, some conclusions will be briefly drawn. In order to have references for comparisons, the results of some studies [16,17] related to the wave energy conditions previously observed in the basin of the Black Sea have been also considered. Thus, it can be first noticed that the general patterns regarding the spatial distribution of the wave energy in the basin of the Black Sea do not substantially change. This means that the western side of the sea still remains its most energetic part. Furthermore, a general enhancement of the wave energy is noticed and it appears that this enhancement is relatively higher in the western side. Although the present study was focused on the high wave energy conditions, the results of the model simulations show that the incidence of the strong storms increases in the last year. At the same time, this feature enhances the incidence of the higher wave energy conditions (that were especially targeted in the present work) followed by the general enhancement of the wave energy, which was already mentioned. Another important observation coming from the analysis of the results coming from the present work, when compared with the previous studies, relates the relatively higher values of the wave power for similar values of the significant wave height. This indicates an enhancement in terms of the wave group velocity, which means that there is a tendency for the waves to travel faster along the Black Sea.

Many previous studies, related to the wind energy potential [4–10], showed that the coastal environment of the Black Sea, and especially its northern and western side, is fully appropriate for the implementation of offshore wind projects. Furthermore, various studies have been also carried out related to the climate change impacts on wind energy potential in the Black Sea [49,50]. According to these, it can be noticed a clear tendency of the wind power enhancement in the western side of the basin. This is expected to be reflected also in an enhancement of the wave power, which will be more significant in the western side of the Black Sea. Furthermore, this western part was also found in previous studies [16,17] to be more energetic. At this point, it has to be also noticed that similar climate change effects are expected also in the Mediterranean Sea [12]. Even so, as regards the waves, it is obvious that the potential in the western side of the Black Sea is lower than in the ocean environment, the total average wave power values being usually lower than 6 kW/m in the western coastal environment of the Black Sea. However, the expected advances in the WEC technologies, coupled with the visible enhancement of wave power, may also give momentum to the wave energy extraction in this part of the sea. This especially concerns the collocation of the wave farms in the vicinity of the future wind projects, so that to benefit from important cost reductions related to the grid connection and to the operational expenditures (OPEX). Furthermore, an important additional argument in favor of marine energy farms is that they may play a significant role in coastal protection [25–27]. At this point, it has to be highlighted that the western side of the Black Sea is currently subjected to high erosion processes that need very expensive periodical investments. From this perspective, the advantage of marine farms is that they diminish the cause (the nearshore wave energy) and not the effect, as most of the other solutions considered for coastal protection.

Funding: This research was funded by the Romanian Executive Agency for Higher Education, Research, Development and Innovation Funding—UEFISCDI, grant number PN-III-P4-IDPCE-2016-0017, associated to the research project REMARC (Renewable Energy extraction in MARine environment and its Coastal impact).

Acknowledgments: The author would like to express his gratitude to the reviewers for their suggestions and observations that helped in improving the present work.

Conflicts of Interest: The author declares no conflicts of interest.

References

1. SET Plan—Declaration of Intent on Strategic Targets in the Context of an Initiative for Global Leadership in Ocean Energy. 2016. Available online: https://setis.ec.europa.eu/system/files/integrated_set-plan/declaration_of_intent_ocean_0.pdf (accessed on 4 April 2018).
2. Jensen, P.H.; Chaviaropoulos, T.; Natarajan, A. Outcomes from the INNWIND.EU Project, LCOE Reduction for the Next Generation Offshore Wind Turbines. 2017. Available online: file:///C:/Users/erusu/Downloads/Innwind-final-printing-version.pdf (accessed on 4 April 2018).
3. Nuclear Power Economics and Project Structuring. April 2016. World Nuclear Association. Available online: http://www.world-nuclear.org/information-library/economic-aspects/economics-of-nuclear-power.aspx (accessed on 4 April 2018).
4. Onea, F.; Rusu, E. Wind energy assessments along the Black Sea basin. *Meteorol. Appl.* **2014**, *21*, 316–329. [CrossRef]
5. Onea, F.; Rusu, E. Evaluation of the wind energy in the north-west of the Black Sea. *Int. J. Green Energy* **2014**, *11*, 465–487. [CrossRef]
6. Global Wind Energy Council (GWEC). *Global Wind Energy Outlook*; GWEC: Brussels, Belgium, 2016.
7. Onea, F.; Raileanu, A.; Rusu, E. Evaluation of the Wind Energy Potential in the Coastal Environment of two Enclosed Seas. *Adv. Meteorol.* **2015**. [CrossRef]
8. Tong, W. Fundamentals of wind energy. In *Wind Power Generation and Wind Turbine Design*; WIT Press: Southampton, UK, 2010; Volume 44, p. 112.
9. Raileanu, A.B.; Onea, F.; Rusu, E. Evaluation of the Offshore Wind Resources in the European Seas Based on Satellite Measurements. In Proceedings of the International Multidisciplinary Scientific GeoConferences SGEM, Albena, Bulgaria, 16–25 June 2015.
10. Onea, F.; Rusu, E. Efficiency assessments for some state of the art wind turbines in the coastal environments of the Black and the Caspian seas. *Energy Explor. Exploit.* **2016**, *34*, 217–234. [CrossRef]
11. Onea, F.; Raileanu, A.; Rusu, E. Evaluation of the wave energy potential in some locations where European offshore wind farms operate. In *Maritime Technology and Engineering 3*; Taylor & Francis Group: London, UK, 2016; pp. 1119–1124.
12. Makris, C.; Galiatsatou, P.; Tolika, K. Climate change effects on the marine characteristics of the Aegean and Ionian Seas. *Ocean Dyn.* **2016**, *66*, 1603–1635. [CrossRef]
13. Ganea, D.; Amortila, V.; Mereuta, E.; Rusu, E. A Joint Evaluation of the Wind and Wave Energy Resources Close to the Greek Islands. *Sustainability* **2017**, *9*, 1025. [CrossRef]
14. Onea, F.; Rusu, L. A long-term assessment of the Black Sea wave climate. *Sustainability* **2017**, *9*, 1875. [CrossRef]
15. Gasparotti, C.; Rusu, E. Methods for the risk assessment in maritime transportation in the Black Sea basin. *J. Environ. Prot. Ecol.* **2012**, *13*, 1751–1759.
16. Rusu, E. Wave energy assessments in the Black Sea. *J. Mar. Sci. Technol.* **2009**, *14*, 359–372. [CrossRef]
17. Rusu, L. Assessment of the wave energy in the Black Sea based on a 15-year hindcast with data assimilation. *Energies* **2015**, *8*, 10370–10388. [CrossRef]

18. Rusu, E.; Soares, C.G. Wave Energy Assessments in the Coastal Environment of Portugal Continental. In Proceedings of the 27th International Conference on Offshore Mechanics and Arctic Engineering, Estoril, Portugal, 15–20 June 2008; Volume 6, pp. 761–772.

19. Zanopol, A.; Onea, F.; Rusu, E. Evaluation of the coastal influence of a generic wave farm operating in the Romanian nearshore. *J. Environ. Prot. Ecol.* **2014**, *5*, 597–605.

20. Zanopol, A.; Onea, F.; Rusu, E. Coastal impact assessment of a generic wave farm operating in the Romanian nearshore. *Energy* **2014**, *72*, 652–670. [CrossRef]

21. Diaconu, S.; Rusu, E. The environmental impact of a Wave Dragon array operating in the Black Sea. *Sci. World J.* **2013**. [CrossRef] [PubMed]

22. Zanopol, A.T.; Onea, F.; Rusu, E. Wave farm influences on the Mangalia nearshore wave pattern. *Int. Multidiscip. Sci. Geoconf.* **2014**, *1*, 621–628.

23. Omer, I.; Mateescu, R.; Vlasceanu, E. Hydrodynamic regime analysis in the shore area taking into account the new master plan implementation for the coastal protection at the Romanian shore, Water Resources, Forest, Marine and Ocean Ecosystems. *SGEM* **2015**, *2*, 651–657.

24. Niculescu, D.M.; Rusu, E.V.C. Evaluation of the new coastal protection scheme at Mamaia Bay in the nearshore of the Black Sea. *Ocean Syst. Eng. Int. J.* **2018**, *8*, 1–20.

25. Bergillos, R.J.; López-Ruiz, A.; Medina-López, E.; Moñino, A.; Ortega-Sánchez, M. The role of wave energy converter farms on coastal protection in eroding deltas, Guadalfeo, southern Spain. *J. Clean. Prod.* **2018**, *171*, 356–367. [CrossRef]

26. Rodríguez-Delgado, C.; Bergillos, R.J.; Ortega-Sánchez, M.; Iglesias, G. Protection of gravel-dominated coasts through wave farms: Layout and shoreline evolution. *Sci. Total Environ* **2018**, *636*, 1541–1552. [CrossRef]

27. Rodríguez-Delgado, C.; Bergillos, R.J.; Ortega-Sánchez, M.; Iglesias, G. Wave farm effects on the coast: The alongshore position. *Sci. Total Environ.* **2018**, *640–641*, 1176–1186. [CrossRef]

28. Booij, N.; Ris, R.C.; Holthuijsen, L.H. A third generation wave model for coastal regions. Part 1: Model description and validation. *J. Geophys. Res.* **1999**, *104*, 7649–7666. [CrossRef]

29. Holthuijsen, H. *Waves in Oceanic and Coastal Waters*; Cambridge University Press: Cambridge, UK, 2007; p. 387.

30. Rusu, E. Strategies in using numerical wave models in ocean/coastal applications. *J. Mar. Sci. Technol. Taiwan* **2011**, *19*, 58–75.

31. SWAN Team. *Scientific and Technical Documentation*; SWAN Cycle III; Delft University of Technology, Department of Civil Engineering: Delft, The Netherlands, 2017.

32. Rusu, L.; Bernardino, M.; Soares, C.G. Wind and wave modeling in the Black Sea. *J. Oper. Oceanogr.* **2014**, *7*, 5–20. [CrossRef]

33. Rusu, L.; Ivan, A. Modelling wind waves in the Romanian coastal environment. *Environ. Eng. Manag. J.* **2010**, *9*, 547–552.

34. Rusu, L.; Butunoiu, D. Evaluation of the wind influence in modeling the Black Sea wave conditions. *Environ. Eng. Manag. J.* **2014**, *13*, 305–314.

35. Rusu, L.; Butunoiu, D.; Rusu, E. Analysis of the extreme storm events in the Black Sea considering the results of a ten-year wave hindcast. *J. Environ. Prot. Ecol.* **2014**, *15*, 445–454.

36. Butunoiu, D.; Rusu, E. Sensitivity tests with two coastal wave models. *J. Environ. Prot. Ecol.* **2012**, *13*, 1332–1349.

37. Rusu, E.; Soares, C.V.; Rusu, L. Computational strategies and visualisation techniques for the wave modeling the Portuguese nearshore. *Marit. Transp. Exp. Ocean Coast.* **2005**, *2*, 1129–1136.

38. Ivan, A.; Gasparotti, C. Influence of the interactions between waves and currents on the navigation at the entrance of the Danube Delta. *J. Environ. Prot. Ecol.* **2012**, *13*, 1673–1682.

39. Ivan, A.; Rusu, E. Assessment of the navigation conditions in the coastal sector at the entrance of the Danube Delta. *Int. Multidiscip. Sci. Geoconf.* **2012**, *3*, 935–942.

40. Rusu, E.; Raileanu, A. A multi parameter data assimilation approach for wave predictions in coastal areas. *J. Oper. Oceanogr.* **2016**, *9*, 13–25. [CrossRef]

41. Butunoiu, D.; Rusu, E. Wave Modeling with Data Assimilation to Support the Navigation in the Black Sea Close to the Romanian Ports. In Proceedings of the 2nd ICTTE Conference, Belgrade, Serbia, 27–28 November 2014; pp. 180–187.

42. Butunoiu, D.; Rusu, E. A Data Assimilation Scheme to Improve the Wave Predictions in the Black Sea. In *OCEANS 2015-Genova*; IEEE Xplore Digital Library: Genoa, Italy, 2015.
43. Rusu, L. A data assimilation scheme to improve the wave predictions in the western side of the Black Sea, Geoconference on Water Resources, Forest, Marine and Ocean Ecosystems. *SGEM* **2014**, *2*, 539–545.
44. Raileanu, A.; Rusu, L.; Rusu, E. Data assimilation Methods to Improve the Wave Predictions in the Romanian Coastal Environment. In Proceedings of the 16th SGEM Conference, Albena, Bulgaria, 30 June–6 July 2016; pp. 855–862.
45. Kalnay, E. *Atmospheric Modeling, Data Assimilation and Predictability*; Cambridge University Press: Cambridge, UK, 2003; p. 341.
46. Janssen, P.A.E.M. Nonlinear four-wave interactions and freak waves. *J. Phys. Oceanogr.* **2003**, *33*, 863–883. [CrossRef]
47. Rusu, E. Numerical Modeling of the Wave Energy Propagation in the Iberian Nearshore. *Energies* **2018**, *11*, 980. [CrossRef]
48. Rusu, E.; Onea, F. Estimation of the wave energy conversion efficiency in the Atlantic Ocean close to the European islands. *Renew. Energy* **2016**, *85*, 687–703. [CrossRef]
49. Davy, R.; Gnatiuk, N.; Pettersson, L.; Bobylev, L. Climate change impacts on wind energy potential in the European domain with a focus on the Black Sea. *Renew. Sustain. Energy Rev.* **2018**, *81*, 1652–1659. [CrossRef]
50. Divinsky, B.V.; Kosyan, R.D. Spatiotemporal variability of the Black Sea wave climate in the last 37 years. *Cont. Shelf Res.* **2017**, *136*, 1–19. [CrossRef]

applied
sciences

MDPI

Article

A Hybrid Fuzzy Analysis Network Process (FANP) and the Technique for Order of Preference by Similarity to Ideal Solution (TOPSIS) Approaches for Solid Waste to Energy Plant Location Selection in Vietnam

Chia-Nan Wang [1,2], Van Thanh Nguyen [1,3,*], Duy Hung Duong [3] and Hoang Tuyet Nhi Thai [3]

1   Department of Industrial Engineering and Management, National Kaohsiung University of Science and Technology, Kaohsiung 80778, Taiwan; cn.wang@nkust.edu.tw
2   Department of Industrial Engineering and Management, Fortune Institute of Technology, Kaohsiung 83160, Taiwan
3   Department of Industrial Systems Engineering, CanTho University of Technology, Can Tho 900000, Vietnam; ddhung.htcn0114@student.ctuet.edu.vn (D.H.D.); thtnhi.htcn0114@student.ctuet.edu.vn (H.T.N.T.)
*   Correspondence: jenny9121989@gmail.com; Tel: +886-906-942-769

Received: 10 June 2018; Accepted: 5 July 2018; Published: 7 July 2018

Abstract: Many research studies have applied the multi-criteria decision making (MCDM) approach to various fields of science and engineering, and this trend has been increasing for many years. One of the fields that the MCDM model has been employed is for location selection, yet very few studies consider this problem under fuzzy environmental conditions. In this research, the authors propose an MCDM approach, including fuzzy analysis network process (FANP), and the technique for order of preference by similarity to ideal solution (TOPSIS), for solid waste to energy plant location selection in Vietnam. In the first stage of this research, the ANP approach with fuzzy logic is applied to determine the weight of criteria. In the FANP model, the value of the criteria is provided by the experts so that the disadvantages of this model are that the input data, expressed in linguistic terms, depends on the experience of experts, and thus involves subjectivity. This is a reason why TOPSIS model was proposed for ranking alternatives in the final stage. Analysis shows that Hau Giang (Decision Making Unit 8 (DMU 8)) is the best location for building solid waste to energy plant, because it has the shortest geometric distance from the positive ideal solution (PIS) and the longest geometric distance from the negative ideal solution (NIS). The contribution of this research is a proposed hybrid FANP and TOPSIS approach for solid waste to energy plant location selection in Vietnam under fuzzy environmental conditions. This paper is also part of an evolution of a new hybrid model that is flexible and practical for decision makers. In addition, the research also provides a special, useful guideline in solid waste to energy plant location selection in many countries, as well as provides a guideline for location selection in other industries. Thus, this research makes significant contributions on both academic and practical fronts.

Keywords: renewable energy; environment; solid waste to energy plant; FANP; TOPSIS; fuzzy logic; MCDM

---

1. Introduction

Presently, the main energy resources in Vietnam are hydropower and thermal power. Vietnam's fossil fuel resources have been exhausted due to overexploitation. Moreover, the plan to develop atomic energy has been halted. Vietnam has been importing materials and primary energy for electricity production, and the development of renewable energy will aid Vietnam to diversify and establish self-reliant power supply and environmental protection. Therefore, the government of Vietnam encourages the development of renewable energy, intelligent grid technology, and new energy technologies, as well as studies on how to exploit renewable energy sources. According to Vietnam's Ministry of Natural Resources and Environment, the total domestic solid waste is approximately 12.8 million tons/year. In Vietnam, solid waste is mainly handled by landfilling and burning, which increases the cost of waste treatments and affects the environment. However, environmental pollution caused by waste incineration may raise concerns over the resident population coming down with the Not-In-My-Backyard (NIMBY) syndrome [1]. Therefore, the construction of solid waste to energy plant is needed in Vietnam.

The Resource Conservation and Recovery Act (RCRA) states that "solid waste" refers to any garbage or refuse, sludge from a wastewater treatment plant, water supply treatment plant, or air pollution control facility and other discarded material, resulting from industrial, commercial, mining, and agricultural operations, and from community activities. Nearly everything we do leaves behind some kind of waste. It is important to note that the definition of solid waste is not limited to waste that is physically solid. Many solid wastes are liquid, semi-solid, or contain gaseous material [2]. Hence, solid waste is one of the main causes of environmental pollution in developing countries [3–5]. There are many technologies for gaseous emission treatment [6,7].

A solid waste to energy plant is a waste management facility that combusts wastes to produce electricity. This type of power plant is sometimes referred to as a trash-to-energy, municipal waste incineration, energy recovery, or resource recovery plant [8]. The advantages of solid waste to energy is the use of closed process, energy production, reduction of greenhouse gases, and saving of land, and it being, essentially, a renewable energy [9].

Solid waste to energy technology is widely used in developing countries. The typical range of net electrical energy that can be produced is approximately 500–600 kWh per ton of waste incinerated.

Many research studies have applied the multi-criteria decision making (MCDM) approaches to various fields of science and engineering, a trend that has been increasing for many years. One of the fields that the MCDM model has been employed is in the location selection problem. Mahmoud A. Hassaan [10] proposed a geographic information systems (GIS) approach for siting a municipal solid waste incineration power plant in Egypt. Tavares et al. [11] used the analytical hierarchy process (AHP) and GIS for siting of an incineration plant for municipal solid waste. H. Y. Yap and J. D. Nixon [12] proposed a multi-criteria analysis of alternatives for energy recovery from municipal solid waste in India and the United Kingdom. Amy H. I. Lee et al. [13] presented a hybrid FAHP – Assurance Region (AR) – Data Envelopment Analysis (DEA) to assess the efficiencies of PV solar plant site candidates. Chia-Nan Wang et al. [14] proposed a hybrid model including DEA–FAHP–TOPSIS for solar power plant location selection in Vietnam. P. Aragonés-Beltrán et al. [15] used ANP for selection of photovoltaic (PV) solar power plant investment projects.

Ali et al. [16] proposed a hybrid GIS and MCDM approach to define the best place for wind farm location. Suh and Brownson [17] proposed GIS and AHP approaches to select PV solar plant sites. Noorollahi et al. [18] combined GIS and fuzzy AHP for land analyses in solar farm site. Audrius Čereška et al. [19] used MCDM for analyzing of steel wire rope diagnostic data.

A generic process of MCDM is shown in Figure 1.

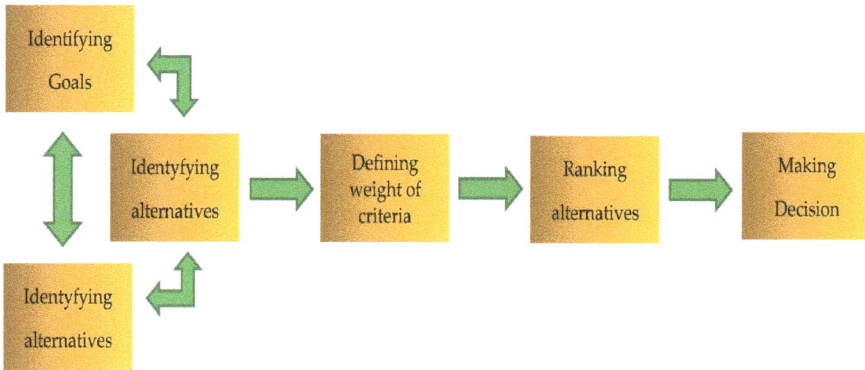

Figure 1. A generic process of multi-criteria decision making (MCDM).

Decision makers should follow a four-stage procedure when making location selection. These steps are as follows and as shown in Figure 2 [20,21].

Step 1: Identify location attributes. In this stage, all attributes that affect the organization will be considered.

Step 2: Identify alternatives. Once decision makers know what attributes affect the organization, they can define alternatives that satisfy the selected attributes.

Step 3: Evaluate alternatives. After a set of alternatives are built, all potential alternatives will be evaluated and ranked by quantitative or qualitative methods.

Step 4: Make a decision. The best alternative will be defined base on the highest-ranking score in the final step.

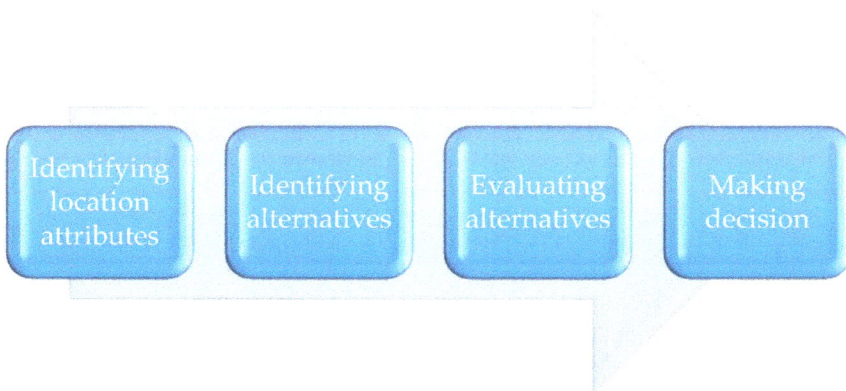

Figure 2. General methodology of location selection problem.

Gan et al. [22] used integrated Triangular Fuzzy Numbers (TFN)-AHP-DEA approaches for renewable energy projects analyzed economic feasibility. Liu et al. [23] proposed a hybrid MCDM model by combining data envelopment analysis (DEA) and the Malmquist model for evaluating the total factor energy efficiency. Asad Asadzadel et al. [24] proposed TOPSIS model for assessing site selection of new towns. Maria Rashidi et al. [25] presented the developed decision support system for asset management of steel bridges within acceptable limits of safety, functionality, and sustainability by using AHP model.

This research proposes a MCDM model including FANP and TOPSIS for solid waste to energy plant location selection in Vietnam under uncertain environmental conditions. In the first stage of this research, FANP is applied to determine the weight of criteria. The steps for implementing the FANP model are as follows:

Step 1: Building FANP model.

Step 2: Set up pair comparison matrix.

Step 3: Calculate maximum individual value.

Step 4: Check consistency. Calculate the vector of the matrix.

Step 5: Form the super matrix.

In the FANP, AHP, or FAHP model, the value of the criteria is provided by the experts, so the disadvantage of the model is that the input data, expressed in linguistic terms, depends on the experience of experts, and thus involves subjectivity. This is a reason why the TOPSIS model is proposed for ranking alternatives in the final stage. The TOPSIS model was employed to rank potential sites. Optimal options have the shortest geometric distance from the positive ideal solution (PIS) and the longest geometric distance from the negative ideal solution (NIS). The advantage of this method is its simplicity and ability to yield an indisputable preference order [26,27]. There are six steps in TOPSIS process, as follows:

Step 1: Determining the normalized decision matrix.

Step 2: Calculating the weight normalized value.

Step 3: Calculating the PIS (D^+) and PIS (D^-).

Step 4: Determining a distance of the PIS (Q_x^+).

Step 5: Determining the relationship proximal.

Step 6: Determining the best option with the maximum value of C_a.

The remainder of the article provides background materials to assist in developing the MCDM model. Then, a hybrid FANP–TOPSIS approach is presented to select the best location for building of a solid waste to energy plant from among eight potential locations in Vietnam. The results, discussion, and the contributions are presented at the end of the paper.

2. Material and Methodology

2.1. Research Development

In this research, we present a hybrid fuzzy ANP and TOPSIS approaches to select the best site for building a solid waste to energy plant in Vietnam. There are three steps in this study, as shown in Figure 3:

Step 1: Identify criteria. In this stage, the criteria for selecting the best location for building solid waste to energy plant will be defined by expert's interviews and literature review from others' research. All of the attributes are shown in Figure 3.

Step 2: Implement fuzzy analysis network process model (FANP). The FANP model is the most effective tool for defining the weight of the criteria. The weight of criteria will be calculated based on economic factors, technical requirements factors, environment factors, and social factors. The weight of all criteria will be used in TOPSIS model.

Step 3: Apply the technique for order of preference by similarity to the ideal solution (TOPSIS) model. The TOPSIS approach is used to rank potential locations. The optimal alternatives will have the shortest geometric distance from the positive ideal solution (PIS) and the longest geometric distance from the negative ideal solution (NIS).

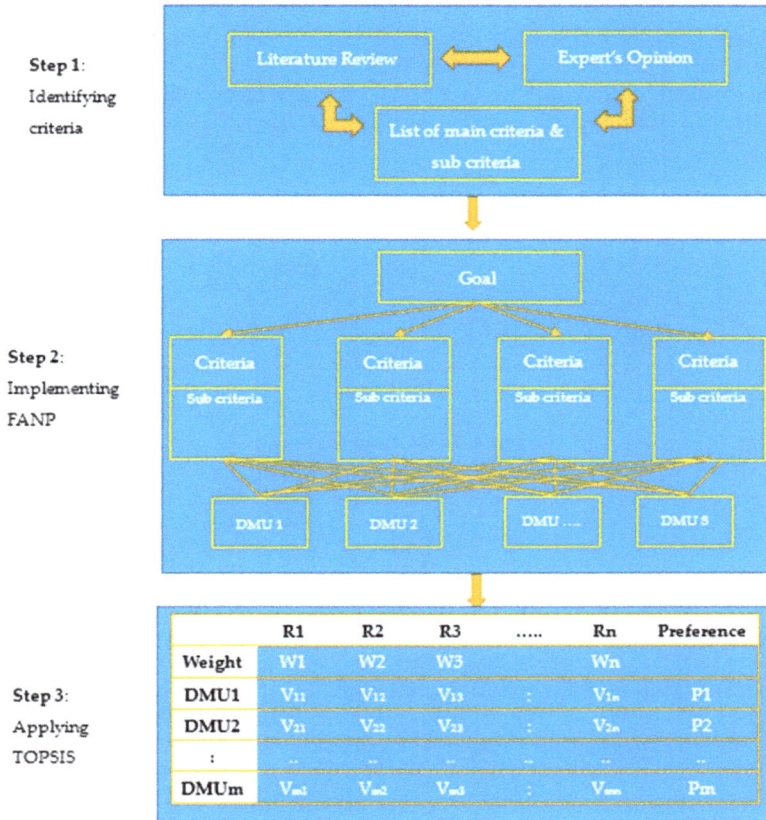

Figure 3. Research methodologies.

2.2. Methodology

2.2.1. Fuzzy Analytic Network Process (FANP)

(1) Analytic Network Process (ANP)

The ANP model was developed to overcome the limitations of AHP, taking into account the hierarchy and interactions between the selection criteria. Moreover, the dynamics and complexity of the majority of decision making environments make ANP an effective tool in addressing such situations. According to Sarkis, ANP is a powerful decision making technique for analyzing key issues related to green supply chain management and environmental business operations, but both AHP and ANP require a resolution. Strategic planning ANP is a combination of two parts [28,29]:

- Primary and secondary criteria that control the interactions.
- The grid effect of elements and clusters.

The goal of the ANP is to use qualitative methods to rank qualitative decisions and to select one or more alternatives that meet the criteria.

(2) Fuzzy Analytic Network Process

Zadeh (1965) [30] introduced the theory to deal with uncertainty due to imprecision and vagueness. There are many forms of fuzzy numbers, such as trapezoidal fuzzy numbers, triangular fuzzy numbers, etc. However triangular fuzzy numbers are often used by efficiency and ease of use. In this study, vendor evaluations were made based on triangular fuzzy numbers, so this fuzzy number was studied [31–35]. The fuzzy triangular numbers are shown in Figure 4.

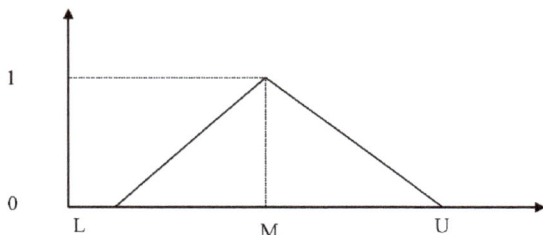

Figure 4. Fuzzy triangular number.

$$\mu\left(x\right) = \begin{cases} \frac{x-L}{M-L}, & L \leq x \leq M \\ \frac{U-x}{U-M}, & M \leq x \leq U \\ 0, & otherwise \end{cases} \tag{1}$$

If $L = M = U$, the fuzzy number A becomes real number. Thus, real numbers are special cases of fuzzy numbers [36].

Fuzzy set theory is a perfect means for modeling uncertainty (or imprecision) arising from mental phenomena, which are neither random nor stochastic. This attitude towards the uncertainty of human behavior led to the study of a relatively new decision analysis field: fuzzy analytical network process [37].

The procedure for implementing the FANP method is as follows:

Step 1: Building FANP model.

Construction of the FANP model structure. Set the hierarchy and define the relationship between the criteria as well as the supplier.

Step 2: Set up pair comparison matrix.

A pairwise comparison of fuzzy numbers is used to perform a pairwise comparison between criteria. The pair comparison matrix is presented as follows:

$$\widetilde{A^k} = \begin{bmatrix} \widetilde{a_{11}^k} & \widetilde{a_{12}^k} & \cdots & \widetilde{a_{1n}^k} \\ \widetilde{a_{21}^k} & \widetilde{a_{22}^k} & \cdots & \widetilde{a_{2n}^k} \\ \cdots & \cdots & \cdots & \cdots \\ \widetilde{a_{n1}^k} & \widetilde{a_{n2}^k} & \cdots & \widetilde{a_{nn}^k} \end{bmatrix}, \tag{2}$$

where

$\widetilde{A^k}$ is called a pairwise comparative matrix of fuzzy elements,

$\widetilde{a_{nn}^k}$ is the triangular fuzzy mean value when comparing the pair of priorities between the elements.

Converting fuzzy numbers to real numbers, triangular fuzzy trigonometric methods are presented as follows [38]:

$$t_{\alpha,\beta}\left(\overline{\alpha}_{ij}\right) = \left[\beta f_\alpha\left(L_{ij}\right) + (1-\beta)f_\alpha\left(U_{ij}\right)\right]; \\ 0 \leq \beta \leq 1, 0 \leq \alpha \leq 1 \tag{3}$$

where

$$f_\alpha\left(L_{ij}\right) = \left(M_{ij} - L_{ij}\right)\alpha + L_{ij} \tag{4}$$

$$f_\alpha(U_{ij}) = U_{ij} - (U_{ij} - M_{ij})\alpha. \tag{5}$$

When matching the diagonal matrix, we have

$$t_{\alpha,\beta}(\bar{\alpha}_{ij}) = \frac{1}{t_{\alpha,\beta}(\bar{\alpha}_{ij})}$$
$$0 \le \beta \le 1, 0 \le \alpha \le 1, i > j. \tag{6}$$

After comparing the fuzzy pairwise comparison matrix, we obtain a matrix that compares the real numbers. This comparison is made between pairs of indicators and is combined into a matrix of n lines and n columns (n: is the number of indicators). Element shows the importance of the indicator i versus the column criteria.

$$A = (m_{ij})_{n \times n} = \begin{bmatrix} 1 & m_{12} & \cdots & m_{1n} \\ m_{21} & 1 & \cdots & m_{2n} \\ \vdots & \vdots & \vdots & \vdots \\ m_{n1} & m_{n2} & \cdots & 1 \end{bmatrix} \tag{7}$$

To evaluate the priority in the FANP model, we use the scale presented in the Table 1 as follows.

Table 1. Fuzzy conversion scale [39].

Intensity of Fuzzy Scale	Linguistic Variables for Relative Weights of Criteria
$\tilde{1} = (1,1,1)$	Equally important
$\tilde{3} = (2,3,4)$	Moderately important
$\tilde{5} = (4,5,6)$	Strongly important
$\tilde{7} = (7,8,9)$	Very strongly important
$\tilde{9} = (9,9,9)$	Extremely strongly important
	Intermediate values between two adjacent judgments;
	$\tilde{2} = (1,2,3);$
$\tilde{2}, \tilde{4}, \tilde{6}, \tilde{8}$	$\tilde{4} = (3,4,5);$
	$\tilde{6} = (5,6,7);$
	$\tilde{8} = (7,8,9);$

After the fuzzy decomposition, in the form of the real comparison matrix. The scale that was suggested by Saaty for AHP and ANP can be used. These scales are shown in Table 2 [40].

Table 2. Priority rating scale.

Priority Level	Number
Equally preferred	1
Moderately preferred	3
Strongly preferred	5
Very strongly preferred	7
Extremely preferred	9
Intermediate judgment values	2, 4, 6, 8

Step 3: Calculate maximum individual value.

To calculate the maximum specific value for the indicator. In particular, the most widely used is lambda max (max) by Saaty's proposition [40]:

$$|A - \lambda_{\max} \cdot I| = 0, \tag{8}$$

where

$\lambda_{\max}$: the maximum value of the matrix.

A: Comparative matrix of pairs of elements.

I: unit matrix of the same level with matrix *A*.

Step 4: Check consistency. Calculate the vector of the matrix.

After calculating maximum individual value, Saaty [40] can use the consistency ratio (CR). This ratio compares the degree of consistency with the (random) objectivity of the data:

$$CR = \frac{CI}{RI}, \tag{9}$$

where

CI: consistency index,

RI: random index.

If $CR \leq 0.1$ is satisfactory, otherwise if $CR \geq 0.1$ then we must conduct a reevaluation of the pair comparison matrix.

$$CI = \frac{\lambda_{max} - n}{n - 1}, \tag{10}$$

where

λ_{max} is the maximum value of the matrix,

n is the number of indicators.

For each n-level comparison matrix, Saaty [40] tested the creation of random matrices and calculated the RI (random index) corresponding to the number of indicators as shown in the Table 3.

Table 3. Randomized index values corresponding to indicators.

N	1	2	3	4	5	6	7	8	9	10
R	0	0	0.52	0.90	1.12	1.24	1.32	1.41	1.45	1.49

Step 5: Form the super matrix

After completing the above steps, a super matrix is formed in Table 4 as follows:

Table 4. Super matrix.

0	U_{12}	0
U_{21}	U_{22}	U_{23}
0	0	0

where

- U_{12} is a matrix formed from the matrix's own vector when comparing the choices for each criterion.
- U_{21} is a matrix formed from its own vector when comparing the criteria for each choice.
- U_{22} is a matrix formed from its own vector when comparing the interaction effect between the criteria.
- U_{23} is a matrix formed from the matrix's own vector when comparing the criteria with each other.

2.2.2. TOPSIS Model

TOPSIS model was proposed Hwang and Yoon [41]. The main concept of TOPSIS is that the best options should have the shortest geometric distance from the positive ideal solution (PIS) and the longest geometric distance from the negative ideal solution (NIS) [42]. There are m alternatives and n criteria, and the result of TOPSIS model shows the score of each option [43]. There are six steps of TOPSIS process, as below:

Step 1: Determining the normalized decision matrix, raw values (a_{ij}) are transferred to normalized values (n_{ij}) by

$$k_{xy} = \frac{b_{xy}}{\sqrt{\sum_x^e b_{xy}^2}}, \ x = 1, \ldots, e; y = 1, \ldots, k. \tag{11}$$

Step 2: Calculating the weight normalized value (v_{ij}), by

$$f_{xy} = P_{xy} h_{xy}, \ x = 1, \ldots, e; y = 1, \ldots, k. \tag{12}$$

where Pj is the weight of the x^{tk} criterion and $\sum_{y=1}^k P_p = 1$.

Step 3: Calculating the PIS (D^+) and PIS (D^-), where l_x^+ indicate the maximum values of f_{xy} and f_x^- indicates the minimum value f_{xy}:

$$D^+ = \{f_1^+, \ldots, f_h^+\} = \left\{ \left(\max_y f_{xy} \middle| x \in A \right), \left(\min_y f_{xy} \middle| x \in A \right) \right\}, \tag{13}$$

$$D^- = \{f_1^-, \ldots, f_n^-\} = \left\{ \left(\min_y f_{xy} \middle| x \in A \right), \left(\max_y f_{xy} \middle| y \in B \right) \right\}, \tag{14}$$

Step 4: Determining a distance of the PIS (Q_x^+) separately by

$$Q_x^+ = \left\{ \sum_{y=1}^k \left(f_{xy} - f_y^+ \right)^2 \right\}^{\frac{1}{2}}, \ x = 1, \ldots, e \tag{15}$$

Similarly, the separation from the NIS (Q_i^-) is given as

$$Q_x^- = \left\{ \sum_{y=1}^k \left(f_{xy} - f_y^- \right)^2 \right\}^{\frac{1}{2}}, \ x = 1, \ldots, e. \tag{16}$$

Step 5: Determining the relationship proximal to the problem-solving model:

$$C_x = \frac{Q_x^-}{Q_x^+ + Q_x^-}, \ x = 1, \ldots, e. \tag{17}$$

Step 6: Determining the best option with the maximum value of C_a.

3. Case Study

The impact of the industrialization and modernization has resulted in ever-increasing solid waste. Vietnam is among five countries in the world with the most solid waste discarded into the ocean. Solid waste in Vietnam includes garbage, construction debris, commercial refuse, hospital waste, etc. A summary of solid waste in Vietnam is shown in Figure 5.

In this research, the authors analyzed eight potential locations for a solid waste to energy plant in Vietnam. The information about eight potential locations is shown in Table 5.

Figure 5. Solid waste in Vietnam.

Table 5. Potential locations for building a solid waste to energy plant.

No	Potential Location	Decision Making Unit (DMU)
1	Long An	DMU 1
2	Tien Giang	DMU 2
3	Can Tho	DMU 3
4	Ben Tre	DMU 4
5	An Giang	DMU 5
6	Vinh Long	DMU 6
7	Dong Thap	DMU 7
8	Hau Giang	DMU 8

The geographical location of eight of potential locations (DMU) are shown in Figure 6.

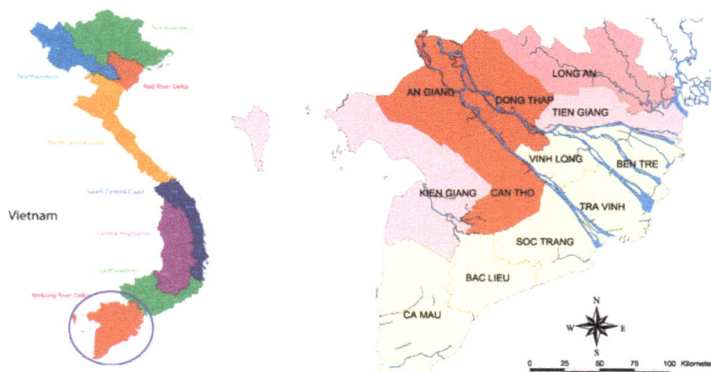

Figure 6. The geographical location of DMU. (Source: UN Development Programme.)

Based on the results of Expert's interviews and literature review, the hierarchical structures of the FANP was contracture, as shown in Figure 7.

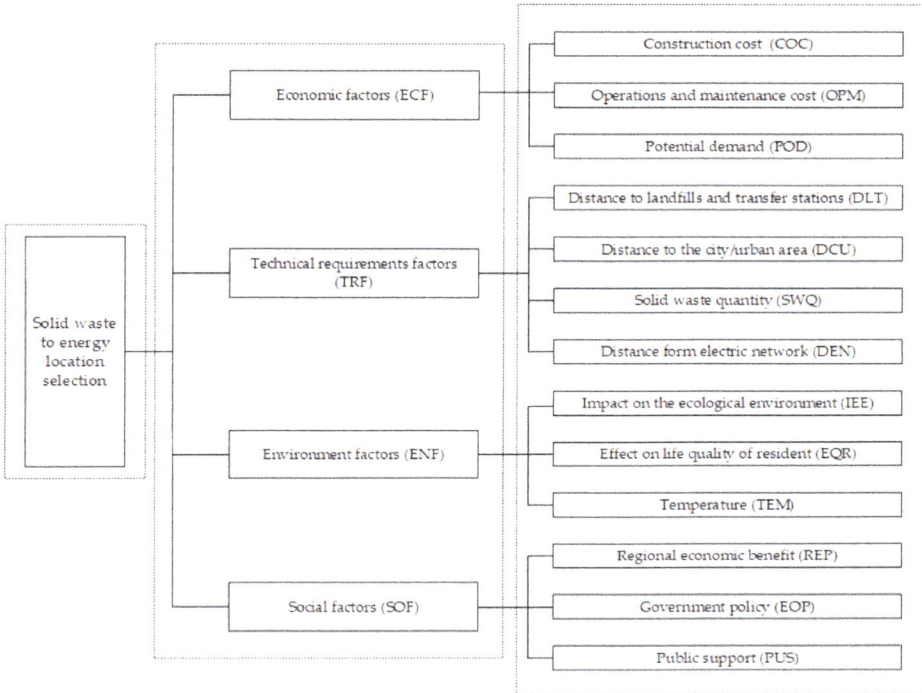

Figure 7. The hierarchical structures of the fuzzy analysis network process (FANP).

A fuzzy comparison matrix for all criteria from FANP model is shown in Table 6.

Table 6. Fuzzy comparison matrices for criteria.

Criteria	EFC	ENF	SOF	TRF
ECF	(1,1,1)	(2,3,4)	(1,2,3)	(3,4,5)
ENF	(1/4,1/3,1/2)	(1,1,1)	(1/4,1/3,1/2)	(1,1,1)
SOF	(1/3,1/2,1)	(2,3,4)	(1,1,1)	(4,5,6)
TRF	(1/5,1/4,1/3)	(1,1,1)	(1/6,1/5,1/4)	(1,1,1)

We used the triangular fuzzy number to convert the fuzzy numbers to real numbers. During the defuzzification, the authors obtain the coefficients $\alpha = 0.5$ and $\beta = 0.5$.

$$g_{0.5,0.5}(\overline{a_{ECF,ENF}}) = [(0.5 \times 2.5) + (1 - 0.5) \times 3.5] = 3$$

$$f_{0.5}(L_{EFC,ENF}) = (3 - 2) \times 0.5 + 2 = 2.5$$

$$f_{0.5}(U_{EFC,ENF}) = 4 - (4 - 3) \times 0.5 = 3.5$$

$$g_{0.5,0.5}(\overline{a_{ENF,EFC}}) = 1/3$$

The remaining calculations are similar to the above calculation, as well as the fuzzy number priority point. The real number priority when comparing the main criteria pairs is presented in Table 7.

Table 7. Real number priority.

Criteria	ECF	ENF	SOF	TRF
ECF	1	3	2	4
ENF	1/3	1	1/3	1
SOF	1/2	3	1	5
TRF	1/4	1	1/5	1

To calculate the maximum individual value as follows:

$$AM1 = (1 \times 3 \times 2 \times 4)^{1/4} = 2.21$$

$$AM2 = (1/3 \times 1 \times 1/3 \times 1)^{1/4} = 0.58$$

$$AM3 = (1/2 \times 3 \times 1 \times 5)^{1/4} = 1.64$$

$$AM4 = (1/4 \times 1 \times 1/5 \times 1)^{1/4} = 0.47$$

$$\sum AM = AM1 + AM2 + AM3 + AM4 + AM5 = 4.9$$

$$\omega_1 = \frac{2.21}{4.9} = 0.45$$

$$\omega_2 = \frac{0.58}{4.9} = 0.12$$

$$\omega_3 = \frac{1.64}{4.9} = 0.33$$

$$\omega_4 = \frac{0.47}{4.9} = 0.1$$

$$\begin{bmatrix} 1 & 3 & 2 & 4 \\ 1/3 & 1 & 1/3 & 1 \\ 1/2 & 3 & 1 & 5 \\ 1/4 & 1 & 1/5 & 1 \end{bmatrix} \times \begin{bmatrix} 0.45 \\ 0.12 \\ 0.33 \\ 0.1 \end{bmatrix} = \begin{bmatrix} 1.87 \\ 0.48 \\ 1.42 \\ 0.4 \end{bmatrix}$$

$$\begin{bmatrix} 1.87 \\ 0.48 \\ 1.42 \\ 0.4 \end{bmatrix} / \begin{bmatrix} 0.45 \\ 0.12 \\ 0.33 \\ 0.1 \end{bmatrix} = \begin{bmatrix} 4.16 \\ 4 \\ 4.30 \\ 4 \end{bmatrix}$$

With the number of criteria is 4, we get $n = 4$, λ_{max} and CI are calculated as follows:

$$\lambda_{max} = \frac{4.16 + 4 + 4.30 + 4}{4} = 4.12$$

$$CI = \frac{\lambda_{max} - n}{n - 1} = \frac{4.12 - 4}{4 - 1} = 0.04$$

For CR, with $n = 4$ we get $RI = 0.9$.

$$CR = \frac{CI}{RI} = \frac{0.04}{0.9} = 0.044$$

We have $CR = 0.09598 \leq 0.1$, so the pairwise comparison data is consistent and need not to be re-evaluated. The results of the pair comparison between the main criteria are presented in Table 8.

Table 8. Fuzzy comparison matrices for criteria.

Criteria	ECF	ENF	SOF	TRF	Weight
ECF	(1,1,1)	(2,3,4)	(1,2,3)	(3,4,5)	0.45
ENF	(1/4,1/3,1/2)	(1,1,1)	(1/4,1/3,1/2)	(1,1,1)	0.12
SOF	(1/3,1/2,1)	(2,3,4)	(1,1,1)	(4,5,6)	0.33
TRF	(1/5,1/4,1/3)	(1,1,1)	(1/6,1/5,1/4)	(1,1,1)	0.1
		Total			1
		$CR = 0.044$			

All the remaining calculation are shown in Appendix A.
The weight of all subcriteria calculated in FANP model are shown in Table 9.

Table 9. The weight of 13 subcriteria.

No	Subcriteria	Weight
1	COC	0.09995
2	DCU	0.07521
3	DEN	0.0473
4	DLT	0.03181
5	EOP	0.03948
6	EQR	0.15249
7	IEE	0.16369
8	OPM	0.10082
9	POD	0.10397
10	PUS	0.03423
11	REP	0.04143
12	SWQ	0.06067
13	TEM	0.04896

TOPSIS model is then applied for ranking all the potential locations. The normalized weight matrix (TOPSIS) are shown in Table 10.

Table 10. Normalized weight matrix (TOPSIS).

Subcriteria	DMUs							
	DMU 1	DMU 2	DMU 3	DMU 4	DMU 5	DMU 6	DMU 7	DMU 8
COC	0.0260	0.0390	0.0520	0.0260	0.0390	0.0260	0.0390	0.0260
OPM	0.0472	0.0236	0.0236	0.0472	0.0472	0.0354	0.0236	0.0236
POD	0.0466	0.0362	0.0310	0.0414	0.0259	0.0414	0.0310	0.0362
DLT	0.0106	0.0106	0.0106	0.0106	0.0141	0.0071	0.0106	0.0141
DCU	0.0272	0.0272	0.0272	0.0272	0.0272	0.0362	0.0181	0.0181
SWQ	0.0182	0.0243	0.0213	0.0243	0.0243	0.0182	0.0213	0.0182
DEN	0.0168	0.0168	0.0112	0.0225	0.0112	0.0225	0.0112	0.0168
IEE	0.0452	0.0226	0.0339	0.0565	0.0565	0.0565	0.0565	0.1017
EQR	0.0490	0.0490	0.0490	0.0588	0.0294	0.0392	0.0490	0.0882
TEM	0.0183	0.0138	0.0183	0.0160	0.0206	0.0160	0.0160	0.0183
REP	0.0174	0.0116	0.0135	0.0155	0.0155	0.0135	0.0174	0.0116
EOP	0.0168	0.0168	0.0131	0.0150	0.0112	0.0131	0.0131	0.0112
PUS	0.0124	0.0124	0.0108	0.0093	0.0139	0.0139	0.0124	0.0108

4. Results and Discussion

Solid waste to energy plant location selection has been identified as an important problem which could affect to the economic and social characteristics of a society. It can be seen that location selection is complicated, in that decision makers must have broad perspectives concerning qualitative and quantitative factors.

As an empirical study, the authors collect data from 8 potential locations in Vietnam. A hierarchical structure to select the best place was built with four main criteria (including 13 subcriteria). Completion of a questionnaire for analyzing in FANP model were done by expert opinion and literature reviews from other research. The ANP model was combined with fuzzy logic to define a priority of each potential location. Then, the TOPSIS model is used for ranking location.

In Figure 8, Hau Giang (DMU 8) has the shortest geometric distance from the PIS and the longest geometric distance from the NIS.

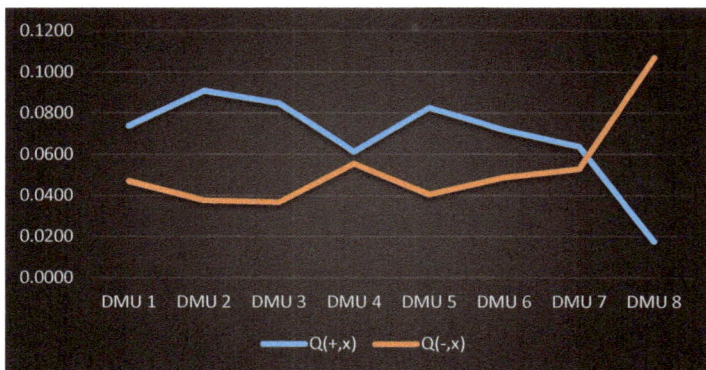

Figure 8. Geometric distance from positive ideal solution (PIS) and negative ideal solution (NIS).

The results of TOPSIS model are shown in Figure 9; based on the final performance score C_x, the final locations ranking list are DMU 8, DMU 4, DMU 7, DMU 6, DMU 1, DMU 5, DMU 3, and DMU 2. The results show that DMU 8 (Hau Giang) is the best location for building a solid waste to energy plant in Vietnam.

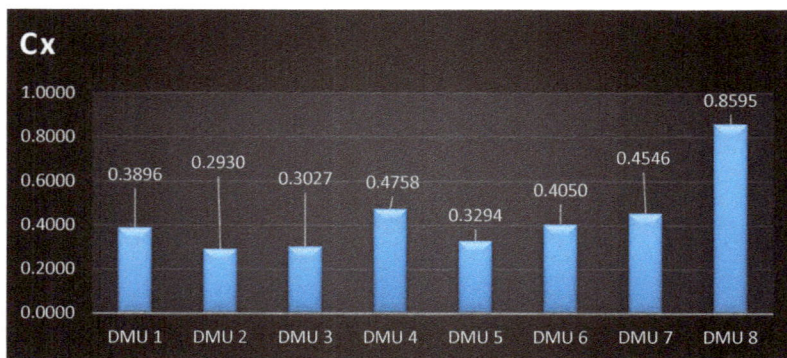

Figure 9. Final performance in technique for order of preference by similarity to ideal solution (TOPSIS) model.

5. Conclusions

Renewable energy plant location selection requires involvement of different decision makers who must evaluate both qualitative and quantitative factors. The fact that economic factors, technical requirements factors, environment factors, and social factors for solid waste to energy plant location selection are considered makes the process more complex. Many studies have applied the MCDM approach to various fields of science and engineering, and this trend has been increasing for many years. One of the fields that the MCDM model has been employed is for location selection, yet very few studies consider this problem under fuzzy environmental conditions. Besides, there is no research that applies the MCDM model for solid waste to energy plant location selection in Vietnam. This is a reason why we proposed a hybrid fuzzy analysis network process (FANP) and the technique for order of preference by similarity to ideal solution (TOPSIS) for solid waste to energy plant location selection in Vietnam.

In the first stage of this research, FANP is applied to determine the weight of criteria. In the FANP model, the value of the criteria is given by the experts, so the disadvantage is that the input data, expressed in linguistic terms, depends on the experience of experts, and is thus subjective. This is a reason why the TOPSIS model was proposed for ranking alternatives in the final stage. The best place for building solid waste to energy plant was concluded to be Hau Giang (DMU 8), because it has the shortest geometric distance from the positive ideal solution (PIS) and the longest geometric distance from the negative ideal solution (NIS).

The contribution of this research is a proposed hybrid FANP and TOPSIS for renewable energy plant location selection in Vietnam under fuzzy environment conditions. Building solid waste to energy plant brings many economic and environmental benefits. Building solid waste to energy plant also creates a new source of renewable energy. This paper also part of the evolution of a new approach that is flexible and practicable to the decision maker. This research provides a useful guideline for solid waste to energy plant location selection in many countries, as well as a guideline for location selection in other industries. Thus, this research has great contributions on academic and practical fronts.

In future research, this hybrid model can be employed to many different countries. In addition, different methods, such as data envelopment analysis (DEA) or the preference ranking organization method for enrichment of evaluations (PROMETHEE), etc., could also be combined for evaluating and selecting locations.

Author Contributions: In this research, C.-N.W. provided the research ideas, designed the theoretical verifications, and reviewed manuscript. V.T.N. contributed the research ideas, designed the frameworks, analyzed the data, and wrote the manuscript. D.H.D. collected data, write a manuscript, H.T.N.T. wrote and format manuscript.

Funding: The authors appreciate the partly funding supported by MOST 107-2622-E-992-012-CC3 from Ministry of Sciences and Technology, and support from National Kaohsiung University of Science and Technology in Taiwan.

Conflicts of Interest: The authors declare no conflict of interest

Appendix A

Table A1. Comparison matrix for ECF.

Criteria	ENF	SOF	TRF	Weight
ENF	(1,1,1)	(3,4,5)	(2,3,4)	0.625013
SOF	(1/5,1/4,1/3)	(1,1,1)	(1/3,1/2,1)	0.1365
TRF	(1/4,1/3,1/2)	(1,2,3)	(1,1,1)	0.238487
Total				1
CR = 0.01759				

Table A2. Comparison matrix for ENF.

Criteria	ECF	SOF	TRF	Weight
ECF	(1,1,1)	(2,3,4)	(1,2,3)	0.527836
SOF	(1/4,1/3,1/2)	(1,1,1)	(1/4,1/3,1/2)	0.139648
TRF	(1/3,1/2,1)	(2,3,4)	(1,1,1)	0.332516
Total				1
$CR = 0.05156$				

Table A3. Comparison matrix for SOF.

Criteria	ECF	ENF	TRF	Weight
ECF	(1,1,1)	(3,4,5)	(1,2,3)	0.58417
ENF	(1/5,1/4,1/3)	(1,1,1)	(1,1,1)	0.184002
TRF	(1/3,1/2,1)	(1,1,1)	(1,1,1)	0.231828
Total				1
$CR = 0.05156$				

Table A4. Comparison matrix for TRF.

Criteria	ECF	ENF	SOF	Weight
ECF	(1,1,1)	(2,3,4)	(4,5,6)	0.636986
ENF	(1/4,1/3,1/2)	(1,1,1)	(2,3,4)	0.258285
SOF	(1/6,1/5,1/4)	(1/4,1/3,1/2)	(1,1,1)	0.104729
Total				1
$CR = 0.03703$				

Table A5. Comparison matrix for EFC based on subcriteria.

Subcriteria	COC	OPM	POD	Weight
COC	(1,1,1)	(2,3,4)	(2,3,4)	0.593634
OPM	(1/4,1/3,1/2)	(1,1,1)	(1,2,3)	0.249311
POD	(1/4,1/3,1/2)	(1/3,1/2,1)	(1,1,1)	0.157056
Total				1
$CR = 0.05156$				

Table A6. Comparison matrix for ENF based on sub-criteria.

Subcriteria	EQR	IEE	TEM	Weight
EQR	(1,1,1)	(2,3,4)	(1,2,3)	0.549946
IEE	(1/4,1/3,1/2)	(1,1,1)	(1,1,1)	0.209843
TEM	(1/3,1/2,1)	(1,1,1)	(1,1,1)	0.240211
Total				1
$CR = 0.01759$				

Table A7. Comparison matrix for SOF based on subcriteria.

Subcriteria	EOP	PUS	REP	Weight
EOP	(1,1,1)	(1/3,1/2,1)	(2,3,4)	0.319618
PUS	(1,2,3)	(1,1,1)	(3,4,5)	0.558425
REP	(1/4,1/3,1/2)	(1/5,1/4,1/3)	(1,1,1)	0.121957
	Total			1
	CR = 0.01759			

Table A8. Comparison matrix for TRF based on subcriteria.

Subcriteria	DCU	DEN	DLT	SWQ	Weight
DCU	(1,1,1)	(2,3,4)	(1/4,1/3,1/2)	(2,3,4)	0.280225
DEN	(1/4,1/3,1/2)	(1,1,1)	(1/5,1/4,1/3)	(1/3,1/2,1)	0.090176
DLT	(2,3,4)	(3,4,5)	(1,1,1)	(1,2,3)	0.471381
SWQ	(1/4,1/3,1/2)	(1,2,3)	(1/3,1/2,1)	(1,1,1)	0.158218
	Total				1
	CR = 0.08237				

Table A9. Comparison matrix for COC.

Subcriteria	OPM	POD	Weight
OPM	(1,1,1)	(3,4,5)	0.8
POD	(1/5,1/4,1/3)	(1,1,1)	0.2
	Total		1
	CR = 0		

Table A10. Comparison matrix for DCU.

Subcriteria	DEN	DLT	SWQ	Weight
DEN	(1,1,1)	(1,2,3)	(1/3,1/2,1)	0.296961294
DLT	(1/3,1/2,1)	(1,1,1)	(1/4,1/3,1/2)	0.163424044
SWQ	(1,2,3)	(2,3,4)	(1,1,1)	0.539614662
	Total			1
	CR = 0.00885			

Table A11. Comparison matrix for DEN.

Subcriteria	DCU	DLT	SWQ	Weight
DCU	(1,1,1)	(3,4,5)	(1,2,3)	0.558424506
DLT	(1/5,1/4,1/3)	(1,1,1)	(1/4,1/3,1/2)	0.121957144
SWQ	(1/3,1/2,1)	(2,3,4)	(1,1,1)	0.319618349
	Total			1
	CR = 0.01759			

Table A12. Comparison matrix for DLT.

Subcriteria	DCU	DEN	SWQ	Weight
DCU	(1,1,1)	(1/4,1/3,1/2)	(1,2,3)	0.249310377
DEN	(2,3,4)	(1,1,1)	(2,3,4)	0.593633926
SWQ	(1/3,1/2,1)	(1/4,1/3,1/2)	(1,1,1)	0.157055696
	Total			1
	$CR = 0.05156$			

Table A13. Comparison matrix for SWQ.

Subcriteria	DCU	DEN	DLT	Weight
DCU	(1,1,1)	(4,5,6)	(3,4,5)	0.673810543
DEN	(1/6,1/5,1/4)	(1,1,1)	(1/4,1/3,1/2)	0.100653892
DLT	(1/5,1/4,1/3)	(2,3,4)	(1,1,1)	0.225535565
	Total			1
	$CR = 0.08247$			

Table A14. Comparison matrix for EOP.

Subcriteria	PUS	REP	Weight
PUS	(1,1,1)	(1/6,1/5,1/4)	0.166667
REP	(4,5,6)	(1,1,1)	0.833333
	Total		1
	$CR = 0$		

Table A15. Comparison matrix for EQR.

Subcriteria	IEE	TEM	Weight
IEE	(1,1,1)	(4,5,6)	0.833333
TEM	(1/6,1/5,1/4)	(1,1,1)	0.166667
	Total		1
	$CR = 0$		

Table A16. Comparison matrix for IEE.

Subcriteria	EQR	TEM	Weight
EQR	(1,1,1)	(5,6,7)	0.857142857
TEM	(1/7,1/6,1/5)	(1,1,1)	0.142857143
	Total		1
	$CR = 0$		

Table A17. Comparison matrix for OPM.

Alternatives	COC	POD	Weight
COC	(1,1,1)	(1/6,1/5,1/4)	0.166666667
POD	(4,5,6)	(1,1,1)	0.833333333
	Total		1
	$CR = 0$		

Table A18. Comparison matrix for POD.

Subcriteria	COC	OPM	Weight
COC	(1,1,1)	(3,4,5)	0.8
OPM	(1/5,1/4,1/3)	(1,1,1)	0.2
	Total		1
	CR = 0		

Table A19. Comparison matrix for PUS.

Subcriteria	EOP	REP	Weight
EOP	(1,1,1)	(2,3,4)	0.75
REP	(1/4,1/3,1/2)	(1,1,1)	0.25
	Total		1
	CR = 0		

Table A20. Comparison matrix for REP.

Subcriteria	EOP	PUS	Weight
EOP	(1,1,1)	(1/3,1/2,1)	0.333333333
PUS	(1,2,3)	(1,1,1)	0.666666667
	Total		1
	CR = 0		

Table A21. Comparison matrix for TEM.

Subcriteria	EQR	IEE	Weight
EQR	(1,1,1)	(1/4,1/3,1/2)	0.249999813
IEE	(2,3,4)	(1,1,1)	0.750000187
	Total		1
	CR = 0		

Table A22. Comparison matrix for COC based on alternatives.

	DMU 8	DMU 1	DMU 2	DMU 3	DMU 4	DMU 5	DMU 6	DMU 7	Weight
DMU 8	(1,1,1)	(1,1,1)	(2,3,4)	(1/4,1/3,1/2)	(3,4,5)	(1,2,3)	(2,3,4)	(3,4,5)	0.1752
DMU 1	(1,1,1)	(1,1,1)	(2,3,4)	(1/3,1/2,1)	(3,4,5)	(2,3,4)	(5,6,7)	(4,5,6)	0.2164
DMU 2	(1/4,1/3,1/2)	(1/4,1/3,1/2)	(1,1,1)	(1/4,1/3,1/2)	(1/4,1/3,1/2)	(1/5,1/4,1/3)	(1/3,1/2,1)	(2,3,4)	0.0506
DMU 3	(2,3,4)	(1,2,3)	(2,3,4)	(1,1,1)	(3,4,5)	(1,2,3)	(2,3,4)	(4,5,6)	0.2617
DMU 4	(1/5,1/4,1/3)	(1/5,1/4,1/3)	(2,3,4)	(1/5,1/4,1/3)	(1,1,1)	(2,3,4)	(1,2,3)	(3,4,5)	0.1067
DMU 5	(1/3,1/2,1)	(1/4,1/3,1/2)	(3,4,5)	(1/3,1/2,1)	(1/4,1/3,1/2)	(1,1,1)	(1,2,3)	(1,2,3)	0.0914
DMU 6	(1/4,1/3,1/2)	(1/7,1/6,1/5)	(1,2,3)	(1/4,1/3,1/2)	(1/3,1/2,1)	(1/3,1/2,1)	(1,1,1)	(4,5,6)	0.0674
DMU 7	(1/5,1/4,1/3)	(1/6,1/5,1/4)	1/3	(1/6,1/5,1/4)	(1/5,1/4,1/3)	(1/3,1/2,1)	(1/6,1/5,1/4)	(1,1,1)	0.0305
Total									1

$CR = 0.09491$

Table A23. Comparison matrix for DCU based on alternatives.

	DMU 8	DMU 1	DMU 2	DMU 3	DMU 4	DMU 5	DMU 6	DMU 7	Weight
DMU 8	(1,1,1)	(3,4,5)	(1,2,3)	(5,6,7)	(4,5,6)	(2,3,4)	(1,2,3)	(3,4,5)	0.2947
DMU 1	(1/5,1/4,1/3)	(1,1,1)	(3,4,5)	(1,2,3)	(4,5,6)	(2,3,4)	(1,1,1)	(2,3,4)	0.1825
DMU 2	(1/3,1/2,1)	(1/5,1/4,1/3)	(1,1,1)	(3,4,5)	(1,2,3)	(1,2,3)	(1/5,1/4,1/3)	(1,2,3)	0.1006
DMU 3	(1/7,1/6,1/5)	(1/3,1/2,1)	(1/5,1/4,1/3)	(1,1,1)	(1/4,1/3,1/2)	(1/3,1/2,1)	(1/5,1/4,1/3)	(1/3,1/2,1)	0.0386
DMU 4	(1/6,1/5,1/4)	(1/6,1/5,1/4)	(1/3,1/2,1)	(2,3,4)	(1,1,1)	(1/4,1/3,1/2)	(1/4,1/3,1/2)	(1/3,1/2,1)	0.0494
DMU 5	(1/5,1/4,1/3)	(1/4,1/3,1/2)	(3,4,5)	(1,2,3)	(2,3,4)	(1,1,1)	(1/5,1/4,1/3)	(1/4,1/3,1/2)	0.0650
DMU 6	(1/4,1/3,1/2)	(1,1,1)	(1/3,1/2,1)	(3,4,5)	(2,3,4)	(3,4,5)	(1,1,1)	(1,2,3)	0.1864
DMU 7	(1/5,1/4,1/3)	(1/4,1/3,1/2)	(1/3,1/2,1)	(1,2,3)	(1,2,3)	(2,3,4)	(1/3,1/2,1)	(1,1,1)	0.0828
Total									1

$CR = 0.08515$

Table A24. Comparison matrix for DEN based on alternatives.

	DMU 8	DMU 1	DMU 2	DMU 3	DMU 4	DMU 5	DMU 6	DMU 7	Weight
DMU 8	(1,1,1)	(3,4,5)	(2,3,4)	(1,2,3)	(4,5,6)	(3,4,5)	(2,3,4)	(1,2,3)	0.2948
DMU 1	(1/5,1/4,1/3)	(1,1,1)	(3,4,5)	(5,6,7)	(4,5,6)	(2,3,4)	(2,3,4)	(1,2,3)	0.2200
DMU 2	(1/4,1/3,1/2)	(1/5,1/4,1/3)	(1,1,1)	(3,4,5)	(2,3,4)	(1,2,3)	(2,3,4)	(1/3,1/2,1)	0.1141
DMU 3	(1/3,1/2,1)	(1/7,1/6,1/5)	(1/5,1/4,1/3)	(1,1,1)	(1,1,1)	(1/3,1/2,1)	(1/4,1/3,1/2)	(1/5,1/4,1/3)	0.0446
DMU 4	(1/6,1/5,1/4)	(1/6,1/5,1/4)	(1/4,1/3,1/2)	(1,1,1)	(1,1,1)	(1/4,1/3,1/2)	(1/4,1/3,1/2)	(1/3,1/2,1)	0.0392
DMU 5	(1/5,1/4,1/3)	(1/4,1/3,1/2)	(1/3,1/2,1)	(1,2,3)	(2,3,4)	(1,1,1)	(1,1,1)	(1/4,1/3,1/2)	0.0687
DMU 6	(1/4,1/3,1/2)	(1/4,1/3,1/2)	(1/4,1/3,1/2)	(2,3,4)	(2,3,4)	(1,1,1)	(1,1,1)	(1/3,1/2,1)	0.0771
DMU 7	(1/3,1/2,1)	(1/3,1/2,1)	(1,2,3)	(3,4,5)	(1,2,3)	(2,3,4)	(1,2,3)	(1,1,1)	0.1415
				Total					1

CR = 0.08044

Table A25. Comparison matrix for DLT based on alternatives.

	DMU 8	DMU 1	DMU 2	DMU 3	DMU 4	DMU 5	DMU 6	DMU 7	Weight
DMU 8	(1,1,1)	(1/4,1/3,1/2)	(1/5,1/4,1/3)	(1/4,1/3,1/2)	(1/4,1/3,1/2)	(1/3,1/2,1)	(1/5,1/4,1/3)	(1/4,1/3,1/2)	0.0398
DMU 1	(2,3,4)	(1,1,1)	(1/3,1/2,1)	(1/4,1/3,1/2)	(1/5,1/4,1/3)	(1/3,1/2,1)	(1/4,1/3,1/2)	(1/5,1/4,1/3)	0.0538
DMU 2	(3,4,5)	(1,2,3)	(1,1,1)	(1/6,1/5,1/4)	(1/3,1/2,1)	(1/4,1/3,1/2)	(1/5,1/4,1/3)	(1/3,1/2,1)	0.0686
DMU 3	(2,3,4)	(2,3,4)	(4,5,6)	(1,1,1)	(2,3,4)	(3,4,5)	(1,2,3)	(3,4,5)	0.2903
DMU 4	(2,3,4)	(3,4,5)	(1,2,3)	(1/4,1/3,1/2)	(1,1,1)	(1,2,3)	(1,1,1)	(2,3,4)	0.1620
DMU 5	(1,2,3)	(1,2,3)	(2,3,4)	(1/5,1/4,1/3)	(1/3,1/2,1)	(1,1,1)	(1/5,1/4,1/3)	(1/4,1/3,1/2)	0.0806
DMU 6	(3,4,5)	(2,3,4)	(3,4,5)	(1/3,1/2,1)	(1,1,1)	(3,4,5)	(1,1,1)	(1,2,3)	0.1862
DMU 7	(2,3,4)	(3,4,5)	(1,2,3)	(1/5,1/4,1/3)	(2,3,4)	(2,3,4)	(1/3,1/2,1)	(1,1,1)	0.1188
				Total					1

CR = 0.08709

Table A26. Comparison matrix for EOP based on alternatives.

	DMU 8	DMU 1	DMU 2	DMU 3	DMU 4	DMU 5	DMU 6	DMU 7	Weight
DMU 8	(1,1,1)	(2,3,4)	(1,2,3)	(1/4,1/3,1/2)	(1/3,1/2,1)	(1,2,3)	(2,3,4)	(3,4,5)	0.1527
DMU 1	(1/4,1/3,1/2)	(1,1,1)	(1,2,3)	(1/3,1/2,1)	(1/5,1/4,1/3)	(1/6,1/5,1/4)	(2,3,4)	(1,2,3)	0.0755
DMU 2	(1/3,1/2,1)	(1/3,1/2,1)	(1,1,1)	(1/4,1/3,1/2)	(1/6,1/5,1/4)	(1/3,1/2,1)	(1,2,3)	(2,3,4)	0.0660
DMU 3	(2,3,4)	(1,2,3)	(2,3,4)	(1,1,1)	(1/5,1/4,1/3)	(1/3,1/2,1)	(2,3,4)	(3,4,5)	0.1610
DMU 4	(1,2,3)	(3,4,5)	(4,5,6)	(3,4,5)	(1,1,1)	(1,2,3)	(2,3,4)	(4,5,6)	0.2891
DMU 5	(1/3,1/2,1)	(4,5,6)	(1,2,3)	(1,2,3)	(1/3,1/2,1)	(1,1,1)	(2,3,4)	(3,4,5)	0.1722
DMU 6	(1/4,1/3,1/2)	(1/4,1/3,1/2)	(1/3,1/2,1)	(1/4,1/3,1/2)	(1/4,1/3,1/2)	(1/4,1/3,1/2)	(1,1,1)	(1,2,3)	0.0495
DMU 7	(1/5,1/4,1/3)	(1/3,1/2,1)	(1/4,1/3,1/2)	(1/5,1/4,1/3)	(1/6,1/5,1/4)	(1/5,1/4,1/3)	(1/3,1/2,1)	(1,1,1)	0.0340
Total									1

$CR = 0.07805$

Table A27. Comparison matrix for EQR based on alternatives.

	DMU 8	DMU 1	DMU 2	DMU 3	DMU 4	DMU 5	DMU 6	DMU 7	Weight
DMU 8	(1,1,1)	(1,2,3)	(3,4,5)	(1,2,3)	(2,3,4)	(1/4,1/3,1/2)	(2,3,4)	(1,2,3)	0.1888
DMU 1	(1/3,1/2,1)	(1,1,1)	(3,4,5)	(2,3,4)	(1,2,3)	(1/4,1/3,1/2)	(1/3,1/2,1)	(2,3,4)	0.1247
DMU 2	(1/5,1/4,1/3)	(1/5,1/4,1/3)	(1,1,1)	(1/3,1/2,1)	(1/4,1/3,1/2)	(1/5,1/4,1/3)	(1/5,1/4,1/3)	(1/3,1/2,1)	0.0362
DMU 3	(1/3,1/2,1)	(1/4,1/3,1/2)	(1,2,3)	(1,1,1)	(1,2,3)	(1/4,1/3,1/2)	(1/5,1/4,1/3)	(1/3,1/2,1)	0.0683
DMU 4	(1/4,1/3,1/2)	(1/3,1/2,1)	(2,3,4)	(1/3,1/2,1)	(1,1,1)	(1/4,1/3,1/2)	(1/4,1/3,1/2)	(2,3,4)	0.0785
DMU 5	(2,3,4)	(2,3,4)	(3,4,5)	(2,3,4)	(2,3,4)	(1,1,1)	(1,2,3)	(4,5,6)	0.2793
DMU 6	(1/4,1/3,1/2)	(1,2,3)	(3,4,5)	(3,4,5)	(2,3,4)	(1/3,1/2,1)	(1,1,1)	(1,2,3)	0.1591
DMU 7	(1/3,1/2,1)	(1/4,1/3,1/2)	(1,2,3)	(1,2,3)	(1/4,1/3,1/2)	(1/6,1/5,1/4)	(1/3,1/2,1)	(1,1,1)	0.0652
Total									1

$CR = 0.07845$

Table A28. Comparison matrix for IEE based on alternatives.

	DMU 8	DMU 1	DMU 2	DMU 3	DMU 4	DMU 5	DMU 6	DMU 7	Weight
DMU 8	(1,1,1)	(1,2,3)	(1,2,3)	(2,3,4)	(1/5,1/4,1/3)	(1,2,3)	(3,4,5)	(1,1,1)	0.1577
DMU 1	(1/3,1/2,1)	(1,1,1)	(2,3,4)	(3,4,5)	(1/3,1/2,1)	(3,4,5)	(2,3,4)	(2,3,4)	0.1843
DMU 2	(1/3,1/2,1)	(1/4,1/3,1/2)	(1,1,1)	(1/3,1/2,1)	(1/5,1/4,1/3)	(1/3,1/2,1)	(1/3,1/2,1)	(1/5,1/4,1/3)	0.0437
DMU 3	(1/4,1/3,1/2)	(1/5,1/4,1/3)	(1,2,3)	(1,1,1)	(1/4,1/3,1/2)	(2,3,4)	(1/3,1/2,1)	(1/4,1/3,1/2)	0.0685
DMU 4	(3,4,5)	(1,2,3)	(3,4,5)	(2,3,4)	(1,1,1)	(4,5,6)	(3,4,5)	(2,3,4)	0.2955
DMU 5	(1/3,1/2,1)	(1/5,1/4,1/3)	(1,2,3)	(1/4,1/3,1/2)	(1/6,1/5,1/4)	(1,1,1)	(1,2,3)	(1/4,1/3,1/2)	0.0592
DMU 6	(1/5,1/4,1/3)	(1/4,1/3,1/2)	(1,2,3)	(1,2,3)	(1/5,1/4,1/3)	(1/3,1/2,1)	(1,1,1)	(1/3,1/2,1)	0.0630
DMU 7	(1,1,1)	(1/4,1/3,1/2)	(3,4,5)	(2,3,4)	(1/4,1/3,1/2)	(2,3,4)	(1,2,3)	(1,1,1)	0.1280
Total									1

CR = 0.0839

Table A29. Comparison matrix for OPM based on alternatives.

	DMU 8	DMU 1	DMU 2	DMU 3	DMU 4	DMU 5	DMU 6	DMU 7	Weight
DMU 8	(1,1,1)	(1,2,3)	(3,4,5)	(1,2,3)	(2,3,4)	(2,3,4)	(1,1,1)	(1,2,3)	0.2218
DMU 1	(1/3,1/2,1)	(1,1,1)	(3,4,5)	(2,3,4)	(1,2,3)	(4,5,6)	(2,3,4)	(1,2,3)	0.2114
DMU 2	(1/5,1/4,1/3)	(1/5,1/4,1/3)	(1,1,1)	(1/4,1/3,1/2)	(1/5,1/4,1/3)	(1/3,1/2,1)	(1/5,1/4,1/3)	(1/4,1/3,1/2)	0.0350
DMU 3	(1/3,1/2,1)	(1/4,1/3,1/2)	(2,3,4)	(1,1,1)	(1/4,1/3,1/2)	(3,4,5)	(1/3,1/2,1)	(1/5,1/4,1/3)	0.0805
DMU 4	(1/4,1/3,1/2)	(1/6,1/5,1/4)	(2,3,4)	(2,3,4)	(1,1,1)	(2,3,4)	(1,2,3)	(1,2,3)	0.1545
DMU 5	(1/4,1/3,1/2)	(1/4,1/3,1/2)	(1,2,3)	(1/5,1/4,1/3)	(1/4,1/3,1/2)	(1,1,1)	(1,1,1)	(1/5,1/4,1/3)	0.0507
DMU 6	(1,1,1)	(1/4,1/3,1/2)	(3,4,5)	(1,2,3)	(1/3,1/2,1)	(1,1,1)	(1,1,1)	(1/3,1/2,1)	0.1017
DMU 7	(1/3,1/2,1)	(1/3,1/2,1)	(2,3,4)	(3,4,5)	(1/3,1/2,1)	(3,4,5)	(1,2,3)	(1,1,1)	0.1445
Total									1

CR = 0.08162

Table A30. Comparison matrix for POD based on alternatives.

	DMU 8	DMU 1	DMU 2	DMU 3	DMU 4	DMU 5	DMU 6	DMU 7	Weight
DMU 8	(1,1,1)	(2,3,4)	(3,4,5)	(1,2,3)	(2,3,4)	(1,2,3)	(1,1,1)	(1/4,1/3,1/2)	0.1646
DMU 1	(1/4,1/3,1/2)	(1,1,1)	(1/3,1/2,1)	(1/5,1/4,1/3)	(1/3,1/2,1)	(1/4,1/3,1/2)	(1/6,1/5,1/4)	(1/5,1/4,1/3)	0.0368
DMU 2	(1/5,1/4,1/3)	(1,2,3)	(1,1,1)	(1/3,1/2,1)	(1/5,1/4,1/3)	(1/4,1/3,1/2)	(1/6,1/5,1/4)	(1/5,1/4,1/3)	0.0410
DMU 3	(1/3,1/2,1)	(3,4,5)	(1,2,3)	(1,1,1)	(1/3,1/2,1)	(1/4,1/3,1/2)	(1/5,1/4,1/3)	(1/3,1/2,1)	0.0729
DMU 4	(1/4,1/3,1/2)	(1,2,3)	(3,4,5)	(1,2,3)	(1,1,1)	(1/4,1/3,1/2)	(1,1,1)	(1/5,1/4,1/3)	0.0985
DMU 5	(1/3,1/2,1)	(2,3,4)	(2,3,4)	(2,3,4)	(2,3,4)	(1,1,1)	(1/3,1/2,1)	(1/3,1/2,1)	0.1344
DMU 6	(1,1,1)	(4,5,6)	(4,5,6)	(3,4,5)	(1,1,1)	(1,2,3)	(1,1,1)	(2,3,4)	0.2366
DMU 7	(2,3,4)	(3,4,5)	(3,4,5)	(1,2,3)	(3,4,5)	(1,2,3)	(1/4,1/3,1/2)	(1,1,1)	0.2154
Total									1

$CR = 0.08708$

Table A31. Comparison matrix for PUS based on alternatives.

Alternatives	DMU 8	DMU 1	DMU 2	DMU 3	DMU 4	DMU 5	DMU 6	DMU 7	Weight
DMU 8	(1,1,1)	(1,2,3)	(1/3,1/2,1)	(2,3,4)	(1/3,1/2,1)	(1/4,1/3,1/2)	(1,2,3)	(1,1,1)	0.0980
DMU 1	(1/3,1/2,1)	(1,1,1)	(1/4,1/3,1/2)	(1,2,3)	(1/5,1/4,1/3)	(1/4,1/3,1/2)	(3,4,5)	(1/5,1/4,1/3)	0.0726
DMU 2	(1,2,3)	(2,3,4)	(1,1,1)	(3,4,5)	(1,2,3)	(2,3,4)	(3,4,5)	(1,2,3)	0.2451
DMU 3	(1/4,1/3,1/2)	(1/3,1/2,1)	(1/5,1/4,1/3)	(1,1,1)	(1/6,1/5,1/4)	(1/3,1/2,1)	(1/4,1/3,1/2)	(1/5,1/4,1/3)	0.0381
DMU 4	(1,2,3)	(3,4,5)	(1/3,1/2,1)	(4,5,6)	(1,1,1)	(1,2,3)	(2,3,4)	(3,4,5)	0.2202
DMU 5	(2,3,4)	(2,3,4)	(1/4,1/3,1/2)	(1,2,3)	(1/3,1/2,1)	(1,1,1)	(1,2,3)	(2,3,4)	0.1564
DMU 6	(1/3,1/2,1)	(1/5,1/4,1/3)	(1/5,1/4,1/3)	(2,3,4)	(1/4,1/3,1/2)	(1/3,1/2,1)	(1,1,1)	(1/3,1/2,1)	0.0572
DMU 7	(1,1,1)	(3,4,5)	(1/3,1/2,1)	(3,4,5)	(1/5,1/4,1/3)	(1/4,1/3,1/2)	(1,2,3)	(1,1,1)	0.1124
Total									1

$CR = 0.08884$

Table A32. Comparison matrix for REP based on alternatives.

Alternatives	DMU 8	DMU 1	DMU 2	DMU 3	DMU 4	DMU 5	DMU 6	DMU 7	Weight
DMU 8	(1,1,1)	(2,3,4)	(3,4,5)	(1,2,3)	(2,3,4)	(2,3,4)	(1,1,1)	(2,3,4)	0.2425
DMU 1	(1/4,1/3,1/2)	(1,1,1)	(4,5,6)	(1,2,3)	(2,3,4)	(3,4,5)	(1,2,3)	(5,6,7)	0.2306
DMU 2	(1/5,1/4,1/3)	(1/6,1/5,1/4)	(1,1,1)	(1/5,1/4,1/3)	(1/3,1/2,1)	(1/4,1/3,1/2)	(1/4,1/3,1/2)	(1/3,1/2,1)	0.0361
DMU 3	(1/3,1/2,1)	(1/3,1/2,1)	(3,4,5)	(1,1,1)	(2,3,4)	(3,4,5)	(1,2,3)	(1,1,1)	0.1478
DMU 4	(1/4,1/3,1/2)	(1/4,1/3,1/2)	(1,2,3)	(1/4,1/3,1/2)	(1,1,1)	(1/4,1/3,1/2)	(1/3,1/2,1)	(1/5,1/4,1/3)	0.0500
DMU 5	(1/4,1/3,1/2)	(1/5,1/4,1/3)	(2,3,4)	(1/5,1/4,1/3)	(2,3,4)	(1,1,1)	(1/4,1/3,1/2)	(1/3,1/2,1)	0.0670
DMU 6	(1,1,1)	(1/3,1/2,1)	(2,3,4)	(1/3,1/2,1)	(1,2,3)	(2,3,4)	(1,1,1)	(1,2,3)	0.1316
DMU 7	(1/4,1/3,1/2)	(1/7,1/6,1/5)	(1,2,3)	(1,1,1)	(3,4,5)	(1,2,3)	(1/3,1/2,1)	(1,1,1)	0.0944
								Total	1

CR = 0.08413

Table A33. Comparison matrix for SWQ based on alternatives.

	DMU 8	DMU 1	DMU 2	DMU 3	DMU 4	DMU 5	DMU 6	DMU 7	Weight
DMU 8	(1,1,1)	(1/3,1/2,1)	(1/5,1/4,1/3)	(1/4,1/3,1/2)	(1/4,1/3,1/2)	(1/5,1/4,1/3)	(1/3,1/2,1)	(1/4,1/3,1/2)	0.0404
DMU 1	(1,2,3)	(1,1,1)	(1/7,1/6,1/5)	(1/5,1/4,1/3)	(1/4,1/3,1/2)	(1/3,1/2,1)	(1/5,1/4,1/3)	(1/3,1/2,1)	0.0446
DMU 2	(3,4,5)	(5,6,7)	(1,1,1)	(1,1,1)	(1,2,3)	(3,4,5)	(2,3,4)	(4,5,6)	0.2613
DMU 3	(2,3,4)	(3,4,5)	(1,1,1)	(1,1,1)	(1,2,3)	(1,2,3)	(3,4,5)	(2,3,4)	0.2183
DMU 4	(2,3,4)	(2,3,4)	(1/3,1/2,1)	(1/3,1/2,1)	(1,1,1)	(2,3,4)	(1,2,3)	(3,4,5)	0.1639
DMU 5	(3,4,5)	(1,2,3)	(1/5,1/4,1/3)	(1/3,1/2,1)	(1/4,1/3,1/2)	(1,1,1)	(1,2,3)	(3,4,5)	0.1167
DMU 6	(1,2,3)	(3,4,5)	(1/4,1/3,1/2)	(1/5,1/4,1/3)	(1/3,1/2,1)	(1/3,1/2,1)	(1,1,1)	(3,4,5)	0.0990
DMU 7	(2,3,4)	(1,2,3)	(1/6,1/5,1/4)	(1/4,1/3,1/2)	(1/5,1/4,1/3)	(1/5,1/4,1/3)	(1/5,1/4,1/3)	(1,1,1)	0.0557
								Total	1

CR = 0.07371

Table A34. Comparison matrix for TEM based on alternatives.

Alternatives	DMU 8	DMU 1	DMU 2	DMU 3	DMU 4	DMU 5	DMU 6	DMU 7	Weight
DMU 8	(1,1,1)	(1,2,3)	(3,4,5)	(2,3,4)	(4,5,6)	(1/4,1/3,1/2)	(1/4,1/3,1/2)	(1,2,3)	0.1594
DMU 1	(1/3,1/2,1)	(1,1,1)	(3,4,5)	(1,2,3)	(3,4,5)	(1/3,1/2,1)	(1,1,1)	(2,3,4)	0.1488
DMU 2	(1/5,1/4,1/3)	(1/5,1/4,1/3)	(1,1,1)	(1/4,1/3,1/2)	(1/3,1/2,1)	(1/5,1/4,1/3)	(1/6,1/5,1/4)	(1/3,1/2,1)	0.0346
DMU 3	(1/4,1/3,1/2)	(1/3,1/2,1)	(2,3,4)	(1,1,1)	(1/4,1/3,1/2)	(1/5,1/4,1/3)	(1/3,1/2,1)	(1/4,1/3,1/2)	0.0565
DMU 4	(1/6,1/5,1/4)	(1/5,1/4,1/3)	(1,2,3)	(2,3,4)	(1,1,1)	(1/3,1/2,1)	(1/4,1/3,1/2)	(1/3,1/2,1)	0.0684
DMU 5	(2,3,4)	(1,2,3)	(3,4,5)	(3,4,5)	(1,2,3)	(1,1,1)	(1,2,3)	(2,3,4)	0.2449
DMU 6	(2,3,4)	(1,1,1)	(4,5,6)	(1,2,3)	(2,3,4)	(1/3,1/2,1)	(1,1,1)	(4,5,6)	0.2078
DMU 7	(1/3,1/2,1)	(1/4,1/3,1/2)	(1,2,3)	(2,3,4)	(1,2,3)	(1/4,1/3,1/2)	(1/6,1/5,1/4)	(1,1,1)	0.0796
				Total					1

CR = 0.0905

References

1. Hou, H.; Li, S.; Lu, Q. Gaseous emission of monocombustion of sewage sludge in a circulating fluidized bed. *Ind. Eng. Chem. Res.* **2013**, *52*, 5556–5562. [CrossRef]
2. United States Environmental Protection Agency. *Criteria for the Definition of Solid Waste and Solid and Hazardous Waste Exclusions*; EPA: Washington, DC, USA, 2018.
3. Costi, P.; Minciardi, R.; Robba, M.; Rovatti, M.; Sacile, R. An environmentally sustainable decision model for urban solid waste management. *Waste Manag.* **2004**, *24*, 277–295. [CrossRef]
4. Fiorucci, P.; Minciardi, R.; Robba, M.; Sacile, R. Solid waste management in urban areas: Development and application of a decision support system. *Resour. Conserv. Recycl.* **2003**, *37*, 301–328. [CrossRef]
5. Nakhla, D.A.; Hassan, M.G.; Haggar, S.E. Impact of biomass in egypt on climate change. *Nat. Sci.* **2013**, *5*, 678–684. [CrossRef]
6. Cheung, W.H.; Lee, V.K.C.; McKay, G. Minimizing dioxin emissions from integrated msw thermal treatment. *Environ. Sci. Technol.* **2007**, *41*, 2001–2007. [CrossRef] [PubMed]
7. Lancia, A.; Karatza, D.; Musmarra, D.; Pepe, F. Adsorption of mercuric chloride from simulated incinerator exhaust gas by means of sorbalittm particles. *J. Chem. Eng. Jpn.* **1996**, *29*, 939–946. [CrossRef]
8. Weinstein, P.E. *Waste-to-Energy as a Key Component of Integrated Solid Waste Management for Santiago, Chile: A Cost-Benefit Analysis*; Fu Foundation School of Engineering and Applied Science, Columbia University: New York, NY, USA, 2006.
9. Psomopoulos, C.; Bourka, A.; Themelis, N.J. Waste-to-energy: A review of the status and benefits in USA. *Waste Manag.* **2009**, *29*, 1718–1724. [CrossRef] [PubMed]
10. Hassaan, M.A. A gis-based suitability analysis for siting a solid waste incineration power plant in an urban area case study: Alexandria governorate, Egypt. *J. Geogr. Inform. Syst.* **2015**, *7*, 643–657. [CrossRef]
11. Tavares, G.; Zsigraiová, Z.; Semiao, V. Multi-criteria gis-based siting of an incineration plant for municipal solid waste. *Waste Manag.* **2011**, *31*, 1960–1972. [CrossRef] [PubMed]
12. Yap, H.Y.; Nixon, J.D. A multi-criteria analysis of options for energy recovery from municipal solid waste in India and the UK. *Waste Manag.* **2015**, *46*, 265–277. [CrossRef] [PubMed]
13. Lee, A.H.I.; Kang, H.-Y.; Lin, C.-Y.; Shen, K.-C. An integrated decision-making model for the location of a pv solar plant. *Sustainability* **2015**, *7*, 13522–13541. [CrossRef]
14. Wang, C.-N.; Nguyen, V.T.; Thai, H.T.N.; Duong, D.H. Multi-criteria decision making (MCDM) approaches for solar power plant location selection in viet nam. *Energies* **2018**, *11*, 1504. [CrossRef]
15. Aragonés-Beltrán, P.; Chaparro-González, F.; Ferrando, J.P.P.; García-Melón, M. Selection of photovoltaic solar power plant investment projects-an anp approach. *Int. J. Environ. Chem. Ecol. Geol. Geophys. Eng.* **2008**, 128–136. [CrossRef]
16. Ali, S.; Lee, S.-M.; Jang, C.-M. Determination of the most optimal on-shore wind farm site location using a GIS-MCDM methodology: Evaluating the case of south korea. *Energies* **2017**, *10*, 2072. [CrossRef]
17. Suh, J.; Brownson, J.R.S. Solar farm suitability using geographic information system fuzzy sets and analytic hierarchy processes: Case study of ulleung island, Korea. *Energies* **2016**, *9*, 648. [CrossRef]
18. Noorollahi, E.; Fadai, D.; Shirazi, M.A.; Ghodsipour, S.H. Land suitability analysis for solar farms exploitation using gis and fuzzy analytic hierarchy process (FAHP)—A case study of Iran. *Energies* **2016**, *9*, 643. [CrossRef]
19. ˇCereška, A.; Zavadskas, E.K.; Bucinskas, V.; Podvezko, V.; Sutinys, E. Analysis of steel wire rope diagnostic data applying multi-criteria methods. *Appl. Sci.* **2018**, *8*, 260. [CrossRef]
20. Reid, R.D.; Sanders, N.R. *Operations Management*; Wiley: Hoboken, NJ, USA, 2009.
21. Belton, V.; Stewart, T. *Multiple Criteria Decision Analysis: An Integrated Approach*; Kluwer Academic: Boston, MA, USA, 2002.
22. Gan, L.; Xu, D.; Hu, L.; Wang, L. Economic feasibility analysis for renewable energy project using an integrated TFN–AHP–DEA approach on the basis of consumer utility. *Energies* **2017**, *10*, 2089. [CrossRef]
23. Liu, J.-P.; Yang, Q.-R.; He, L. Total-factor energy efficiency (TFEE) evaluation on thermal power industry with DEA, malmquist and multiple regression techniques. *Energies* **2017**, *10*, 1039. [CrossRef]
24. Asadzadeh, A.; Sikder, S.K.; Mahmoudi, F.; Kötter, T. Assessing site selection of new towns using topsis method under entropy logic: A case study: New towns of tehran metropolitan region (TMR). *Environ. Manag. Sustain. Dev.* **2014**, *3*, 123–137. [CrossRef]

25. Rashidi, M.; Ghodrat, M.; Samali, B.; Kendall, B.; Zhang, C. Remedial modelling of steel bridges through application of analytical hierarchy process (AHP). *Appl. Sci.* **2017**, *7*, 168. [CrossRef]

26. Feng, C.-M.; Wang, R.-T. Performance evaluation for airlines including the consideration of financial ratios. *J. Air. Transp. Manag.* **2000**, *6*, 133–142. [CrossRef]

27. Sen, P.; Yang, J.-B. *Multiple Criteria Decision Support in Engineering Design*; Springer: London, UK, 1998.

28. Sarkis, J. A strategic decision framework for green supply chain management. *J. Clean. Prod.* **2003**, *11*, 397–409. [CrossRef]

29. Saaty, T.L. *Decision Making with Dependence and Feedback: The Analytic Network Process*; Rws Publications: Pittsburgh, PA, USA, 1996.

30. Zadeh, L.A. Fuzzy sets. *Inform. Control.* **1965**, *8*, 338–353. [CrossRef]

31. Lee, A.H.I. A fuzzy supplier selection model with the consideration of benefits, opportunities, costs and risks. *Expert. Syst. Appl.* **2009**, *36*, 2879–2893. [CrossRef]

32. Lee, A.H.I.; Kang, H.-Y.; Hsu, C.-F.; Hung, H.-C. A green supplier selection model for high-tech industry. *Expert. Syst. Appl.* **2009**, *36*, 7917–7927. [CrossRef]

33. Lee, A.H.I.; Kang, H.-Y.; Chang, C.-T. Fuzzy multiple goal programming applied to tft-lcd supplier selection by downstream manufacturers. *Expert. Syst. Appl.* **2009**, *36*, 6318–6325. [CrossRef]

34. Lee, A.H.I.; Kang, H.-Y.; Wang, W.-P. Analysis of priority mix planning for the fabrication of semiconductors under uncertainty. *Int. J. Adv. Manuf. Technol.* **2006**, *28*, 351–361. [CrossRef]

35. Cheng, C.-H. Evaluating weapon systems using ranking fuzzy numbers. *Fuzzy Sets Syst.* **1999**, *107*, 25–35. [CrossRef]

36. Dehghani, M.; Esmaeilian, M.; Tavakkoli-Moghaddam, R. Employing fuzzy anp for green supplier selection and order allocations: A case study. *Int. J. Econ. Manag. Soc. Sci.* **2013**, *2*, 565–575.

37. Kahraman, C.; Ertay, T.; Büyüközkan, G. A fuzzy optimization model for QFD planning process using analytic network approach. *Eur. J. Oper. Res.* **2006**, *171*, 390–411. [CrossRef]

38. Lin, R.; Lin, J.S.-J.; Chang, J.; Tang, D.; Chao, H.; Julian, P.C. Note on group consistency in Analytic Hierarchy Process. *Eur. J. Oper. Res.* **2008**, *190*, 672–678. [CrossRef]

39. Kuswandari, R. *Assessment of Different Methods for Measuring the Sustainability of Forest Management Retno Kuswandari*; University of Twente: Enschede, The Netherlands, 2004.

40. Saaty, T.L. *The Analytic Hierarchy Process: Planning, Priority Setting, Resources Allocation*; McGraw-Hill: New York, NY, USA, 1980.

41. Hwang, C.-L.; Yoon, K. *Multiple Attribute Decision Making: Methods and Applications a State-of-the-Art Survey*; Springer: Berlin/Heidelberg, Germany, 1981.

42. Assari, A.; Maheshand, T.M.; Assari, E. Role of public participation in sustainability of historical city: Usage of topsis method. *Indian J. Sci. Technol.* **2012**, *5*, 2289–2294.

43. Jahanshahloo, G.R.; Lotfi, F.H.; Izadikhah, M. Extension of the topsis method for decision-making problems with fuzzy data. *Appl. Math. Comput.* **2006**, *181*, 1544–1551. [CrossRef]

applied
sciences

MDPI

Article

Smart System for the Optimization of Logistics Performance of the Pruning Biomass Value Chain

Girma Gebresenbet [1], Techane Bosona [1,*], Sven-Olof Olsson [2] and Daniel Garcia [3]

[1] Department of Energy and Technology, Swedish university of agricultural sciences,
   P.O. Box 75651 Uppsala, Sweden; girma.gebresenbet@slu.se
[2] Mobitron AB, P.O. Box 56146 Huskvarna, Sweden; soo@mobitron.se
[3] Research Centre for Energy Resources and Consumption-CIRCE, Mariano Esquillor Gomez,
   15 50018 Zaragoza, Spain; daniel.garcia@fcirce.es
* Correspondence: techane.bosona@slu.se; Tel.: +46-18671851

Received: 19 June 2018; Accepted: 10 July 2018; Published: 18 July 2018

Featured Application: The specific application of this work is that it improves management of solid biofuel supply chain.   Its potential application is for integrated product distribution management where product quality control and traceability can be integrated to increase customer satisfaction and resource utilization, and reduce logistics cost as well as environmental impact.

Abstract: Agricultural pruning biomass is one of the important resources in Europe for generating renewable energy. However, utilization of the agricultural residues requires development of efficient and effective logistics systems. The objective of this study was to develop smart logistics system (SLS) appropriate for the management of the pruning biomass supply chain. The paper describes the users' requirement of SLS, defines the technical and functional requirements and specifications for the development of SLS, and determines relevant information/data to be documented and managed by the SLS. This SLS has four major components: (a) Smart box, a sensor unit that enables measurement of data such as relative humidity, temperature, geographic positions; (b) On-board control unit, a unit that performs route planning and monitors the recordings by the smart box; (c) Information platform, a centralized platform for data storing and sharing, and management of pruning supply chain and traceability; and (d) Central control unit, an interface linking the Information platform and On-board control unit that serves as a point of administration for the whole pruning biomass supply chain from harvesting to end user. The SLS enables the improvement of performance of pruning biomass supply chain management and product traceability leading to a reduction of product loss, increased coordination of resources utilisation and quality of solid biofuel supply, increased pruning marketing opportunity, and reduction of logistics cost. This SLS was designed for pruning biomass, but could also be adapted for any type of biomass-to-energy initiatives.

Keywords: smart logistics system; smart box; information platform; renewable biomass energy; agricultural pruning

1. Introduction

Biomass is an important source of renewable energy sources. Biomass comprises among others, vegetation, energy crops, biosolids, forestry and agricultural residues, organic municipal waste, as well as some industrial waste [1]. Agricultural residues from the pruning operations of fruit plantations can be used as a renewable energy source. However, such utilization of pruning residues has been limited due to logistic-related constraints in harvesting, processing and transporting activities [2–4]. Although there is a potential source of renewable energy from agricultural pruning residues, the challenges and opportunities of the logistics operations have not been assessed well considering the entire chains from pruning harvesting to transport to end users.

It is therefore important to develop new improved logistics for pruning residues, which covers harvesting, transport and storage for agricultural pruning (fruit tree, vineyards and olive groove prunings and branches from up-rooted trees). This study aims to develop an integrated concept where location and quality allow a wise decision tool to support decisions for logistic operators. This paper describes the users' need for a smart logistics system (SLS), defines the components of SLS with technical and functional requirements and specifications, and determines relevant information/data to be documented and managed by the SLS.

1.1. Telematics Information System

Telematics is modern technology of receiving, storing, and sending of information in a wireless form. In a broad sense, it means, the use of informatics and telecommunication in an integrated form. In a telematics system, technologies such as the Global Positioning System (GPS), General Packet Radio Service (GPRS), and GSM (Global System for Mobile Communications) are used. GPS is a space-based global navigation satellite system (GNSS) that provides reliable location and time information in all weather and at all times unless the receiver is underground or in unobstructed line of sight to four or more GPS satellites. The GPRS is a packet-oriented mobile data service on the 2G and 3G cellular communication systems. The GSM, originally from Groupe Spécial Mobile, is the world's most popular standard for mobile telephony systems.

According to the European Telecommunication Standards Institute (ETSI), GSM was originally designed to operate in the 900 MHz band and gradually adapted for operation in bands up to 1900 MHz [5]. GPRS was commercially introduced first in early 2000s. It is enabler for always-on data connection with essential applications for web browsing and offers faster data rates than GSM i.e., up to 171 kbit/s, theoretically [6].

Telematics can be used for providing real-time information in fleet management, vehicle tracking, and monitoring freight and driver data. Mobile telematics are emerging as new models connecting smartphones to a car's computer system and exchange data using the wireless network. In logistics companies, systems developed to improve fleet management mainly apply GPS and GSM.

Telematics refers to the combination of telecommunication and informatics and it is of key importance in an intelligent transport system (ITS) [7]. Telematics can be expressed as "the use of wireless devices and "black box" technologies to transmit data in real time back to an organization" [8]. Mobile telematics are emerging as new models connecting smartphones to a car's computer system and exchange data using the wireless network. In logistics companies, systems developed to improve fleet management mainly apply GPS and GSM to retrieve real-time information.

The management of goods during transport requires tracking (ability to locate goods at any time), tracing (ability to know the product movement from source to end user), and monitoring (ability to know and control product quality status during transportation) information and its analysis. This management approach is relevant for agricultural goods and it can be performed based on technologies such as wireless sensors for climatic conditions (such as temperature and humidity) which can be installed in a truck trailer [7]. Oliveria et al. [9] proposed a logistics management system for cargo transport based on geofencing algorithms and radio-frequency technology, in order to improve services, reduce costs and ensure the safety in cargo transportation. Xiao et al. [10] used a wireless sensor network to monitor temperature and relative humidity in order to improve cold chain logistics for table grapes and increase traceability and product quality information.

Some of the important communication technologies applicable in goods transport systems and implemented in the current SLS include QR codes, Bluetooth wireless communications devices, GPS and GPRS/GSM receivers. A QR Code is a matrix symbology where the symbols consist of an array of square modules arranged in a square pattern [11]. Bluetooth is a technology for wireless communication between different components. In the current Smart system, Bluetooth is used for communication between the Smart box code and scanner for reading the QR code, and IR (infrared) communication is used for communication between a PC and the Smart box. IR communication is used to load the required parameters, start data collection period, and download data from the Smart box.

1.2. Electron Trade and Its Advantages

In any trading system, market participants interact in various ways [12]. Historically, this interaction has been based on face-to-face communication. Modern electronic communication networks (ECNs) could eliminate the face-to-face communication at physical locations. This could increase competition and reduce transaction costs [13]. Referring to United States Securities and Exchange Commission, Hendershott [13] defined ENCs as "electronic trading systems that automatically match with sell orders at specified prices". ENCs bypass human intermediaries (e.g., dealers) and offer faster trade execution.

The Committee on the Global Financial System (CGFS) broadly defined [12] the electronic trading system (ETS) as "a facility that provides some or all of the following services: electronic order routing (the delivery of orders from users to the execution system), automated trade execution (the transformation of orders into trades) and electronic dissemination of pre-trade (bid/offer quotes and depth) and post-trade information (transaction price and volume data)". In ETS, the trading processes and trading relationship are automated using computers. ETS differs from traditional trading systems in many ways [12]:

➢ it is location-neutral and allows continuous multilateral interaction;
➢ it can be scaled up to handle more trades by increasing the computer network capacity;
➢ it integrates the different parts of the trading process from start to end.

The trading system can be structured as decentralized markets (relies largely on bilateral interaction of market participants) or centralized markets (participants interaction is fully multilateral). Multilateral interaction is realized when trading activities are pooled on a single platform.

1.3. Basic Concept of Web-Based Information Platform

The motivations why enterprises invest in information systems include cutting cost, increasing productivity (without increasing cost) and increasing product and/or service quality [14]. Although information systems are introduced with sizable investment costs, they may be underutilized or abandoned unless accepted by intended users [14,15]. The usage of information systems depend on the perceived usefulness and how user friendly the systems are. Perceived usefulness may include improvement of work quality, working time, work performance and control over the work. Similarly, the perceived ease of use of information system can be conceived if it is easy to operate, enables users to do what they want to do, and if it is flexible and requires less mental effort to interact with it [14]. In the SLS to be developed, a web-based Information platform should be user friendly and have the perceived usefulness described above.

The Information platform, as part of this new SLS, has integrated the QR code generation and printing options to enhance a continuous flow of information associated with the pruning biomass to be traded using the SLS. According to ISO/IEC 18004:2015 [11], there are four technically different, but closely related QR Codes: QR Code Model 1 (with a capacity to store up to 1167 numerals), QR Code Model 2 (an enhanced form of model 1 with a capacity to store up to 7089 numerals and used as the basis of the first edition of ISO/IEC 18004), QR Code (closely similar to QR Code Model 2 with additional facility for a mirror image oriented appearance of symbols and this code was used as basis of the second edition of ISO/IEC 18004:2006), and Micro QR Code format (a variant of QR Code which enables a small to moderate amount of data to be represented by a small symbol) [11,16].

2. Biomass Supply Chain and Its Actors

2.1. Biomass Supply Chain

Sources of pruning wood could be different from fruit trees, for example Almond, Fruit, Vineyard, And Olive, Apple and Cherry prunings. As indicated in Figure 1, the chain includes pruning, collection, off-farm storage and energy production. Pruning activities are carried out mainly manually by farmers. Appropriate machinery can be used to collect (harvest) the pruning and bale or chip it as required based on biomass quality requirements by energy plants (end users). In order to monitor the quality characteristics of pruning (bales and chips) as it moves from farm to end users (see Figure 1), the SLS should create an application platform where different actors in the supply chain can interact.

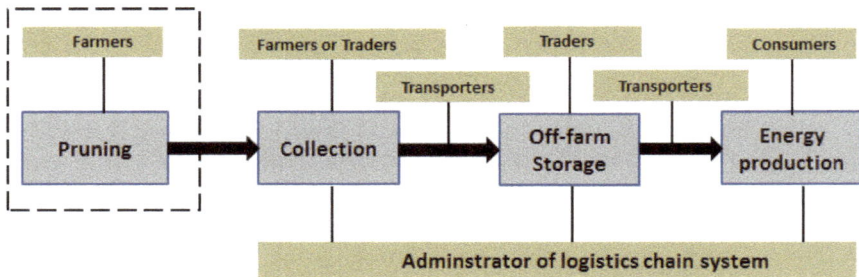

Figure 1. Different actors along the pruning biomass logistics chain.

2.2. Actors in the Pruning Biomass Supply Chain and Their Role in the Smart System

For the pruning supply chain, four actors of the biomass supply chain were considered:

Producers: Farmers are primary producers of biomass. They produce, harvest, process (chipping, baling) and sell biomass. An account for access to the Information platform could be created for each farmer participating in the system, so that they can put all the necessary information regarding their products.

Traders: The biomass traders buy, store, and sell the biomass. Therefore, traders who own storage and fulfil quality control criteria could be registered and get an ID number from system to gain access to the platform for information sharing.

Transporters: Biomass transporters or logistics operators are those who handle biomass procurement, source identification, and a proper way of delivering to end users with the desired characteristics. These logistics operators can play complex roles by interacting with different actors.

Consumers: Consumers are end-users who can order biomass product needed. They can have their own account to access the Information platform perform online order and receive information. They specify the quality and quantity of biomass required at power plant. They also determine and control the state of deliveries. Consumers can also follow information flow continuity regarding their orders and deliveries.

3. Components of Smart Logistics System (SLS) Prototype

The Smart Logistics System (SLS) has four units (see Figure 2): (a) a Sensor unit (Smart box); (b) an On-board control unit; (c) a web-based Information platform, and (d) Central control unit. The Smart box tool is closely controlled by the On-board control unit. The Central control unit manages also the web-based Information platform and monitors information and material flow as well as facilitating interaction among different actors of biomass supply chain (see Figure 3). In this Smart system, the GPS, GPRS and GSM modules are integrated to monitor the information from the Smart box to the platform and Central control system.

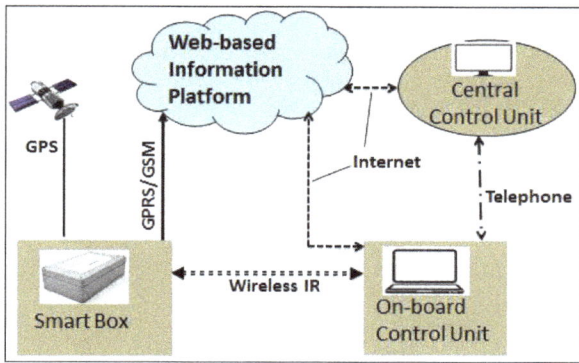

Figure 2. The conceptual organization of smart logistics system (SLS) components (Adapted from [17] with slight modification).

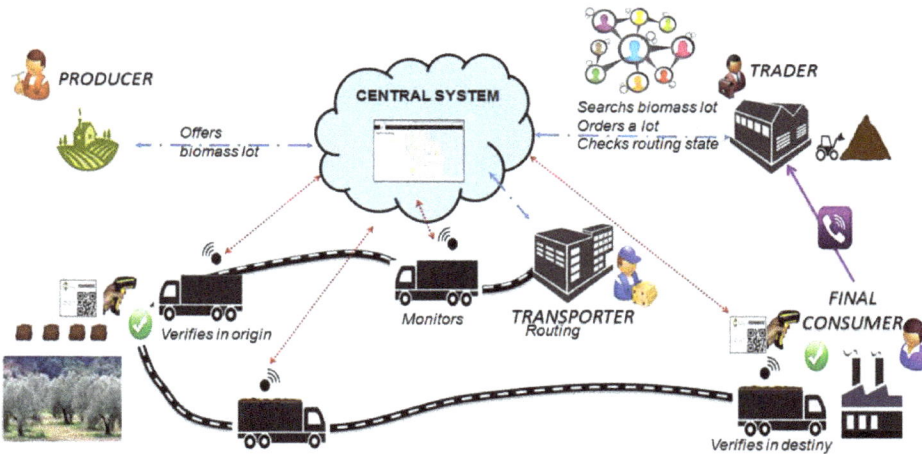

Figure 3. Linkage between different activities of actors along biomass supply chain with application of SLS. Source: Created by authors using free Icons obtained from FREEPIK [18].

3.1. Sensor Unit (Smart Box)

The sensor unit measures continuously the parameters such as relative humidity, temperature (with possibility of measuring at different positions at the same time), geographical position and tracking, and vehicle speed. It has accessories such as a barcode or QR (Quick Reader) and sensors for temperature and moisture content measurements at remote distance from sensor unit. The extension of the sensor unit should be placed in the loading compartment (trailer) of the vehicle and able to communicate wirelessly with the On-board control system, with the ability to transfer and receive data. The sensor unit (Smart box) is assembled with its accessories, GPS and GPRS/GSM modules connected with antennas. Sensors for temperature and relative humidity are also connected to it so that continuous measurement can be done during transport from producers to the required destinations.

3.2. Components of Smart Box Tool (Sensor Unit)

The Smart box uses a sensor probe (see Table 1) which could be placed in the loading compartment (trailer) of the vehicle and measures temperature and relative humidity of the pruning biomass during transport with time interval determined prior to start of the measurement. The start and end of measurements and the real-time performance of the sensor unit is monitored by the transport operator using the On-board control unit. The Smart box registers the GPS coordinates during transport. It also captures the QR code readings (data on biomass quality parameters) on the labels during scanning and retrieves the quality parameters of the transported prunings.

Table 1. Characteristics and functions of components of Smart box tool.

Component	Characteristics Description	Function as Applied in SLS
Registration unit	Registration Unit — Holder for Registration Unit. Protected by a holder and Metal Case. Model of the registration unit CS 0940-6	The Smart box which is based on the platform for the Cargolog Impact Recorder System is a measuring unit that records the temperature and humidity of the pruning biomass (using the sensor probe), QR readings (using scanner), and Global Positioning System (GPS) coordinates (using GPS signal receivers) during biomass transport. It also transmits the recorded data to the central server using General Packet Radio Service/Global System for Mobile Communications (GPRS/GSM) tools. It is provided with a holder so that it can be installed with appropriate protection against mechanical damage. The smart box could either be equipped with a battery or receive a power supply of 12 or 24 volt from the vehicle's battery.
Metal case	Metal Case with four magnets	The metal case is used for protecting the smart box from mechanical damage, aggressive gases, liquids and humidity. The Smart box is installed in the metal case with its holder.
Power cable	Power supply cable	The power supply cable is used to connect the Smart box to external power source of 12 or 24 V.
Sensor, sensor probe, and cable	Sensor probe inserted into biomass	Temperature and relative humidity sensors are installed in the probe for protection. The probe is connected to the Smart box and long enough to measure at any distance required within 15 m. The sensors are protected by a hard cover. The hard probe enables the sensors to be inserted into biomass to measure temperature and humidity during transport. The measurement interval can be adjusted as required.
GPS antenna	GPS antenna with magnetic foot. Cable length: 5 m. Size in mm (L × W × H): 50 × 40 × 10	The GPS antenna receives signals from satellites and enables the GPS system to determine the geographical coordinates. The GPS antenna is directly connected to the GPS module in the Smart box.
GPRS/GSM antenna	Antenna with magnetic foot. Cable length: 2.5 m	The GPRS/GSM antenna transmits data from the Smart box to the Central control unit. The GPRS/GSM antenna is directly connected to the Smart box by 2.5 m long cable. The antenna has a magnetic foot so that it can be placed at appropriate position with a clear view for optimal transmission of data/information.

Table 1. *Cont.*

Component	Characteristics Description	Function as Applied in SLS
Scanner	Type: PowerScan™ BT9500 family Model: PBT9500 Capability: Single scanner and omni-directional reading; improved ergonomic and user comfort Holder: HLD-P080	The Power™ 9500 is used to read QR codes on labels of the pruning biomass. It can read QR codes with a distance of over 1.0 m/3.3 ft, depending on the barcode's resolution. Its features support 1D, stacked and 2D codes, postal codes and image capture. The Power 9500 scanner enables to perform a reliable, and long-term operation. PowerScan™ 9500 scanner combines omnidirectional reading capabilities with an optical characteristics and it is able to read any kind of bar code, regardless of the orientation. It has a HLD-P080 plastic holder designed to be mounted on a wall or fixed on a desk.
Infrared (IR) Interface	Type: Cargolog FAT90 IR Interface Version: 60944 TopCom version with USB Size in mm (L × W × H): 150 × 48 × 35	The IR interface is used for communication between the Smart box and On-board control using Bluetooth. The CargoLog PC software enables parameters to be to downloaded and sampling sequences to be initiated, and reads out the collected data after or during a record period.

Three new Smart box prototypes were developed using an up-to-date data log system to test their performance. These three Smart box tools could be identified with their serial numbers when these are connected to the central control and monitoring system. The prototypes use Smart box and associated accessories (see Table 1). In order to perform measurements, the Smart system has 6 different components: Smart box for recording required data and transferring data; **sensors** for measuring temperature and humidity; GPS module and its antenna for detecting geographical coordinates; GPRS/GSM module and its antenna for wireless communication with the Central control unit; a Scanner for scanning QR code from pruning biomass label.

In the Smart System, PowerScan™ 9500 scanner [19] was implemented to read QR codes. PowerScan™ PBT9500 has powerful algorithm that enables to read the QR code. The QR code implemented in this SLS has standard QR code Specification (ISO/IEC 18004:2000 bzw. ISO/IEC 18004:2006) which has been replaced by ISO/IEC 18004:2015 [11]. The scanner communicates with the Smart box via wireless communication using Bluetooth.

The functions of the Smart box were programmed using the CargoLog PC software, and connected to the web address of the Information platform [20]. In order to use the geographical information systems with up-to-date information on facility locations and roads as integrated in Google Maps, the web-based Information platform is linked with Google Maps using Google Maps Engine API version ct 06.2015. This capability of the smart system enables the smart tool to interact with updated Google Maps with detailed information for the geographical position.

3.3. Assembly of Smart Box (Sensor Unit) Components

Once the Smart box is connected to its accessories such as power supply, GPS and GPRS/GSM antennas and temperature and relative humidity sensor probe, it should be installed in the metal case that has four power magnets (see Table 1 and Figure 4). Using magnets, the metal case can be mounted on the vehicle at the suitable position. The sensor probe should be inserted into the biomass loaded in the loading compartment (trailer) of the vehicle. The GPS antenna and GPRS/GSM antenna are connected to the Smart box with 5 m- and 2.5 m-long cables respectively. The antennas have magnetic feet and can be affixed on top of a vehicle. For the best signal reception, the GPS antenna should be placed in a horizontal unshielded line of sight to establish contact with the satellites. The GPRS/GSM antenna should be placed in a way it can get a clear view. The scanner was provided with a plastic holder for better handling and protection (see Figure 5).

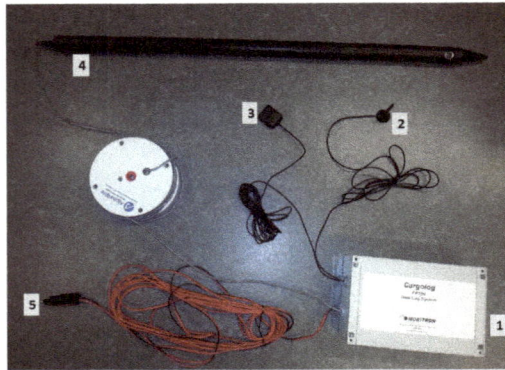

Figure 4. Smart box system assembly: (**1**) Smart box protected with metal case; (**2**) GPRS/GSM antenna; (**3**) GPS antenna; (**4**) temperature and humidity measuring sensor probe; (**5**) power cable.

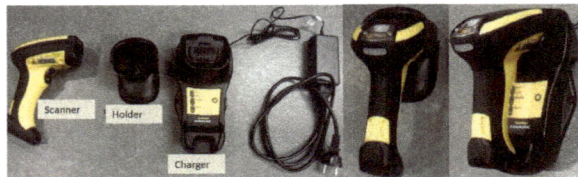

Figure 5. Power^{TM} 9500 scanner with its components.

3.4. On-Board Control System

The On-board control system is an important unit which should be linked to the sensor unit via wireless connection (Bluetooth, or IR) and to the Information platform and Central control unit (Figure 2).

This unit is always on the vehicle and is monitored by the transporter. The main functions are to: monitor the sensor unit (Smart box); receive and transfer data from the sensor unit to the central platform and to the Central control unit; receive and transfer data from and to web-based Information platform and Central control system; and perform route planning.

The SLS has an ability to consider the pick-up and delivery locations on Google Map-based geographic information system (GIS) and generate optimum biomass delivery routes. The transporters of biomass use this functionality of the smart system to plan their optimum driving routes based on locations of pruning (on farm or at storage site) and order locations (consumers' locations). This functionality leads to economic and environmental efficiency of the pruning supply chain.

The On-board control unit uses CargoLog [21] software designed for a PC running on Windows XP/Windows 7/Windows 8, with minimum 20 MB free disk space and 1 free USB port. From the On-board control unit, the transport operator can monitor the performance of Smart box tool (parameters to be measured, measurement intervals and data transfer intervals to the Central control system).

Using the Web-based information platform [20], the SLS has the ability to consider the pick-up and delivery locations on Google Map integrated into the platform. Using the google based geographic information systems, the Smart System enables the transporter to generate optimum pruning delivery routes.

3.5. Information Platform

The web-based platform has the characteristics of an on-line marketing system. The system has been built in such a way that all actors such as farmers/producers, users/customers, transport companies and the on-board control system have access to the platform. The platform will be managed by the Central control system. The platform has two components:

Component 1: where the producers and processors put in information about their products; and

Component 2: where the product request and billing system is included.

The basic information which the farmers and processors should put in the platform includes quality parameters of biomass and the amount of biomass so that the end-users (for example a power plant) could order as required.

The system allows the actors to access the platform and provide and receive the required data. Special software was developed to enable the actors such as biomass producers and traders to be able to print a QR code and label their products. It enables them to print a label with QR code and lot number according to the pre-established labelling characteristics. The detailed information on traceability system linked to this SLS is reported by Bosona et al. [17].

The platform provides an e-trading platform for effective marketing of pruning biomass. Figure 6 depicts an example of a first page display where different actors can login and use it.

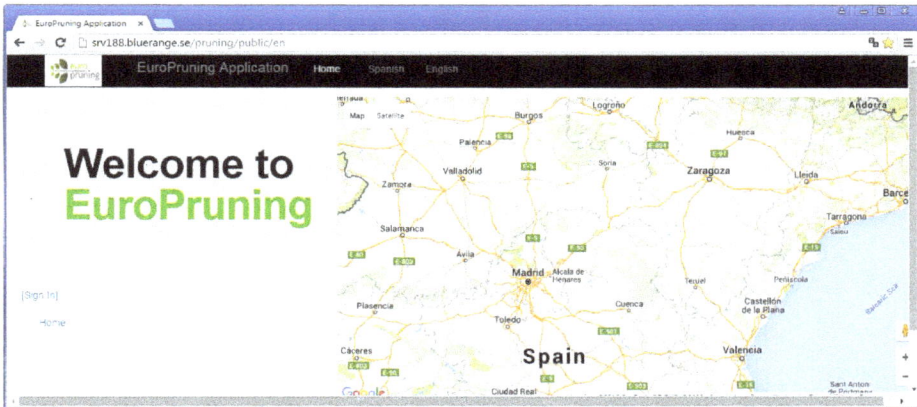

Figure 6. SLS Information platform displaying home page for signing in. http://europruning.mobitron. se/public/en.

3.5.1. Activities by Different Actors

The information platform provides support functionality to different actors such as farmers (raw material producer), solid biofuel traders, transporters, and final consumers (the power plant). Figure 7 summarises the major activities of the actors using the platform while Figure 8 depicts the network of actions indicating how the platform enhances interaction among actors of the pruning supply chain.

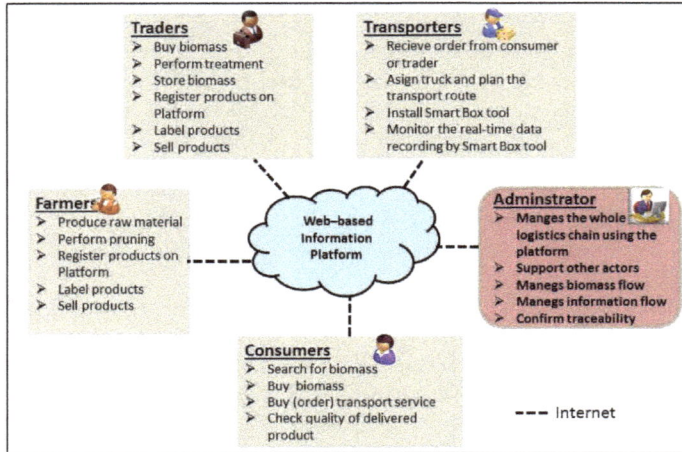

Figure 7. Major activities of different users of the platform. In order to allow the actors to perform the activities indicated, the platform has modules and links to defined actions making the platform more user friendly.

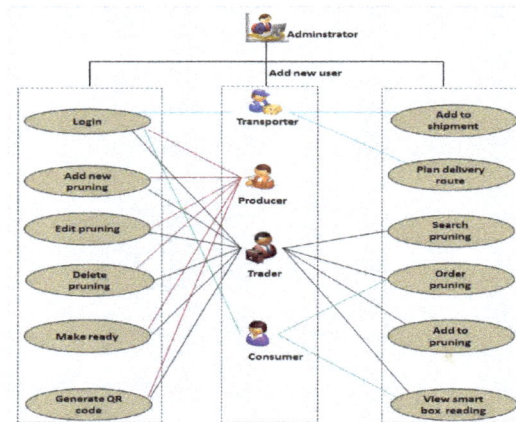

Figure 8. Network of actions by different users of the platform.

During development of the platform, firstly, the biomass quality parameters and related information required to be gathered and stored were organized in a manner convenient to be implemented in the online platform. For this, a spreadsheet database was created for storing the data entered into the platform. Based on the defined activities, user-friendly interfaces have been created for each category of actors and administrators. Programing languages such as C+; php, java, and Delphi were used during developing the platform.

The platform [20] has been integrated also with Google Maps using Google Maps Engine API version ct 06.2015. This enables the smart system to interact with updated Google Maps with detailed information for the geographical position and to display (on the platform) the locations of pruning farms, biomass storage and processing sites, energy plants, the planned routes, and eventually the delivery routes. Using geographic information provided by the user (e.g., latitude–longitude coordinates of a facility under consideration) and GPS coordinates recorded and transmitted by the Smart box, the platform provides online mapping and display services.

The platform has been integrated with a QR code generation option for identifying biomass products entering and leaving the pruning supply chain. The standard specification of this QR code (ISO/IEC 18004:2000 bzw ISO/IEC 18004:2006) has been replaced by ISO/IEC 18004:2015 in 2015 (ISO, 2015).

This information platform can be upgraded and improved to increase its robustness and/or add additional functionality based on the feedback of users and experience of the logistics manager who administrates the Central control unit.

3.5.2. Interfaces of the SLS Platform for Different Actors

The platform has the characteristics of an on-line marketing system. For instance, the interface for farmers has a format for entering biomass quality parameters and generating lot numbers while the interface for transporters has features for receiving orders (to provide transport service for consumers or traders) and performing route planning. All registered actors such as farmers (producers), traders, transporters, and consumers have to login the web-based application platform before performing any action.

Interface for administrator: the platform has been designed to have additional functional features for administrator. For instance, the administrator can add new users, edit the profile of registered users, or delete users who want to unsubscribe from the system. This enables the logistics manager of the logistics smart system to administer the whole material and information flow along the biomass logistics chain.

Interface for biomass producers: the new pruning biomass enters into the trading system when the farmer registers it in the application platform entering all required information about the product. The platform allows the producers (farmers or traders) to enter and edit the information regarding their pruning products (see Figure 9). The tool allows a producer to enter location information for each pruning batch, if its location differs from the location of a company (producer). It also enables the producers to generate a QR code of biomass products with a specific lot number. The lot number generation has been linked to the actual processing date i.e., baling or chipping date.

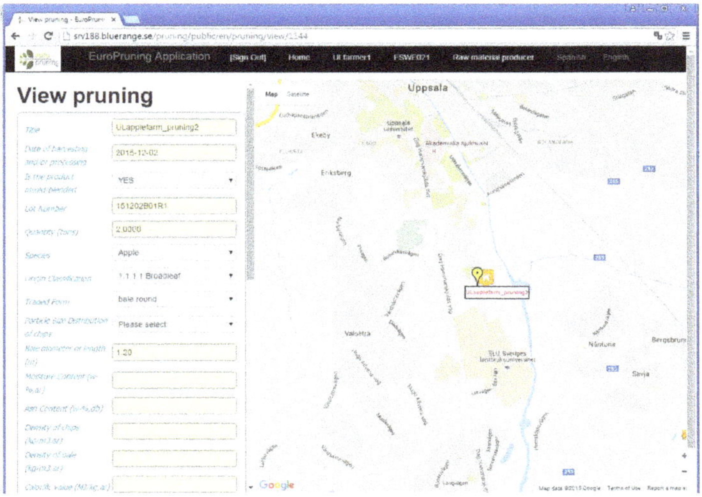

Figure 9. *Cont.*

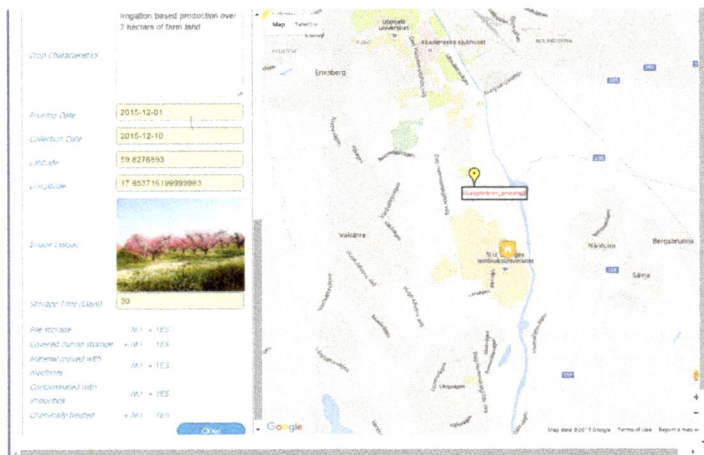

Figure 9. Farmer's interface as signed in by a raw material producer. The page displays a data entering format where some of the parameters will be added.

The interface allows the biomass providers (farmers and traders) to print the label associated to their products. Figure 10 presents label with company name (label holder), product lot number, and QR code.

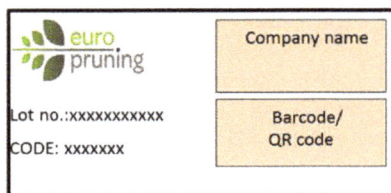

Figure 10. Sample of label indicating appearance of company name with its identification code, lot number of ordered biomass product, and QR code.

Interface for traders: the platform supports the traders during procurement and selling the pruning biomass. Traders can search for available products in the platform and order it for purchasing. Traders can use also the platform to register their biomass products (after storing and/or processing to add value) on the platform to be available for sale.

Interface for consumers: the consumers have access to search for available products that are ready for sale, read quality parameter information provided by producers, make orders (to purchase products) and select an appropriate transport company to obtain a transport service for purchased biomass product.

Interface for transporters: the interface allows a transporter to enter a different start, storage, and end location for each shipment route. The platform allows the transporters to receive different orders from different consumers and add the accepted orders into planned shipments. It enables transporters to identify the location of products and the location of the customers' store as well as the quantities required at each store. The location information attached to each user profile will be used to determine the source and destination locations for each order as well as in planning the route for each shipment.

3.6. Central Control Unit

The central control unit is the governing unit linked to the Information platform and On-board control unit (see Figure 2). It could be able to receive and transfer information within the decided intervals of time, and saves information. It provides a human–machine interface to administer and enable the monitoring of the system.

The Central control unit was designed to enable the administrator of the pruning trading system to monitor the SLS performance as whole. In general, it receives, controls, and transfers information from one actor to other actors along the pruning supply chain. This is an important unit that enables evaluation of the performance of the SLS and seeks continuous improvement measures that could lead to an economically effective and environmentally sustainable supply chain of renewable biomass energy generation.

4. Implementation and Management of the SLS

4.1. Material Flow Along the Logistics Chain

In the smart system, the web-based Information platform plays an important role to run and manage the whole pruning trading system. The tool allows each actor to understand the types of data to share and receive using a user-friendly interface as explained above.

Once the new pruning biomass enters the system, it passes different stages along the logistics chain. On the platform, the pruning position can be pinpointed under the link 'Status' which could be displayed as 'Pending', 'Ready', 'Ordered', or 'Routed'. These terms indicating the status of recorded pruning biomass are defined in Table 2.

Table 2. Pruning record status.

Status	Definition
Pending	The pruning has been entered into the system; the provider can still edit or delete that pruning.
Ready	The pruning has been flagged as ready; all consumers can access the system to view its quality parameters and order it.
Ordered	The pruning has been ordered by a buyer (consumer or trader) and assigned to a specific transporter. That transporter can access the system to view its source and destination locations and add it to a shipment.
Routed	The pruning has been selected by a transporter to be included in one of the shipment routes. The transporter plans its best delivery route

In general, integrating the Smart box tool and web-based Information platform, the SLS enables users to perform tracking, tracing, and monitoring activities [8] for improved management of the pruning biomass supply chain.

4.2. Data Acquisition and Management

The platform enables users to have an effective data acquisition, monitoring, and utilization system. Yu et al. [22] discussed the importance of data-driven supply chains and indicated that coordination and supply chain responsiveness increases the financial performance of firms. The developed platform receives data from two sources:

(i)   Actors (farmers, traders, etc.) can enter data (regarding biomass quality parameters and biomass supply related activities) directly into the platform.

(ii) The Smart box measures some parameters such as relative humidity, temperature, GPS coordinates during biomass transport and transmits these data directly to the platform via GPRS/GSM devices. The platform can receive data from different Smart box tools and store the data along with the identification number (Cargolog serial number) of the respective Smart box.

The collected data will be stored in the server of the platform and monitored by the administrator of the system. The platform has a spreadsheet-based database on the server. This data can be used for different purposes, such as analysis of biomass quality against required quality standards, performance analyses of biomass distribution and traceability systems, and identifying improvement strategies.

The data gathered and stored in the data base can be categorized as:

(i)   information that should be kept for documentation, but non-traceable; and
(ii)   information that should be included in the labelling/traceability system and follows the product.

The information that documented in the database and needed to be public could be printed as a barcode or QR code. The farmers/producers will then deliver the barcode/QR code with the pruning to the transport company. The information from the barcode or QR code can also be used for traceability purposes. The information from the barcode/QR code will be first transferred to the sensor unit and then transferred to the web-based platform. Tables 3 and 4 present data source, destination and transfer intervals.

Table 3. Data type, source, destination and utilization.

Data Type	Data Source	Device (to Transfer Data to Smart Box)	Data Destination	Main Data User
Temperature	New measurement by sensor placed in the prunings	Sensor probe with cable	Information platform server (and On-board control unit PC)	Consumer, producer, Administrator (Central control unit)
Relative humidity	New measurement by sensor placed in the prunings	Sensor probe with cable	Information platform server (and On-board control unit PC)	Consumer, producer, Administrator
Pruning characteristics	QR code on labels of pruning transported	Scanner	Information platform server (and On-board control unit PC)	Consumer, Trader, Transporter, Administrator
GPS coordinates	Earth orbiting satellites	GPS antenna and GPS receiver	Information platform server (and On-board control unit PC)	Transporter, Administrator

Table 4. Smart box data recording and transferring intervals.

Parameter	Unit	Recommended Recording Interval	Recommended Transfer Interval	Comment
Temperature	°C	15 min	30 min	The intervals can be adjusted by On-board control unit
Relative humidity	%	15 min	30 min	The intervals can be adjusted by On-board control unit
GPS coordinates (latitude, longitude)	Decimal degrees	5 min	30 min	-
QR reading	Text	Twice	Twice	QR reading is done during loading the product and after end of delivery (unloading)

4.3. Order Management Using the Platform

Order management is one of important contributions of the information platform. This functionality enables online pruning biomass trading (i.e., pruning e-trade). Two types of orders are handled in this SLS: Pruning biomass order and Transport service order.

Pruning biomass order management: in this smart system, the contact could be directly between farmers and consumers or between farmers and traders as well as between traders and consumers (see Table 5).

Table 5. Order tracking using smart logistics system.

Pruning Title (or Lot No.)	Offer (to Sale)	Order (to Purchase)	Offer (As Purchased or after Additional Processing)	Order
To be assigned by pruning provider	by producer	by consumer	-	-
To be assigned by pruning provider	by producer	by trader	by trader	by consumer

Transport service order management: traders and consumers order the transport service. Transporters are order receivers. The transporter has the right to accept or reject the service ordered by traders or consumers while the traders and consumers have the right to choose transporters from a list of transporters registered in the SLS.

In both cases of order management (product and service orders), there should be a predefined contractual agreement between the concerned actors using the SLS platform. Such a contract number regarding the concerned actor can be registered (by the administrator using 'Add New User' link) within the profile of the respective user. The financial transaction between the actors will be based on this procurement agreement. There is no financial transaction to be done on this platform and no financial related data to be documented in the spreadsheet-based database.

In this study, three Smart box prototypes have been developed; the SLS system has been tested in Sweden, Germany, and Spain with effective results and improvement feedbacks. Detailed performance evaluation results of this SLS will be reported separately. This SLS has the ability to generate optimum delivery routes for biomass products. It was found to be an important tool to improve the performance of pruning biomass supply chain management and product traceability by reducing biomass loss (in terms of quality and quantity) and increasing the quality of solid fuel and delivery service, creating effective pruning marketing channels, and reducing logistics and transaction costs.

As a final remark, it must be noted that the improvements in logistics that the system facilitates are applicable to any type of biomass. Therefore, the whole system can be easily adapted to improve the management of biomass like forestry woodchips, agrarian residues, or energy crop production and marketing.

5. Conclusions

The smart logistics system (SLS) has been designed with the objective to effectively and efficiently manage the pruning biomass supply chain leading to more sustainable biomass utilization. The smart system allows all recognized actors such as farmers, traders, transporters, and end users (power plants) to have access to the web-based application platform and provide and receive data, order, and activity reports. The SLS has four major components:

I. **Smart box (Sensor unit)**: for performing measurements of relative humidity, temperature, location records, truck speed and routes, and information associated with QR codes.

II. **On-board control unit**: for performing route planning, monitoring measurements by the sensor unit, and controlling data/information flow between Smart box tools and the web-based Information platform.

III. **Information platform**: for performing documentation and data sharing, facilitating interaction between different actors and biomass trading, and facilitating the management of pruning supply chain and traceability.

IV. **Central control unit**: for providing human-computer interface (linked to the Information platform and On-board control unit) and administering the whole pruning trading and delivery system.

Three Smart box prototypes have been developed and the whole SLS system has been tested in Sweden, Germany, and Spain with effective results and improvement feedback. This smart logistics system has the ability to generate optimum delivery routes for biomass products, and its application leads to economic and environmental efficiency of the pruning supply chain. It also describes how biomass producers, traders, transporters, and end users use the SLS and play their roles in promoting an economically effective and environmentally more sustainable logistics system for a pruning biomass trading system.

Author Contributions: G.G. conceived the SLS. G.G., D.G., T.B. and S.-O.O. and designed the SLS. S.-O.O. led the development of prototypes. G.G., D.G. and T.B. participated in monitoring and providing feedback for the development of SLS. T.B. and G.G. wrote the paper.

Funding: This research was funded by [European Union Seventh Framework Programme (FP7/2007-2013)] grant number [312078]. This work was part of international project EuroPruning: "Development and implementation of a new and non-existent logistics chain for biomass from pruning".

Conflicts of Interest: The authors declare no conflict of interest.

References

1. Iakovou, E.; Karagiannidis, A.; Vlachos, D.; Toka, A.; Malamakis, A. Waste biomass-to-energy supply chain management: A critical synthesis. *Waste Manag.* **2010**, *30*, 1860–1870. [CrossRef] [PubMed]
2. Vela'zquez-Marti', B.; Ferna'ndez-Gonza'lez, E.; Lo'pez-Corte's, I.; Salazar-Herna'ndez, D.M. Quantification of the residual biomass obtained from pruning of vineyards in Mediterranean area. *Biomass Bioenergy* **2011**, *35*, 3453–3464. [CrossRef]
3. Vela'zquez-Marti', B.; Ferna'ndez-Gonza'lez, E.; Lo'pez-Corte's, I.; Salazar-Herna'ndez, D.M. Quantification of the residual bioma ss obtained from pruning of trees in Mediterranean olive groves. *Biomass Bioenergy* **2011**, *35*, 3208–3217. [CrossRef]
4. Esteban, L.S.; Carrasco, J.E. Biomass resources and costs: Assessment in different EU countries. *Biomass Bioenergy* **2011**, *35*, S21–S30. [CrossRef]
5. Mobile Technologies GSM. Available online: http://www.etsi.org/technologies-clusters/technologies/mobile/gsm?highlight=YToxOntpOjA7czozOiJnc20iO30 (accessed on 10 July 2018).
6. General Packet Radio Service, GPRS. Available online: http://www.etsi.org/index.php/technologies-clusters/technologies/mobile/gprs (accessed on 9 July 2018).
7. Santa, J.; Zamora-Izquierdo, M.A.; Jara, A.J.; Gómez-Skarmeta, A.F. Telematic platform for integral management of agricultural/perishable goods in terrestrial logistics. *Comput. Electron. Agric.* **2012**, *80*, 31–40. [CrossRef]
8. IT Glossary. Available online: https://www.gartner.com/it-glossary/telematics (accessed on 10 July 2018).
9. Oliveira, R.R.; Cardoso, I.M.G.; Barbosa, J.L.V.; da Costa, C.A.; Prado, M.P. An intelligent model for logistics management based on geofencing algorithms and RFID technology. *Expert Syst. Appl.* **2015**, *42*, 6082–6097. [CrossRef]
10. Xiao, X.; Wang, X.; Zhang, X.; Chen, E.; Li, J. Effect of the quality property of table grapes in cold chain logistics-integrated WSN and AOW. *App. Sci.* **2015**, *5*, 747–760. [CrossRef]
11. Information Technology—Automatic Identification and Data Capture Techniques—QR Code Bar Code Symbology Specification. Available online: https://www.iso.org/obp/ui/#iso:std:iso-iec:18004:ed-3:v1:en (accessed on 10 July 2018).
12. The Implications of Electronic Trading in Financial Markets. Committee on the Global Financial System (CGFS). Bank for International Settlements. Available online: http://www.bis.org/publ/cgfs16.pdf (accessed on 18 June 2018).
13. Hendershott, T. Electronic trading in financial markets. *IT Prof.* **2003**, *4*, 10–14. [CrossRef]
14. Legris, P.; Ingham, J.; Collerette, P. Why do people use information technology? A critical review of the technology acceptance model. *Inf. Manag.* **2003**, *40*, 191–204. [CrossRef]
15. Yi, M.Y.; Hwang, Y. Predicting the use of web-based information systems: Self-efficiency, enjoyment, learning goal orientation, and the technology acceptance model. *Int. J. Hum.-Comput. Stud.* **2003**, *59*, 431–449. [CrossRef]
16. Types of QR Codes. Available online: http://www.qrcode.com/en/codes/ (accessed on 18 June 2018).

17. Bosona, T.; Gebresenbet, G.; Olsson, S. Traceability System for Improved Utilization of Solid Biofuel from Agricultural Prunings. *Sustainability* **2018**, *10*, 258. [CrossRef]
18. Recursos Gráficos Para Todos. Available online: https://www.freepik.es/ (accessed on 18 June 2018).
19. Cargolog Impact Record System. Available online: http://mobitron.se/en/cargolog-impact-recorder-system/power-9500-2/ (accessed on 16 July 2018).
20. Official Website of Europruning Application Platform. Available online: http://europruning.mobitron.se/public/en (accessed on 16 July 2018).
21. Cargolog Impact Record System. Available online: http://mobitron.se/en/cargolog-impact-recorder-system/ (accessed on 16 July 2018).
22. Yu, W.; Chavez, R.; Jacobs, M.A.; Feng, M. Data-driven supply chain capabilities and performance: A resource-based view. *Transp. Res. Part E Logist. Transp. Rev.* **2018**, *114*, 371–385. [CrossRef]

applied
sciences

MDPI

Article

Evaluation of a Smart System for the Optimization of Logistics Performance of a Pruning Biomass Value Chain

Techane Bosona [1,*], Girma Gebresenbet [1], Sven-Olof Olsson [2], Daniel Garcia [3] and Sonja Germer [4]

[1] Department of Energy and Technology, Swedish University of Agricultural Sciences, P.O. Box 75651 Uppsala, Sweden; girma.gebresenbet@slu.se
[2] Mobitron AB, 561 46 Huskvarna, Sweden; soo@mobitron.se
[3] Fundacion Circe Centro de Investigacion de Recursos Yconsumos Energeticos, CIRCE, Mariano Esquillor Gomez, 15 50018 Zaragoza, Spain; daniel.garcia@fcirce.es
[4] Leibniz-Institut für Agrartechnik und Bioökonomie e.V. (ATB), Max-Eyth-Allee 100, 14469 Potsdam, Germany; sgermer@atb-potsdam.de
* Correspondence: techane.bosona@slu.se; Tel.: +46-18671851

Received: 24 July 2018; Accepted: 16 October 2018; Published: 19 October 2018

Featured Application: This work can be used as benchmark for performance evaluation of tools to improve supply chain management in general and solid biofuel supply chain in particular. The potential application is for integrated goods distribution management that promotes efficient resource utilization and sustainability.

Abstract: The paper presents a report on the performance evaluation of a newly developed smart logistics system (SLS). Field tests were conducted in Spain, Germany, and Sweden. The evaluation focused on the performance of a smart box tool (used to capture information during biomass transport) and a web-based information platform (used to monitor the flow of agricultural pruning from farms to end users and associated information flow). The tests were performed following a product usability testing approach, considering both qualitative and quantitative parameters. The detailed performance evaluation included the following: systematic analysis of 41 recordable parameters (stored in a spreadsheet database), analysis of feedback and problems encountered during the tests, and overall quality analysis applying the product quality model adapted from ISO/IEC FDIS 9126-1 standard. The data recording and storage and the capability to support product traceability and supply chain management were found to be very satisfactory, while assembly of smart box components (mainly the associated cables), data transferring intervals, and manageability could be improved. From the data retrieved during test activities, in more than 95% of the parameters within 41 columns, the expected values were displayed correctly. Some errors were observed, which might have been caused mainly by barriers that could hinder proper data recording and transfer from the smart box to the central database. These problems can be counteracted and the performance of the SLS can be improved so that it can be upgraded to be a marketable tool that can promote sustainable biomass-to-energy value chains.

Keywords: smart logistics system; information platform; pruning biomass; performance evaluation; product quality model; product usability testing

---

1. Introduction

The use of pruning biomass for renewable energy production is one of the renewable energy uses being promoted in Europe. This study was part of an EU EuroPruning project, "Development and implementation of a new and non-existent logistics chain for biomass from pruning", which was aimed at developing improved logistics chains for biomass from agricultural pruning residues [1]. This includes the development of a new decision tool for a pruning biomass trading system and logistics management (see Figures 1 and 2). To fulfill this, a smart logistics system (SLS) was developed as described by Gebresenbet et al. [2]. Computer technology has changed the historical face-to-face communication in trading systems [3], and the SLS enables utilization of recent technology.

Understanding and managing the material and information flow (see Figure 1) enables efficient and effective utilization of resources. The SLS developed for this purpose has been reported in detail in [2]. But there are associated research questions: What is the performance of the SLS tool in real conditions? What is the feedback of end users of the tool?

The objective of this study was to evaluate the performance of the SLS under field conditions. This is important, as the SLS is a newly developed tool and evaluating its performance is important. This report describes the methodology applied and the results obtained from applying the SLS tool in real conditions. This is important for replication of the study and improvement of the SLS tool.

1.1. Pruning Biomass Logistics Chain

Figures 1 and 2 describe activities at different stages of a logistics chain. Pruning fruit trees (removing top or unwanted branches) (see Figure 2a) is done by farmers on an annual or biennial basis to maintain the desired tree form and structure and to increase the productivity of fruit trees. Pruning harvesting is often integrated with chipping or baling (see Figure 2b). The harvested product can be stored in the form of bales or chips (see Figure 2c).

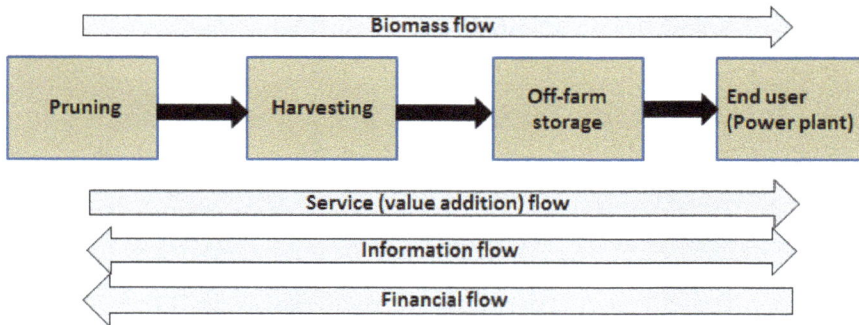

Figure 1. Typical pruning biomass logistics chain.

Storage of biomass is related to seasonal variability and logistics operation cost, and storage locations should be efficient to reduce transportation and operation costs. The location of biomass storage can vary depending on the nature of the biomass supply chain [4]: on-field storage, intermediate storage (between the fields and power plants), and storage facility at the biomass power plant. Biomass transportation is an important activity in biomass-to-energy systems. This includes both on-farm transportation and main transportation (from farm to storage or power plant).

a. Pruning activities

b. Harvesting of pruni chipping (left) and baling (right,

c. Storage of biomass in the form of bales and chips

d. Boiler for biomass with combustion chamber (at power plant)

Figure 2. Different stages of logistics chain in the process of biomass-to-energy conversion.

1.2. Smart Logistics System and Its Utilization by Different Actors

The SLS has four major components, as indicated in Figure 3 and presented in detail by Gebresenbet et al. [2]. The smart box is a sensor unit for measuring parameters such as relative humidity, temperature, geographic position, and route tracking, and information associated with quick read (QR) codes. The onboard control unit is used for planning transport routes and monitoring recordings by Cargolog. The information platform is used for documentation and data sharing, and to facilitate biomass trading and management of the pruning supply chain and traceability. The central control unit is a point of administration of the biomass trading and logistics system and links the information platform and onboard control unit.

The two components of the SLS (the smart box and information platform) have been integrated so that the smart box is functionally connected with the web address of the information platform [5]. The smart box records and transmits data to the central information platform. The information platform is designed to facilitate interactions among biomass supply chain actors and data collection, as well as the management of the entire logistics of the pruning biomass supply chain (see Figure 2 and Table 1). Table 1 presents how different actors use the platform to interact regarding the flow of materials and related information along the pruning biomass supply chain.

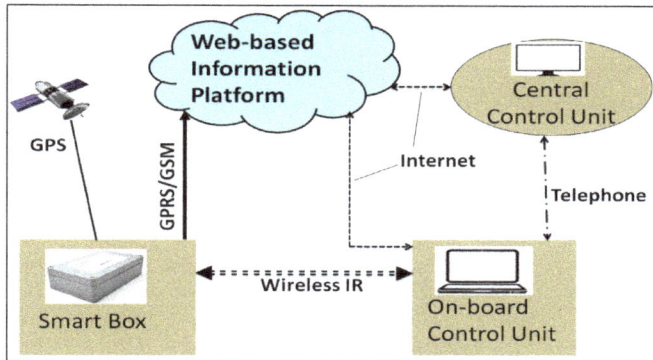

Figure 3. Major components of the smart logistics system. Reproduced from permission of [2] (MDPI, 2018).

Table 1. Actions to be performed by main actors of the pruning supply chain.

Action	Actor	Description of Actions to Be Performed
Add new user	Administrator	Logistics manager who administers the smart system can add new users to the system. New label codes will be assigned to new actors.
Add new pruning	Farmers	Farmers have to input pruning quality parameters required by end users.
Edit pruning	Farmers	Farmers can update input information such as pruning quality parameters while it is in "pending" status.
Delete pruning	Farmers	Farmers can delete pruning as long as it is still pending and has not been flagged as "ready".
Make ready	Farmers	Farmers can flag pruning as "ready" so that consumers can view that pruning and order it.
Generate QR code	Farmers	Farmers can generate a quick read (QR) code for "pending", "ready", and "ordered" pruning.
Order pruning	Consumers	Consumers order pruning based on quality parameters. Consumers select a transporter from a list registered in the central system.
Add to shipment	Transporters	Transporters select orders to be included in one shipment in order to plan the route between locations of chosen orders.
Plan route	Transporters	Transporters generate driving instructions for source and destination locations for all orders included in one shipment.
Add to pruning	Traders	Traders use the tool to select which orders are part of a new treated pruning offer. Trader can add new pruning quality parameters, e.g., if bales are chipped.

The SLS will be used to improve the performance of the biomass supply chain focusing on pruning biomass. In evaluating the performance of supply chain management, it is important to identify the performance measurement metrics within the context of the supply chain under evaluation [6]. The overall performance of a supply chain depends on the role of each stakeholder in the chain [6,7]. Shashi et al. [6] discussed that stakeholders' interest, value addition, and partners' performance play important roles in the overall performance of a supply chain.

The rest of this paper is structured as follows. Section 2 describes the testing of the SLS tool and evaluation methodology. Section 3 presents the performance evaluation results and discussion, while Section 4 presents the conclusion.

2. Testing and Evaluation Methodology

2.1. Usability Testing

Usability testing is an evaluation method that has the largest impact on product [8]. It is one method of evaluating the learning and use of new products. In this study, the usability testing approach was used considering six basic characteristics: system to evaluate, focus, participants, tasks, data, and results [8]. These characteristics were applied as explained in Table 2. Both qualitative and quantitative data analyses were used in the evaluation of SLS.

Table 2. Usability testing characteristics.

Basic Characteristics	Description as Applied in SLS Testing
System to evaluate	Used to evaluate the SLS. Usability testing is an evaluation approach that can be applied to evaluate almost any product or technology; for example, software for database management, network management tools, early-stage prototypes, and related help manuals.
Focus	The focus was on usability of the SLS. The test was intended to validate the first release of the smart system (smart box and information platform) prototype, with less consideration of marketability of the product.
Participants	The intended main participants were (potential) end users of the smart system who were actors of the pruning supply chain (farmers, pruning traders, transporters, power plants). Test participants and administrators interacted during tests. Training and instruction guides were provided before starting. During this testing, research centers such as Leibniz Institute for Agricultural Engineering Potsdam-Bornim (ATB) in Germany, Research Centre for Energy Resources and Consumption (CIRCE) in Spain, and the Swedish University of Agricultural Sciences (SLU) as well as Gruyser (transport dealer) participated actively.
Tasks	The smart system has different functional features. Each participant considers the specific functionality of the system that best fits him/her. Test participants who act as farmers perform tasks that the pruning producers perform, while traders perform what processors and traders of biomass perform using the system. Similarly, transporters and consumers test features of the system that serve the transport company and end consumers of the pruning biomass.
Data	Pruning biomass–related data were recorded, stored in a database, and analyzed. Both problems and positive aspects noticed during tests were analyzed. Data collected with three prototypes were used to evaluate the functionality of similar features of the system but with different smart box prototypes. Accordingly, a data triangulation test approach was used, especially to test data storage and display capability of the central information platform by analyzing data sourced from testing with three different prototypes.
Results	Test results were used to improve the smart system and communicate the research outputs. These analysis results are documented, archived, and used to identify what problems surfaced and how to solve them.

2.2. Metrics and Product Quality Model for Evaluation of SLS

2.2.1. Metrics for Performance Evaluation

The SLS deals with data recording, transfer, and storage in a centralized database through the integrated action of a smart box and a web-based information platform (see Figure 3). The functional features of the smart box tool and information platform were integrated using Cargolog PC software, Cargolog FAT90 V2 (Mobitron, Huskvarna, Sweden). In this report, as part of the evaluation of the smart system, metrics were selected and adapted from international standards for systems and software quality requirements and evaluation (ISO/IEC 25010:2011) as described in [9] and Table 3.

Table 3. Quality attributes considered for performance evaluation of the SLS.

Attribute	Description
Functionality	How easy it is for the system to integrate its functional units while maintaining the security and accuracy of service provided
Reliability	How well the smart system can provide service with required precision
Usability	How easy it is for the smart system to learn, operate, and analyze the data and make decisions
Efficiency	Amount of resources and time required by the system to perform its intended function
Maintainability	How easy it is for the system to identify and fix an error
Portability	How easy it is to move the smart box from one place to another and to move the application platform from one server to another

2.2.2. Product Quality Model

The general product quality model for internal and external quality evaluation of information technology (software product) described in ISO/IEC FDIS 9126-1 (see Figure 4) was adapted for evaluation of the smart system [10]. The evaluation is based on the definitions given in international standards for information technology–based product quality (ISO/IEC FDIS 9126-1) as indicated in Table 3 and Figure 4. The detailed analysis results are presented in Table 9.

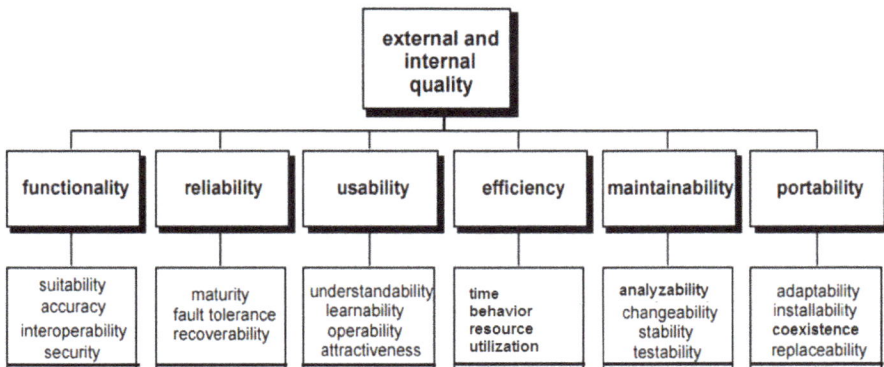

Figure 4. Product quality model for internal and external quality assessment (adapted with modification from ISO/IEC FDIS 9126-1).

2.3. Smart System Testing

2.3.1. Smart Box Installation and Training

Before starting the actual testing in Sweden, Spain, and Germany, training/guidance was provided on proper installation and operation of the smart system. In addition to this training, two user guide documents were prepared and provided to actors involved in testing the tool. The first user guide enables smart box users to install Cargolog PC software on computers for onboard monitoring purposes and to mount the smart box tool (with metal case) appropriately on the truck (see Figure 5). The second guide enables users to use the information platform efficiently.

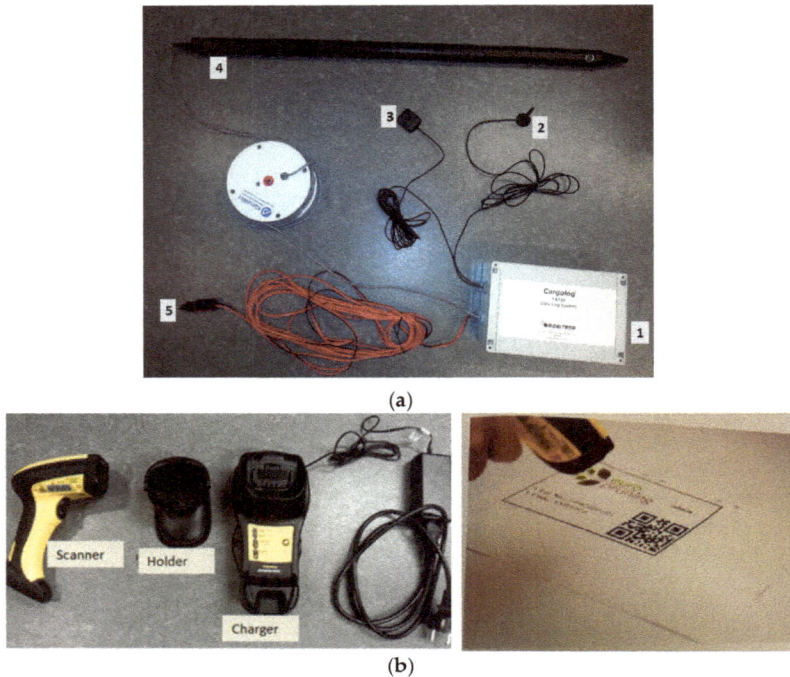

Figure 5. (**a**) Smart Box assembly: (1) smart box protected with metal case; (2) GPRS/GSM antenna; (3) Global Positioning System (GPS) antenna; (4) temperature and humidity measuring sensor probe; (5) power cable. (**b**) Power^TM 9500 scanner with its components.

2.3.2. Smart System Field Testing

In this study, three prototypes were developed. The testing and evaluation activities were done with all the three prototypes during 2015–2016. Although testing of the smart system was done during different stages of the development of the system, this report presents only the testing and evaluation results of the final smart system prototype used by end users during final testing activities. The smart boxes were used at farm-to-storage and storage-to-consumer transport stages.

3. Evaluation Results and Discussion

3.1. Data Retrieved from Central Database of the Smart System

In the spreadsheet-based central database, the data are stored in 41 columns (see Table 4 and Appendix A). Data entered by actors directly on the platform and data transferred from the smart box are stored mainly in the 41 columns, while some additional information can be displayed and visualized on the platform. It is important to test the performance of the smart system by considering how correctly the recorded values and information are displayed in each of the 41 columns. This enables us to understand the performance of the smart system and identify the columns where incorrect information might be displayed. In general, during this test, about 104 records of product lines (with specific lot numbers) were retrieved from the SLS database. This in turn enabled us to identify functional features of the SLS linked to errors in displayed values and to figure out what further improvement would be needed to upgrade the system.

Table 4. Parameters and how they are implemented in the spreadsheet-based database on the platform server under 41 columns.

Column No.	Column Header as Implemented in Database	Description of Parameters under Each Column Header
1	title	Name given to the biomass registered with a unique lot number
2	status	Status of the registered biomass along logistics chain (e.g., delivered or not)
3	providerID	Label code assigned to the provider (i.e., farmer or trader) of the biomass
4	transporterID	Label code assigned to the transport company delivering the biomass
5	consumerID	Label code assigned to the end user that ordered (purchased) the biomass
6	shipmentID	Number to identify the delivery route of the biomass with the indicated lot number
7	Cargolog SerialNo	Identification number of smart box used by the transporter while transporting this specific biomass
8	deliveryCode	Product delivery identification number to monitor the traceability of product movement along the chain
9	latitude	Latitude of the location where the biomass is to be picked up
10	longitude	Longitude of the location where the biomass is to be picked up
11	destinationLongitude	Longitude of the location where the biomass is to be delivered
12	destinationLatitude	Latitude of the location where the biomass is to be delivered
13	lotNumber	LotNumber associated with the specific biomass product under consideration
14	quantity	Quantity of biomass associated with each lot number
15	species	Sources of agricultural pruning where biomass product is produced
16	originClassification	Identification of origin and source according to classification in standard EN-ISO 17225-1:2014
17	tradedForm	Information on traded form of biomass
18	particleSize	Particle size distribution of chips
19	baleDiameter	Diameter of traded bale
20	moistureContent	Moisture content of biomass
21	ashContent	Ash content of biomass
22	densityChips	Density of traded chips
23	densityBale	Density of traded bales
24	calorificValue	Caloric value of traded biomass
25	cropCharacteristics	Additional information regarding biomass production and source
26	pruningDate	Date when farmer pruned fruit or other trees
27	collectionDate	Date when pruning is gathered (on farm) to be transported or processed (e.g., chipping)
28	storageDays	Duration of storage at storage site, in days
29	piled	Information to identify whether biomass is stored as large pile or spread as small heaps
30	chemicallyTreated	Information regarding whether biomass product is chemically treated or not
31	covered	Information indicating whether biomass storage is covered or not
32	moved	Information to identify whether machinery is used to move biomass at storage site
33	contaminated	Information to identify whether biomass has been contaminated with impurities
34	useStorage	Information indicating whether storage is used and if it is located at a different position than the address of farm or trader
35	storageLongitude	Longitude value of storage
36	storageLatitude	Latitude value of storage
37	pickupDate	Date and time when the product is picked up for delivery to intended destination

Table 4. Cont.

Column No.	Column Header as Implemented in Database	Description of Parameters under Each Column Header
38	deliveryDate	Date and time when the product is delivered to intended destination
39	createdDate	Date and time indicating when the product is registered on the platform server
40	modifiedDate	Date and time indicating when the status (see column 2) of the product along the delivery process is changed by next actor
41	modifiedByUserID	Label code assigned to the actor who changed the status of product (see column 40)

3.2. Data on Product Delivery Information

Using the SLS tool, the system administrator can easily identify products entered into the system, products for which transport routes are planned, and products that have been delivered to end users. Table 5 indicates that, out of the recorded 104 products, route planning was tested for 54%, while actual product delivery of 16% was confirmed according to the data captured by the smart system.

Table 5. Number of products registered in platform during field testing.

Country	Recorded Product with Specific Lot Number (N)	Route Planned for Product Delivery		Product Delivered to Consumer *	
		Number	%	Number	%
Sweden	28	21	75	14	50
Spain	65	31	48	NA	NA
Germany	11	4	36	3	27
Total	104	56	54	17	16

* Percentage of delivered products does not necessarily indicate efficiency, because some registered products may not be ordered by traders/consumers; for some products, route may be planned but they may not be delivered to consumers. NA, not available.

3.3. Data Recorded and Visualized Directly on Platform

The feedback from users regarding the attractiveness and user-friendliness of the functional features of the smart system is very important to improve the system. The feedback may be on utilization of functional features to perform specific actions and/or visualize the results.

Samples of visualizations of recorded data are presented in Figures 6–9, which describe information gathered with smart box 2015085119 (see Figure 6) at a test site in Germany during transport of pruning products registered as Field3, lot number 151008B03R1.

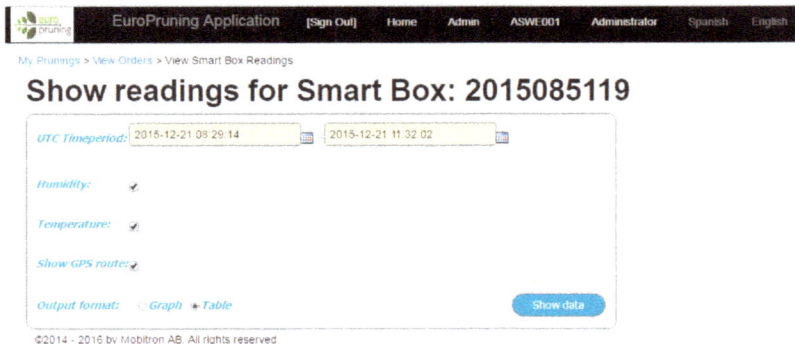

Figure 6. Visualization of smart box reading link indicating its Cargolog serial number.

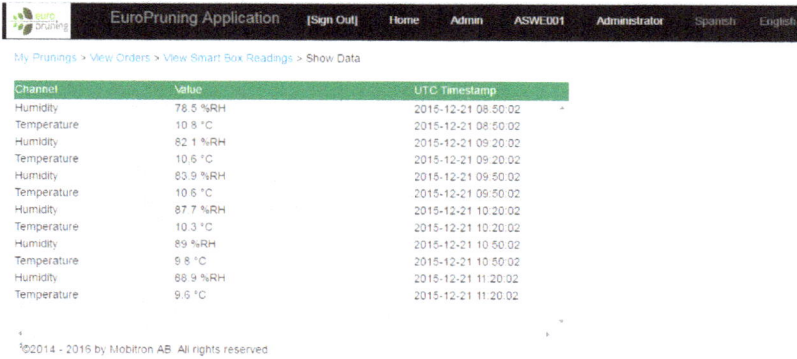

Channel	Value	UTC Timestamp
Humidity	78.5 %RH	2015-12-21 08:50:02
Temperature	10.8 °C	2015-12-21 08:50:02
Humidity	82.1 %RH	2015-12-21 09:20:02
Temperature	10.6 °C	2015-12-21 09:20:02
Humidity	83.9 %RH	2015-12-21 09:50:02
Temperature	10.6 °C	2015-12-21 09:50:02
Humidity	87.7 %RH	2015-12-21 10:20:02
Temperature	10.3 °C	2015-12-21 10:20:02
Humidity	89 %RH	2015-12-21 10:50:02
Temperature	9.8 °C	2015-12-21 10:50:02
Humidity	88.9 %RH	2015-12-21 11:20:02
Temperature	9.6 °C	2015-12-21 11:20:02

Figure 7. Visualization of measured parameter values (temperature and humidity as recorded on 21 December 2015, from 08:29:14 to 11:32:02 and displayed as a table).

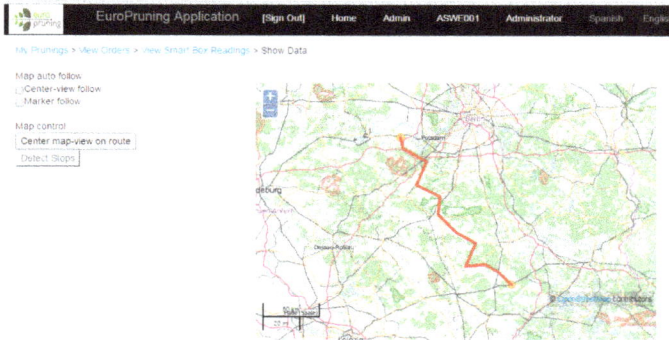

Figure 8. Delivery route based on GPS coordinates recorded by the smart box.

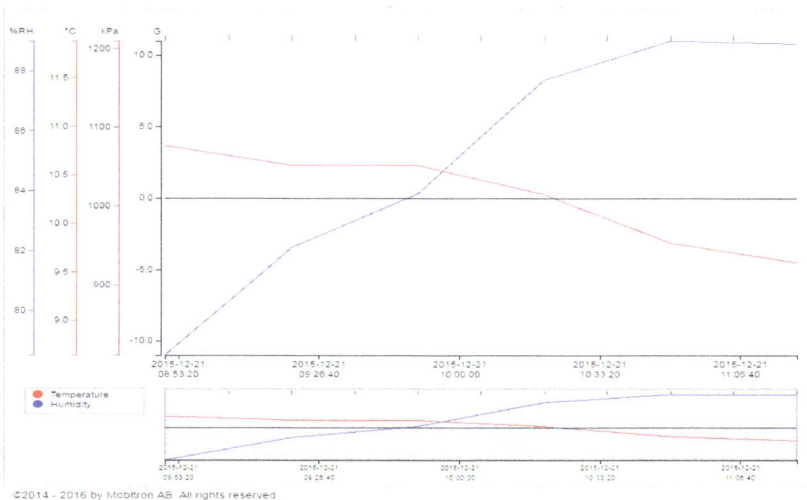

Figure 9. Example of smart box reading visualization: graphic presentation of measured parameters (along with displayed values of temperature, humidity, and GPS coordinates as recorded on 21 December 2015, from 08:29:14 to 11:32:02).

The relative humidity data retrieved from the testing activity in Germany are presented in Table 6, together with the moisture content of the biomass at the storage site. It must be noted that relative humidity indicates the humidity in the air inside the truck box after loading the woodchips. In the initial moment, the moisture measured was just the air moisture at the site. Once the probe (see Figure 5a) was inserted into the truck box, the air in the internal void volume among the woodchips increased its moisture content until it reached equilibrium with the moisture of the woodchips.

Table 6. Example of moisture content data compiled from test in Germany.

Product (Lot Number)	Recorded Moisture Content (%)			Remark
	Storage (start)	Storage (end)	Relative Humidity (RH) Measured by Smart Tool	
Field1 (151007B01R1)	31.68	31.68	NA	Data were not transferred properly
Field3 (151008B03R1)	31.68	13.36	78.5–88.9	RH measured during transport
Field7 (151008B05R1)	31.68	13.36	91.8–93.2	RH measured during transport
Field8 (151008B04R1)	31.68	31.68	NA	Route planned, but pruning was not transported

3.4. Information Flow and Product Traceability Performance of the Smart System

The information gathered using the smart system was rich enough to construct an effective traceability map of each registered product along the pruning supply chain [11]. Detailed pruning quality parameters and the product traceability system are provided in [12,13]. The SLS was integrated with the biomass traceability system, with pruning quality and related traceability parameters. The usability testing indicated that all of the traceability-related parameters were implemented on the information platform of the SLS (see Figure 3). The identification of producers (provider ID), transporters (transporter ID), and consumers (consumer ID), and the specific name (title) and production lot number assigned to each registered product are mandatory parameters to trace product movement along the pruning supply chain (see Table 4). Table 7 indicates that except for smart box number (Cargolog serial no.), delivery code, pickup date, and delivery date, all important parameters are recorded correctly. The errors in records of Cargolog serial no., pickup date, and delivery date could be mainly due to barriers in data transfer from the smart box to the central database. In the case of Spain, the Cargolog serial no., pickup date, and delivery date were not recorded at all in the central database, indicating that the information was not properly transferred from the smart box tool to the central database.

For illustration purposes, a product from the test in Germany is considered here. For a product registered with the title "Field7" and lot number 151008B05R1, the producer, transporter, and consumer are identified as FDEU020, DDEU012, and CDEU013, respectively (see Appendix A). The corresponding complete product identification code can be constructed as FDEU020-DDEU012-CDEU013. Whenever a product traceability issue is initiated, further detailed traceability-related information can be retrieved from gathered data for each stage of the logistics chain. This can be done by an experienced administrator of the smart system who can interpret all recorded values in each of the 41 columns of the spreadsheet-based central database. The data barriers in delivery date (column 38) and Cargolog serial number (column 7) may hinder the system administrator from confirming the final delivery of the product to the intended consumer. In such cases, the administrator should find additional information by telephone or internet conversation, for instance, by introducing an option (on the platform) for the end consumer to confirm the product delivery. This enables the system administrator to easily construct the product delivery identification code and strengthen the traceability and management of the entire pruning supply [10].

Table 7. Major parameters for traceability information continuity and their recording performance

Column Header (Indicating Parameter) as Implemented in Central Database	Correctly Recorded Values of Registered Products for Which Route Planning Was Done (%)			
	Sweden N = 21	Spain N = 31	Germany N = 4	Relevance Level of the Parameter for Traceability Information
title	100	100	100	Mandatory
Status	100	100	100	Supporting information
providerID	100	100	100	Mandatory
transporterID	100	100	100	Mandatory
consumerID	100	100	100	Mandatory
cargologSerialNo	67	0	75	Supporting information
deliveryCode	91	68	100	Very important
lotNumber	100	100	100	Mandatory
pickupDate	67	0	75	Supporting information
deliveryDate	67	0	50	Very important
modifiedByUserID	100	100	100	Supporting information

N, number of routed chains as indicated.

The smart system was developed to effectively gather, monitor, store, and analyze data to improve pruning logistics management (even though the tool could be utilized for any biomass supply chain). As indicated in Table 8, this integrated system enables improved performance of pruning supply chain management and product traceability by reducing biomass loss (in terms of quality and quantity), increasing the quality of solid fuel and delivery service, and reducing logistics and transaction costs [2,11,14]. Therefore, evaluation results indicate that the SLS is important tool for management of the biomass value chain and trading system. It enables tracing and tracking of the product, controls product quality, facilitates information flow, reduces management cost, and enables managers to manage the entire supply chain. Table 8 indicates that most of the major functional features performed satisfactorily.

Table 8. Analysis results of overall satisfactory level of major functionalities.

Functional Features	Main Actor Responsible for the Activity	Satisfaction Level (Ease of Use of the Tool and Precision of Results) *				
		1	2	3	4	5
Data entering (uploading) onto web-based platform	Biomass producer				√	
Data recording by smart box	Transporter				√	
Data transfer from smart box to central database	Transporter			√		
Searching for available product and ordering for purchase	Trader and consumer					√
Whole pruning supply chain management support	System administrator				√	
Product quality monitoring and traceability capability	System administrator, consumer				√	

* This analysis was done assuming that the system will be used by trained actors and system managers. 1 = poor; 2 = fair; 3 = good; 4 = satisfactory; 5 = very satisfactory.

3.5. Evaluation of Smart System Using Product Quality Model

In this section, the product quality model is used to analyze the performance of the smart system. The feedback from the testing was systematically used as input information for the product quality model. This input includes observations of the experts participating in testing the tool (see Table 9).

Table 9. Evaluation of the SLS using product quality model for internal and external quality assessment (adapted with modification from ISO/IEC FDIS 9126-1).

Attribute	Smart Logistics System Performance Description
Functionality	**Suitability**: The smart system is designed for management of pruning biomass logistics activities. It is suitable for collecting and managing pruning-related data and provides service for all registered actors of the pruning supply chain. **Completeness**: The system enables complete data to be obtained regarding the biomass quality and quantity as well as its flow along the supply chain. **Accuracy**: For more than 95% of data parameters included in the spreadsheet-based database, the system accurately displays the recorded data and information. In some cases, errors were noticed, mainly due to problems encountered during data transfer from the smart box. **Interoperability**: The smart system effectively integrates different functional units such as GPS and GPRS/GSM devices, temperature and humidity recording tools, and the information platform. The data stored in the central database can be downloaded and easily analyzed using spreadsheets, facilitating further interoperability of the system.
Efficiency	**Time saving**: The tool enables transporters to plan their best transport routes, reducing driving time and distance. Each producer can use the online platform to announce its products while end users and/or traders can buy the ready products and transport services easily online, where the system is managed by an administrator. **Resource utilization**: Once the pruning biomass-to-energy value chain is initiated, the smart system facilities the coordinated utilization of available resources owned by actors in the chain. **Capacity**: The smart system has functional capacity to record, transfer, and store adequate data along the pruning biomass logistics chain from producer to end user. It facilities the traceability of pruning quality and logistics management, leading to economic efficiency.
Usability	**Appropriateness**: The smart system is appropriate, as it enables recording, documentation, and having adequate data centrally, which in turn facilitates the performance analysis and traceability of pruning biomass. **Learnability**: The tool has a guiding manual to facilitate training and learning to use the system. Once registered by the administrator, each actor can easily practice and use the information platform. **User error protection**: Once training is provided, there are fewer user errors when using the smart box. Once the power cable is plugged in, the recording will be triggered (started and ended) by a separate portable scanner used to read QR codes on biomass labels. While using the information platform, each registered actor can record and edit only the data he/she provides but not those of other users. Therefore, user error is minimized and can be easily corrected by the respective responsible actor. **Accessibility**: The smart system will be accessible to all interested actors involved in the pruning-to-energy value chain. However, each user should be registered by the system administrator first and get a specific identification code (labelling code as described in [2,10]). **Understandability**: The system can be used effectively if initial training and user guide documents are provided. However, how to interpret some displayed values and how to analyze the gathered data can be difficult for many users and should be handled only by experienced (well-trained) system administrators. **Attractiveness**: The web-based user interface information platform has integrated up-to-date Google Maps, which increases the aesthetics and attractiveness of the online platform. **Operability**: Components of the smart box are connected to a recording unit by cables [2] (see Figure 5). It has no plug-in and plug-out system for the cables increasing difficulty during operation, indicating that improvement is required. There should also be power on and off buttons for user-friendly operation.
Reliability	**Maturity**: The smart system is newly developed and tested for the first time. Further repeated tests and evaluation are recommended to increase its maturity level. **Availability**: The tool is newly developed and not available on the market. From feedback during testing activities, there is a potential market for the smart system, and the tool could be made available for marketing. **Fault tolerance**: The test results indicate that all registered pruning products were successfully recorded and stored in the central database with associated product quality characteristics. This indicates that a trader or consumer can confidently order any product registered on the information platform and made ready for sale. Other faults, if any, in relation to real-time data during transport may be tolerable. **Recoverability**: Once the product is registered in the central platform, much associated information will be generated as the product moves downward along the pruning logistics chain. This increases the recoverability of some missed data, if any.

Table 9. *Cont.*

Attribute	Smart Logistics System Performance Description
Security	**Confidentiality**: Personal information of registered actors is confidential, as it can only be accessed by system administrators (of the information platform). **Integrity**: The system well integrates data recorded by the smart box and related data provided by pruning biomass producers. **Accountability**: All registered users are accountable for the information they provide to intended users of the information platform. The system enables pinpointing damages that could happen at any stage along the logistics chain through the pruning traceability system integrated in the smart system. **Authenticity**: All registered actors have specific codes and access to and recognition of data and information they provided.
Maintainability	**Modularity**: The smart box components can be disassembled and reassembled by the prototype developer. This facilitates maintenance service. **Reusability**: The system can be maintained and reused. Maintenance service may be required on average once a year, and some parts of the smart box may need to be replaced. **Modifiability**: The information platform can be modified based on the interest of users or if additional service is required. **Testability**: Both the smart box and the information platform are testable. After appropriate maintenance, the system will be well tested before it is used.
Portability	**Adaptability**: The possibility of adapting to logistics of biomass other than pruning residues, such as forest wood products, was taken into consideration during the development of the smart logistics system. **Installability**: The smart box has a metal case with a magnetic foot for easy and appropriate mounting on a truck. The metal case is used to protect the smart box from mechanical damage, aggressive gases, liquids, and humidity. **Packaging**: The smart box needs to be shipped from its manufacturing place to end users. The packaging system for this should be hard enough for protection from damage due to impact. Improved packaging is recommended for this purpose.

3.6. Major Recommended Improvements

When the SLS was conceived, it was designed for use in supporting and improving biomass logistics, not only for pruning biomass, but for any type of biomass, such as forestry woodchips and herbaceous agrarian residues. For this purpose, further improvements are suggested so that the SLS can be effectively adapted for biomass-to-energy businesses. The performance test results indicate that the SLS performs data collection and storage in the central database satisfactorily. The stored data and other relevant information enable the tool to support product traceability and supply chain management very effectively. For further improvement and service quality, assembly of the smart box components (mainly the associated cables) and data transfer intervals as well as column arrangement in the spreadsheet format for data storage could be improved more.

In general, each batch of pruning biomass data entered into the online platform directly by actors (farmers, traders, etc.) was found to be well documented in the central database and easily visualized by intended users. However, some errors were noticed regarding data on parameters measured by the smart box (such as relative humidity, temperature, Global Positioning System (GPS) coordinates during biomass transport) and transmitted to the database. These errors may be caused by improper initialization of the smart box, unplugging the power cables before data transmission is complete, or unexpected barriers that could interrupt data transmission from the smart box to the online platform.

The smart system enables users to determine the transport distance and truck speed at the route planning stage. However, the real routes followed during transport are displayed only on maps. The actual driving distance and time (or speed) are not determined and displayed along the route maps. Through further improvement, these values can be either displayed on the maps or documented in the spreadsheet-based central database. To increase effective utilization of the information platform, additional functional options should be created (on the platform interface) for transporters and end consumers where they can confirm product delivery and acceptance, respectively. In addition, a billing system could be included as a component to facilitate trading of biomass products.

4. Conclusions

The testing and evaluation of a smart logistics system was carried out in order to identify functional limitations and important improvements to be implemented. The functional features of the smart box and centralized information platform were considered during testing. The features were designed to enable users to have an effective information acquisition, monitoring, and utilization system while promoting sustainable pruning for the energy value chain. All three smart box prototypes were tested together with the information platform. Based on tests done in Spain, Germany, and Sweden, the performance evaluation of the smart system was carried out by systematically analyzing:

> ➤ values of parameters recorded and stored in the spreadsheet database with 41 columns (for each pruning product registered with a specific lot number),
> ➤ feedback and problems encountered during the test, and
> ➤ selected performance metrics in relation to product quality model adapted from ISO/IEC FDIS 9126-1 standard.

The performance test results indicate that the smart system satisfactorily performed data collection and storage in the central database. The stored data and other relevant information enabled the tool to support product traceability and supply chain management very effectively.

In general, for each batch of pruning biomass (with specific lot number) entering and leaving the biomass supply chain, data entered into the online platform directly by actors (farmers, traders, etc.) were well documented in the central database and easily visualized by intended users. The information platform enabled users to display the locations of pruning farms, biomass storage and processing sites, energy plants, and the planned and/or actual delivery routes according to the needs of different actors, such as raw materials producers, traders, and consumers. However, some errors were noticed regarding the smart box identification numbers (expected to be transferred from the smart box to the central database during product pickup and delivery), product delivery codes (expected to be generated by the smart system), and product pickup and delivery times. These errors might have been caused by improper initialization (while triggering the smart box with a power scan), unplugging the power cables before data transmission was complete, or unexpected barriers that interrupted data transmission from the smart box to the online platform. To counteract these problems and increase information continuity along the pruning supply chain as well as effective utilization of the smart system, additional functional features should be created so that transporters and consumers can confirm product delivery and acceptance, respectively. This option enables retrieval of important information that could be missed due to problems encountered by the smart box during data transfer to the central database and increases system performance so that it can be upgraded to a marketable tool promoting sustainable biomass development for renewable energy generation.

Author Contributions: G.G. and S.-O.O. conceptualized the validation the SLS. G.G., D.G., T.B. and S.-O.O. designed the methodology of performance evaluation of SLS. S.-O.O., D.G., T.B., and S.G. participated in field test activities, monitoring and providing feedback for the development of SLS. T.B. wrote the paper while G.G. reviewed it.

Funding: This research was funded by the European Union Seventh Framework Programme (FP7/2007-2013), grant number 312078. This work was part of international EuroPruning project, "Development and implementation of a new and non-existent logistics chain for biomass from pruning".

Conflicts of Interest: The authors declare no conflict of interest.

Appendix A

Table A1. Sample of data recorded and stored in Spreadsheet data base during Smart System testing in Germany.

1	2	3	4	5	6	7	8	9	10	11
title	status	Provider ID	Transporter ID	Consumer ID	Shipment ID	Cargolog Serial No	Delivery Code	latitude	longitude	Destination Longitude
ATBTest	ordered	FSWE001	DDEU012	CSWE001		2015085119	FSWE001-CSWE001	52.403554	13.063443	13.97707995
ATBTest2	ordered	FSWE001	DDEU012	CSWE001		2015085119	FSWE001-CSWE001	52.305625	12.836196	13.9770799
ATBTest3	ordered	FSWE001	DDEU012	CSWE001		2015085119	FSWE001-CSWE001	52.484537	12.838255	13.9770799
ATBTest4	ordered	FSWE001	DDEU012	CSWE001		2015085119	FSWE001-CSWE001	52.277906	13.141066	13.9770799
ATBTest5	ordered	FSWE001	DDEU012	CSWE001		2015085119	FSWE001-CSWE001	52.514216	12.904860	13.9770799
Delivery 02	routed	FDEU004	DSWE001	CSWE001			FDEU004-CSWE001	51.1656	10.45152600	13.9770799
Field 1	routed	FDEU020	DDEU012	CDEU014	21		FDEU020-CDEU014	52.373172	12.879107	12.9964099
Field 2	ordered	FDEU020	DDEU012	CDEU011	78		FDEU020-CDEU011	52.390351	12.732488	13.3759440
Field 4	ready	FDEU020						52.450598	12.833617	
Field 3	routed	FDEU020	DDEU012	CDEU015	80	2015085119	FDEU020-CDEU015	52.373889	12.854585	13.7257835
Field 8	routed	FDEU020	DDEU012	CDEU013	67		FDEU020-CDEU013	52.36694	12.87653	12.6542489

12	13	14	15	16	17	18	19	20	21
title	Lot Number	quantity	species	Origin Classification	Traded Form	Particle Size	Bale Diameter	Moisture Content	Ash Content
Destination Latitude									
ATBTest	57.74525490	151012B01R1	3.0000	Lemon	1.1.1.1	chips			
ATBTest2	57.74525490	151016B01R1	6.0000	Orange	1.1.1.1	chips	P16A		
ATBTest3	57.74525490	151016B02R1	10.0000	Tangerine	1.1.1.1	branches	P16A		
ATBTest4	57.74525490	151016B03R1	50.0000	Apple	1.1.1.1	bale_round		1.20	
ATBTest5	57.74525490	151016B04R1	60.0000	Plum	1.1.1.1	chips			
Delivery 02	57.7452549	141127N01R1	3.0000	Cherry	1.1.1.4	chips	P45A	1.00	60.00	2.00
Field 1	52.47596	151007B01R1	1.9000	Apple	1.1.1.1	bale_round		1.20	
Field 2	52.509649	151008B01R1	13.0000	Apple	1.1.1.1	bale_round		1.20	
Field 4		151008B02R1	1.0000	Cherry	1.1.1.1	bale_round		1.00	
Field 3	51.6350451	151008B03R1	0.6000	Cherry	1.1.1.1	bale_round		1.00	
Field 8	53.5605655	151008B04R1	2.6000	Apple	1.1.1.1	bale_round		1.20	
Field 7	53.5605655	151008B05R1	8.3000	Apple	1.1.1.1	bale_round		1.20	

Table A1. Cont.

title	22 Density Chips	23 Density Bale	24 Calorific Value	25 Crop Characteristics	26 Pruning Date	27 Collection Date	28 Storage Days	29 piled	30 Chemically Treated	31 covered
ATBTest					14 April 2015	4 May 2015	150	0	0	0
ATBTest2					4 October 2015	5 October 2015	10	0	0	0
ATBTest3					1 September 2015	5 October 2015	20	0	0	0
ATBTest4					9 August 2015	14 August 2015	30	0	0	0
ATBTest5					8 June 2015	17 August 2015	60	0	0	0
Delivery 02	2.00	1.00	1.00		27 November 2015	28 November 2014	2	1	1	1
Field 1					23 February 2014	9 March 2014	180	1	0	0
Field 2					4 March 2015	16 October 2014	180	1	0	0
Field 4					25 February 2014	18 March 2014	180	1	0	0
Field 3					23 March 2014	26 March 2015	180	1	0	0
Field 8					26 March 2014	30 March 2015	180		0	0
Field 7					27 March 2014	14 November 2015	180	1	0	0

title	32 moved	33 contaminated	34 Use Storage	35 Storage Longitude	36 Storage Latitude	37 Pickup Date	38 Delivery Date	39 Created Date	40 Modified Date	41 Modified By User ID
ATBTest	0	0	1	13.012569	52.43828	15 October 2015 08:32	15 October 2015 11:56	12 October 2015 09:22	18 November 2015 11:04	DDEU012
ATBTest2	0	0	0			16 October 2015 07:59	16 October 2015 09:11	16 October 2015 08:27	16 November 2015 10:02	DDEU012
ATBTest3	0	0	0			16 October 2015 12:09		16 October 2015 08:32	18 November 2015 11:04	DDEU012
ATBTest4	0	0	0			16 October 2015 09:38	16 October 2015 12:09	16 October 2015 08:33	18 November 2015 11:04	DDEU012
ATBTest5	0	0	0			16 October 2015 10:20	16 October 2015 12:09	16 October 2015 09:27	18 November 2015 11:04	DDEU012
Delivery 02	1	0	0					27 November 2015 12:01	31 March 2015 08:01	DSWE001
Field 1	1	1	0			18 December 2015 08:56		8 October 2015 14:59	21 December 2015 08:14	DDEU012
Field 2	1	1	0					8 October 2015 15:02	21 December 2015 08:08	DDEU012
Field 4	1	1	0					8 October 2015 15:04	8 October 2015 15:19	FDEU020
Field 3	1	0	0			21 December 2015 08:29	21 December 2015 11:32	8 October 2015 15:11	8 February 2016 10:18	DDEU012
Field 8	1	1	0					8 October 2015 15:15	18 November 2015 11:04	DDEU012
Field 7	1	0	0			17 November 2015 11:17	19 November 2015 11:23	8 October 2015 15:17	19 November 2015 11:39	DDEU012

References

1. EuroPruning. Available online: http://www.europruning.eu (accessed on 28 December 2017).
2. Gebresenbet, G.; Bosona, T.; Olsson, S.-O.; Garcia, D. Smart System for the Optimization of Logistics Performance of the Pruning Biomass Value Chain. *Appl. Sci* **2018**, *8*, 1162. [CrossRef]
3. Hendershott, T. Electronic trading in financial markets. *IT Prof.* **2003**, *4*, 10–14. [CrossRef]
4. Iakovou, E.; Karagiannidis, A.; Vlachos, D.; Toka, A.; Malamakis, A. Waste biomass-to-energy supply chain management: A critical synthesis. *Waste Manag.* **2010**, *30*, 1860–1870. [CrossRef] [PubMed]
5. Europruning Application Platform. Available online: http://europruning.mobitron.se/public/en (accessed on 16 July 2018).
6. Cerchione, S.; Cerchione, R.; Singh, R.; Centobelli, P.; Shabani, A. Food cold chain management: From a structured literature review to a conceptual framework and research agenda. *Int. J. Logist. Manag.* **2018**, *29*, 792–821. [CrossRef]
7. Seuring, S.; Muller, M. From a literature review to a conceptual framework for sustainable supply chain management. *J. Clean Prod.* **2008**, *16*, 1699–1710. [CrossRef]
8. Dumas, J.S.; Fox, J.E. Usability testing: Current practice and future directions. In *Human-Computer Interaction: Development Process*; Book Chapter; CRC Press: Boca Raton, FL, USA, 2009; pp. 232–253. Available online: https://www.taylorfrancis.com/books/e/9781420088892/chapters/10.1201%2F9781420088892-18 (accessed on 18 October 2018).
9. ISO/IEC 25010:2011(en). Systems and Software Engineering-Systems and Software Quality Requirements and Evaluation (Square)-System and Software Quality Models. Available online: https://www.iso.org/obp/ui/#iso:std:iso-iec:25010:ed-1:v1:en (accessed on 23 July 2018).
10. ISO/IEC (2000). Information Technology—Software Product Quality—Part 1: Quality Model. ISO/IEC FDIS 9126-1. Available online: http://www.cse.unsw.edu.au/~cs3710/PMmaterials/Resources/9126-1%20Standard.pdf (accessed on 24 March 2016).
11. Bosona, T.; Gebresenbet, G.; Olsson, S.-O. Traceability System for Improved Utilization of Solid Biofuel from Agricultural Prunings. *Sustainability* **2018**, *10*, 258. [CrossRef]
12. CIRCE (2014). Description of the Biomass Specifications for the Value Chain. Deliverable 2.1 of EuroPruning Project. Available online: http://www.europruning.eu/web/lists/pubfiles.aspx?type=pubdeliverables (accessed on 29 May 2017).
13. Smart Logistics and Cost-Effective Transportation. Available online: http://www.europruning.eu/web/data/category.aspx?id=smartlogistics (accessed on 30 December 2017).
14. Santa, J.; Zamora-Izquierdo, M.A.; Jara, A.J.; Gómez-Skarmeta, A.F. Telematic platform for integral management of agricultural/perishable goods in terrestrial logistics. *Comput. Electron. Agric.* **2012**, *80*, 31–40. [CrossRef]

applied
sciences

MDPI

Article

A New Hybrid Approach for Wind Speed Forecasting Applying Support Vector Machine with Ensemble Empirical Mode Decomposition and Cuckoo Search Algorithm

Tongxiang Liu [1], Shenzhong Liu [2,*], Jiani Heng [2] and Yuyang Gao [2]

[1] Faculty of Professions, University of Adelaide, Adelaide 5000, Australia;
    tongxiang.liu@student.adelaide.edu.au
[2] School of Statistics, Dongbei University of Finance and Economics, Dalian 116025, China;
    hengjn13@lzu.edu.cn (J.H.); gaoyuyang0315@gmail.com (Y.G.)
*   Correspondence: liusz@dufe.edu.cn; Tel.: +86-13804258916

Received: 12 July 2018; Accepted: 20 September 2018; Published: 28 September 2018

Featured Application: Wind speed forecasting.

Abstract: Wind speed forecasting plays a crucial role in improving the efficiency of wind farms, and increases the competitive advantage of wind power in the global electricity market. Many forecasting models have been proposed, aiming to enhance the forecast performance. However, some traditional models used in our experiment have the drawback of ignoring the importance of data preprocessing and the necessity of parameter optimization, which often results in poor forecasting performance. Therefore, in order to achieve a more satisfying performance in forecasting wind speed data, a new short-term wind speed forecasting method which consists of Ensemble Empirical Mode Decomposition (EEMD) for data preprocessing, and the Support Vector Machine (SVM)—whose key parameters are optimized by the Cuckoo Search Algorithm (CSO)—is developed in this paper. This method avoids the shortcomings of some traditional models and effectively enhances the forecasting ability. To test the prediction ability of the proposed model, 10 min wind speed data from wind farms in Shandong Province, China, are used for conducting experiments. The experimental results indicate that the proposed model cannot only improve the forecasting accuracy, but can also be an effective tool in assisting the management of wind power plants.

Keywords: cuckoo search algorithm; support vector machine; ensemble empirical mode decomposition; wind speed forecasting; forecasting validity

---

1. Introduction

1.1. Background and Motivation

In recent years, a demand for clean and renewable energy resources has increased significantly because of the air pollution caused by traditional fossil fuels. Wind power, which is one of the most promising renewable resources, proved to be an ideal alternative. Currently, wind energy has been successfully employed in many countries, representing approximately 10% of energy consumption in Europe and more than 15% in countries like America and Spain [1]. However, wind speed, which is one of the most essential factors in wind power generation, is difficult to forecast due to many natural factors such as pressure, temperature, and the rotation of the planet.

Therefore, in the development of wind energy, wind speed forecasting is rather important. The accuracy of the forecasting result can influence the wind rotating equipment, operation cost, and the limitations of wind power penetration. With a precise prediction of wind speed, the dispatching department is able to make adjustments to the program efficiently, so that the influence of wind energy itself on the power station and the adverse effect of the wind farm on the power system can be minimized, making wind power more competitive in the global electricity market.

1.2. Existing Models

Currently, developing an accurate and effective wind speed forecasting model is a priority in many countries. These models can be divided into four categories based on the specific time horizon: Very short-term, short-term, medium-term, and long-term forecasting [2]. Very short-term and short-term forecasting can provide aids and references when making economic load dispatch plans, while medium and long-term forecasting are employed for the maintenance of the grid, wind farm planning, and providing information for site selection [3,4]. Using computational methods, these forecasting methods can be further categorized into four different types: Physical, statistical, intelligent, and hybrid methods. Physical methods are usually based on numerical weather prediction (NWP), which simulates the physical characteristics of the atmosphere through applying physical rules and geographical conditions, though there are still many difficulties in employing this model for wind speed forecasting directly—for example, the forecasting accuracy, resolutions of space and time, and the importance of the physical procedures. Statistical methods are used for wind speed forecasting through setting mathematical models, which are similar to the direct random time-series models [5]. As one of the most widely employed statistical methods, the Autoregressive Integrated Moving Average (ARIMA) model is adopted to predict wind speed. Moreover, an Autoregressive Fractional Integrated Moving Average (ARFIMA) model is used in wind speed prediction because of its superior ability to select valid information from the past time series more effectively, compared with the traditional ARIMA model. In recent years, intelligent approaches such as artificial neural networks (ANN) and back propagation neural networks (BP), with the aim of reducing errors by employing past time series, are being applied to forecast future wind speed because of the rising popularity of artificial intelligence [6–12].

Due to the complexity of the raw wind speed series, forecasting the traits in the time series accurately is very difficult to achieve with individual models. Therefore, hybrid models which adopt multiple approaches to increase the prediction ability of the raw time series, or which assemble multiple forecasting methods to obtain the traits of the raw data to forecast wind speed, have been proposed in many studies. Guo et al. combined the Seasonal Autoregressive Intergrated Moving Average model (SARIMA) and the Least Square Support Vector Machine model (LSSVM) to obtain a more accurate forecasting model [13]. This new approach can effectively catch the seasonal and nonlinear features in the input data. Similarly, Guo et al. used a back-propagation neural network, and relied on the theory of using a seasonal exponential adjustment to reduce seasonal influence from the raw data, to create a new hybrid approach [14]. Experimental results showed that this new model is able to effectively increase the forecasting precision, as compared to the results of an independent back propagation neural network (BPNN) with no adjustment of the seasonal exponent. From the studies of Pourmousavi et al., a new and very short-term wind speed hybrid model was proposed and proven to significantly increase the prediction range [15,16]. Liu et al. combined Empirical Model Decomposition to create a new hybrid Empirical Mode Decomposition-Artificial Neural Network (EMD-ANN) model. The proposed approach was effective in capturing jumping samples in raw data with high noise [17]. Mohammadi et al. proposed a Stackelberg game technique to improve the efficiency of electric grid [18]. Kianoosh et al. introduced a new multi-time-scale approach modeled by historical time-series for electric data forecasting [19]. Haque et al. put forward the idea of combining multiple soft computing models with a data preprocessing technique to perform short-term wind speed forecasting [20]. Li and Shi conducted an experiment to make a comprehensive comparison among three ANN using one h ahead of forecasting; namely, ADLINE (adaptive linear element), RBFNN (radial basis function neural network), and BPNN (back-propagation neural network) [21].

Moreover, to reduce the influence of data diversity, Ortiz-García et al. introduced several novel SVM structures, which achieved a better performance than a similar multilayer perception [22].

1.3. Introduction of the Proposed Model

However, these individual models mentioned above have certain drawbacks. The disadvantages are summarized as followed: (1) They usually require a huge amount of wind speed data in order to build a model for a precise prediction, because of the inner irregularity and instability within the raw data. If the raw data suddenly changes due to environmental factors, the error of the forecasting result can be relatively high [23]; (2) Some models only try to match the fitness of the model closer to the original data. As a result, when processing the wind speed data with high noise and irregularity, it is difficult to fit the individual model to it by using only traditional physical or statistical methods, causing poor forecasting accuracy and low efficiency; (3) Individual forecasting methods ignore the importance of data preprocessing and the necessity of model parameter optimization. Because of this, the accuracy of the result is not satisfying; (4) While some new models take advantage of artificial intelligence to enhance their forecasting ability, they still have the problem of over-fitting and a low convergence rate [24].

Therefore, in order to solve the problems mentioned above, a new hybrid model for more precise wind speed forecasting and better evaluation, achieved by adapting the Ensemble Empirical Mode Decomposition (EEMD), the Cuckoo Search Optimization (CSO) algorithm, and the Support Vector Machine (SVM) model, is introduced in this paper.

The main contributions and innovations of this paper compared to other studies in the field of wind speed forecasting are summed up as follows:

- The newly proposed approach in this paper takes advantage of the data preprocessing method and the algorithm of parameter optimization to enhance the performance of the SVM model. In this paper, the raw time series is first decomposed into several sub-signals, among which signals with high frequency ones are removed and the rest are restructured to obtain a stationary time series, with which the intrinsic characteristics of the wind speed data can be better captured and analyzed so that the forecasting accuracy can be greatly improved.

- This paper employs the Cuckoo Search algorithm to optimize the parameters of the SVM before training. The CS algorithm, which has the advantage of a powerful capability in terms of global optimization, requires few parameters and has strong multi-objective problem solving ability, and therefore can significantly improve the accuracy of the forecasting. The Support Vector Machine can overcome the difficulties of traditional models, such as the curse of dimensionality, falling into local optima easily, and over-learning.

- To verify the forecasting ability of the proposed approach, conventional models like BPNN, RBFNN, and ARIMA are used for comparison. A more comprehensive evaluation is conducted, including multi-step forecasting experiments and performance evaluation metrics such as six indexes and a DM(Diebold-Mariano) test, to assess and analyze the performance of the newly developed method.

The model used the raw data from wind turbines in a large wind farm. The result of this paper shows that the newly developed approach effectively improves the forecasting accuracy and can be successfully applied in wind grids to provide statistical support for making operation plans and managing the power station.

1.4. Structure of the Paper

This paper is organized as follows. Section 2 concisely introduces the required techniques. Forecasting performance evaluation criterion and numerical experimentation are introduced in Sections 3 and 4. In Sections 5 and 6, the results and the superiority of the proposed approach are discussed by using comparisons with other conventional methods. Finally, Section 7 concludes this paper and introduces the direction for future studies.

2. Materials and Methods

In this section, main theories about the proposed model will be introduced first. Then, the proposed hybrid model is introduced. DM tests and forecasting effectiveness are also introduced.

2.1. Empirical Mode Decomposition (EMD)

EMD, which has been proved to be an effective data preprocessing method, is usually applied to reduce the noise of a time series, which is non-stationary [25]. The main theory of EMD is that by local characteristic timescale filtering, the original series is decomposed into oscillatory modes (IMFs). The IMFs follow the rules as shown below: (1) The extremes and the zeros must have the same quantity or the difference of quantity must be within one in the whole time series; (2) The upper and lower bounds' mean must be equal to zero. Let the raw data be $s(t)(t = 1, 2, \ldots, l)$. The main procedures of EMD are listed below:

(a) Set all local maxima and minima of time series.
(b) Connect all local extreme to produce the upper bound $e_{up}(t)$ and the lower bound $e_{low}(t)$ by applying a cubic spline.
(c) Compute the mean value from the upper and lower bounds $m(t) = [e_{up}(t) + e_{low}(t)]/2$.
(d) Compute the difference value between the raw data and the mean value $h(t) = s(t) - m(t)$.
(e) Inspect if $h(t)$ fits characteristics of IMF. If yes, $h(t)$ is defined as the ith IMF and the residual $r(t) = s(t) - h(t)$ will replace $s(t)$. If no, $s(t)$ will be replaced by $h(t)$.
(f) Repeat the above-mentioned procedures. Stop when the value of the two successive siftings' standard deviation is lower than the threshold set earlier.

By utilizing mentioned procedures, a group of IMFs is settled from the raw data, which is arranged based on the frequency from high to low.

Definition 1. *The raw data, which is used for decomposition and consists of n IMFs and one residual is obtained as:*

$$s(t) = \sum_{i=1}^{n} c_i(t) + r_n(t) \tag{1}$$

where n is the number of IMFs, $r_n(t)$ is the residuals representing a trend in $s(t)(t = 1, 2, \ldots, l)$, and $c_i(t)(i = 1, 2, \ldots, n)$ represents the IMFs. When describing the local characters of a raw time series, each IMF is independent.

2.2. Ensemble Empirical Mode Decomposition (EEMD)

EEMD, which can successfully overcome the shortcomings of EMD, is first introduced by Wu and Huang [26]. The main theory of EEMD is that by using the features of the noise, the problem of mode mixing can be effectively solved. The original time series are combined with true signals and noise. Thus, to extract the true signals in the raw data, a white noise is added to the original data. Procedures of EEMD can be described as follows.

- Step (a). Add a white noise to the raw time series.
- Step (b). Based on the method of EMD, decompose the time series with the added white noise to nIMFs.
- Step (c). Repeat the mentioned two steps, but add the white noise at different scales each time.
- Step (d). Calculate the means of each IMF of decomposition to constitute the final IMFs.

Through the mentioned process, the whole series of white noise, which are added into the raw data, can provide a unified reference range to assist the process of noise reduction. As a result, the true IMFs can be obtained from the original data.

Definition 2. *The relationship among the ensemble number, the error tolerance, and the added noise level is described based on the research of Wu and Huang:*

$$N_\varepsilon = \frac{\varepsilon^2}{\varepsilon_n^2} \tag{2}$$

where ε represents the amplitude of the added noise, ε_n is the error's standard deviation, and N_ε is value of ensemble members. Generally, it is suggested that an amplitude fixed at 0.2 will result in an accurate result [26]. In this study, the value of ensemble members are set to 100 and the optimal standard deviation of white noise series are settled from 0.1 to 0.2.

2.3. Cuckoo Search Optimization (CSO) Algorithm

The CSO, which was recently introduced by Yang et al., is a meta-heuristic search algorithm [27]. The main principle of the algorithm is originated from the application of cuckoo birds' obligating brood parasitic behavior and the behavior of Lévy flight. Generally, by the switching/discovery probability (pa), Cuckoo Search algorithm is able to perform more powerful capability of optimization [28]. In order to successfully adopt the CS algorithm, three basic assumptions are illustrated as follows:

- The egg which is generated by each cuckoo bird represents a solution in a time period, and it is dumped randomly in the nest.
- The nests which contain better eggs (better solution) are described as the best nests and they will be passed to the next generation.
- The available host nests' number is restricted to n, and each host bird is able to recognize the cuckoo bird's egg with a probability $Pa \in [0,1]$. As a result, the host bird has two possible choices, which are either throwing away the egg or giving up the whole nest and finding a new location to build a new nest.

Definition 3. *Based on the Lévy flight behavior of the cuckoo bird's nest-seeking nature, set the current solution as x_i^t for ith cuckoo, then new solution x_i^{t+1} is generated as follows:*

$$x_i^{t+1} = x_i^t + \alpha \oplus L(\lambda), i = 1, 2, 3 \cdots n$$
$$L(\lambda) = t^{-\lambda}, (1 < \lambda \leq 3) \tag{3}$$

where $\alpha > 0$ is the size of each step which relates to the optimization. $\oplus$ is the entry wise multiplications. $L(\lambda)$ stands for a Lévy distribution which has an infinite mean and variance. Lévy flight essentially provides a random walk while the random step length is drawn from $L(\lambda)$, which can produce a random walk process. Around the best solution, the local search process can be faster because of the new solution generated by Lévy walk [29]. The flowchart of CSO is shown in Figure 1.

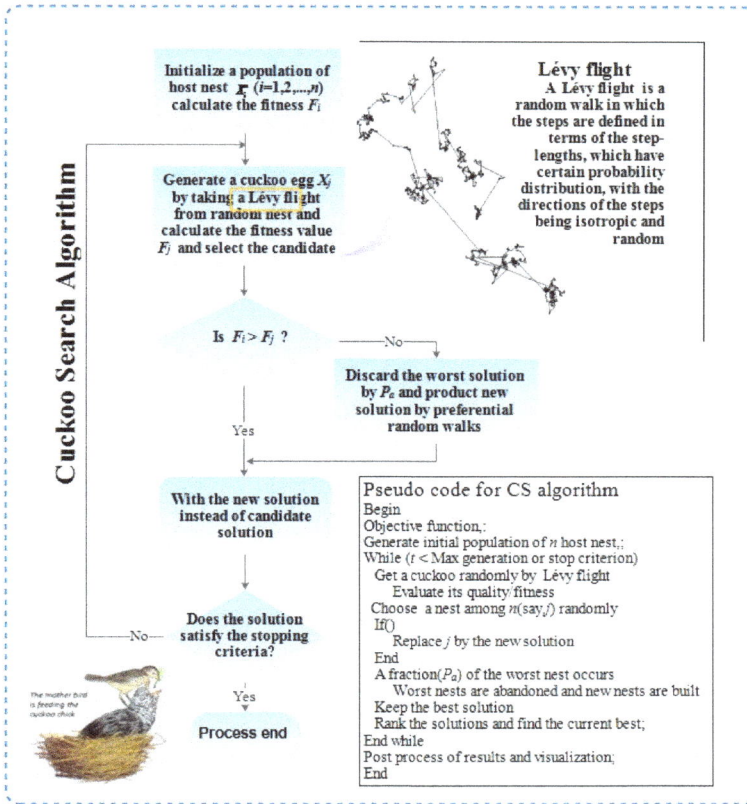

Figure 1. The flowchart of the Cuckoo Search Algorithm.

2.4. Support Vector Machine (SVM)

The Support vector machine, which is introduced by Vapnik, is one of the newest algorithms [30]. Different from the conventional models, SVM follows the rule of statistical machine learning process and structural risk minimization to obtain the minimal upper bound generalization error, which is the biggest advantage of SVM compared with other models. Due to this, SVM has a great popularity and a wide use in the area of pattern recognition, classification, and regression analysis and forecasting [31,32].

Definition 4. *Suppose a set of data* $\{x_i d_i\}_i^n$, *the n-dimensional input vector is represented as* x_i, *and the output is expressed as* d_i. *The estimating function of SVM can be represented as:*

$$f(x) = w \cdot \phi(x) + b \tag{4}$$

where $\phi(x)$ *is a nonlinear mapping while w, b is the weight vector and scalar, separately, and they are estimated through the equation:*

$$R_{SVMs}(C) = \frac{1}{2}\|w^2\| + C\frac{1}{n}\sum_{i=1}^{n} L(x_i, d_i) \tag{5}$$

where C represents the penalty factor of the error. The $L(x_i, d_i)$ is the loss function $C\frac{n}{2}\sum_{i=1}^{n} L(x_i, d_i)$ represents the empirical error.

Definition 5. *Setting the upper and lower excess deviation ξ_i and ξ_i^* as the positive slack variables, the optimization problem is obtained as the follows:*

$$MinimizeR_{SVMs}(w, \xi^{(*)}) = \frac{1}{2}\|w^2\| + C\frac{1}{n}\sum_{i=1}^{n}(\xi_i + \xi_i^*) \tag{6}$$

$$Subject\ to \begin{cases} d_i - w\phi(x_i) - b_i \leq \varepsilon + \xi_i \\ w\phi(x_i) + b_i - d_i \leq \varepsilon + \xi_i^* \\ \xi_i, \xi_i^* \geq 0, i = 1, \ldots, l \end{cases} \tag{7}$$

where $w^2/2$ represents the regularization term. ε stands for the loss factor, and its value is related to the approximate accuracy of the input sample. l represents the quantity of elements in the sample data series.

Equation (4) can be explained through the Lagrange function, which is described as follows:

$$f(x) = \sum_{i=1}^{n}(\alpha_i - \alpha_i^*)K(x, x_i) + b \tag{8}$$

where α_i and α_i^* are the multipliers of Lagrange function. $K(x, x_i)$ represents the kernel function and $K(x_i, x_j) = \phi(x_i)\phi(x_j)$, which is the dot product of the two inner vectors x_i, x_j in $\phi(x_i)$ and $\phi(x_j)$.

Gaussian function, which is known for its features of simplicity, efficiency, and reliable computing ability, has proved to be one of the most effective core functions [30]. SVM model, which employs it as the core function, can effectively obtain the complex features of nonlinearity of the original data and result sample by matching input data into a higher-dimensional feature space.

Definition 6. *Gaussian function, which is employed as core function of SVM model, is described as follows:*

$$K(x_i, x_j) = \exp\left(-\gamma\|x_i - x_j\|^2\right) \tag{9}$$

where the γ stands for the parameter of the kernel function. x_i, x_j are vector quantities in the input space.

In this paper, parameters (γ, C), which are considered as the two most valuable figures that influence the performance of wind speed prediction. The CSO is used to optimize them.

2.5. Introduction of the EEMD-CSO-SVM Model

Based on the above-mentioned methods, the EEMD-CSO-SVM model is proposed for wind speed forecasting. Taking one of the datasets as an example, the main procedures of the proposed model is shown in Figure 2. Firstly, the EEMD noise reduction technique is applied to the raw wind speed data in order to obtain a stationary time series. In this paper, the raw data is decomposed into 10 IMFs, among which the first IMF is considered as noise and removed. The rest are recombined for the forecasting. Then, each set of data from the four wind turbines are used to perform three-step forecasting to testify the validity of the proposed model. The de-noised wind speed data is applied to train the SVM model whose key parameters (γ, C) are optimized by the Cuckoo Search algorithm. The results of forecasting are recorded for further analysis.

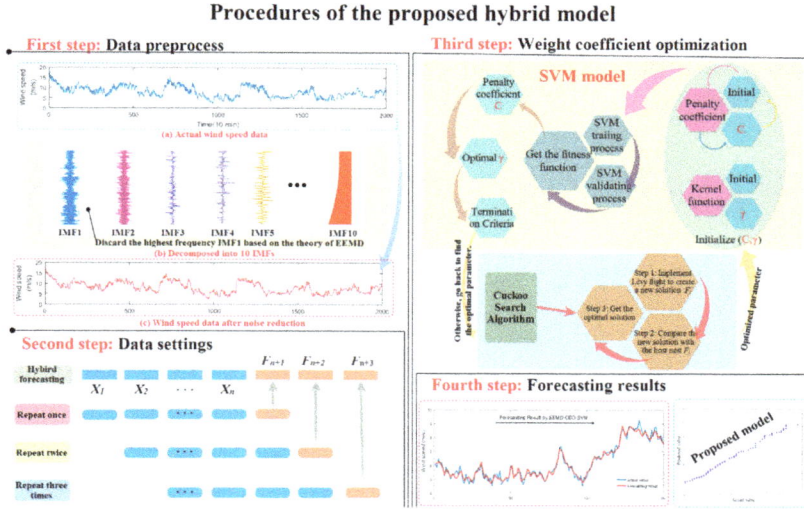

Figure 2. The flowchart of the proposed forecasting method.

2.6. DM Test and Forecasting Effectiveness

The DM test, which focuses on the difference in the forecasting precision between the proposed hybrid model and other traditional methods, is described as follows:

$$H_0 : E(d_n) = 0, \forall n$$
$$H_1 : E(d_n) \neq 0, \exists n \tag{10}$$

The value of the DM test is described as follows:

$$DM = \frac{\sum_{h=1}^{k} \left(L\left(\varepsilon_{t+h}^{(i)} \right) - L\left(\varepsilon_{t+h}^{(j)} \right) \right) / k}{\sqrt{S^2/k}} s^2 \tag{11}$$

where ε_{t+h} is the forecasting error.

S^2 represents the estimation value for the variance of $d_h = L\left(\varepsilon_{t+h}^{(i)} \right) - L\left(\varepsilon_{t+h}^{(j)} \right)$. L is the loss function, which represents the accuracy of the forecasting.

Among the loss functions, absolute deviation error loss and square error loss are widely applied. They are described as follows:

Absolute deviation error loss:

$$L\left(\varepsilon_{t+h}^{(i)} \right) = \left| \varepsilon_{t+h}^{(i)} \right| \tag{12}$$

Square error loss:

$$L\left(\varepsilon_{t+h}^{(i)} \right) = \left(\varepsilon_{t+h}^{(i)} \right)^2 \tag{13}$$

If no significant differences are found between the performances of the included methods, the null hypothesis will be rejected.

The null hypothesis is described as follows:

$$|DM| > z_{\alpha/2} \tag{14}$$

where $z_{\alpha/2}$ represents the critical value of the standard value distribution when the value of significance is α.

Forecasting effectiveness, which evaluates the performance of the proposed model using the sum of the squared errors and the mean squared deviation of the forecasting results is also applied. The forecasting effectiveness is described as follows [33].

The kth order forecasting effectiveness unit is described as:

$$m^k = \sum_{n=1}^{N} Q_n A_n^k$$
$$\sum_{n=1}^{N} Q_n = 1$$
(15)

where Q_n stands for the discrete probability distribution when the time is n. A_n represents the forecasting accuracy.

The k-order forecasting effectiveness is presented as:

$$H\left(m^1, m^2, \cdots, m^k\right)$$
(16)

The first order forecasting effectiveness is defined as $H(m^1) = m^1$ while the second-order forecasting effectiveness is the difference between the standard deviation and expectation, which is shown as:

$$H\left(m^1, m^2\right) = m^1\left(1 - \sqrt{m^2 - (m^1)^2}\right)$$
(17)

3. Performance Evaluation Criterion

In this study, to test the accuracy of the proposed approach, six evaluation indexes including MAE, MAPE, RMSE, WI, E_{NS}, and E_{LM} are applied for evaluating forecasting accuracy. These indexes are shown as follows in Table 1:

Table 1. The description of the error evaluation indexes.

Metric	Definition	Equation
MAE	The mean absolute error of N forecasting results	$\text{MAE} = \frac{1}{N} \sum_{i=1}^{N} |yp_i - y_i|$
MAPE	The average of N absolute percentage error	$\text{MAPE} = \frac{1}{N} \sum_{i=1}^{N} \left|\frac{yp_i - y_i}{y_i}\right| \times 100\%$
RMSE	The square root of the average of the error square	$\text{RMSE} = \sqrt{\frac{1}{N} \sum_{i=1}^{N} (yp_i - y_i)^2}$
WI	Willmott's Index	$\text{WI} = 1 - \left[\frac{\sum_{i=1}^{N} (yp_i - y_i)^2}{\sum_{i=1}^{N} (|yp_i - \bar{y}| + |y_i - \bar{y}|)^2}\right], 0 \le \text{WI} \le 1$
E_{NS}	Nash-Sutcliffe coefficient	$E_{NS} = 1 - \left[\frac{\sum_{i=1}^{N} (yp_i - y_i)^2}{\sum_{i=1}^{N} (y_i - \bar{y})^2}\right], \infty \le E_{NS} \le 1$
E_{LM}	Legates and McCabe Index	$E_{LM} = 1 - \left[\frac{\sum_{i=1}^{N} |y_i - yp_i|}{\sum_{i=1}^{N} |y_i - \bar{y}|}\right], \infty \le E_{LM} \le 1$

Respectively, where N is the total output samples; y_i represents the actual series; and yp_i stands for the prediction results. For WI, E_{NS}, and E_{LM}, when the values of them are closer to 1, the model achieves a higher performance.

4. Numerical Experimentation

The Windows 7 Professional operating system was used to perform the experiments. The specific version of software, which was used to conduct the proposed model, is Matlab2016a. The details of the hardware are: Intel Core i5-3230M2.60 GHz CPU, and 4 GB RAM.

4.1. Introduction of Datasets

Penglai is located in the east coast of Shandong in China. It has a 3100 km coastline. Thus, although the whole region is not large, it is famous for its abundant wind power due to this unique geographical feature. The approximate wind power capacity of the region is 67 million KW. In this paper, datasets 1, 2, 3, and 4 were chosen from a wind farm in Penglai, with the latitude from 120°43′ N to 120°47′ N and longitude from 37°50′ E to 37°37′ E. These datasets are located in mountain and hilly areas whose altitude is from 100 m to 240 m. The features of the wind power generator are provided as follows: Rated power: 1500 KW. Height of measurement: 70 m. Sampling time period: 10 min. Scanning frequency: 144 times per day. All four datasets were included in the experiment, to help analyze the differences in the results. Multistep forecasting was also conducted in this paper.

Each group is divided into a training group and a testing group. The size ratio of the training and testing group is set to 9:1. The training sample included a total of 1350 10-min wind speed series, and the testing sample contains 150 wind speed data points. The ratio of input and output data is set to 4:1. Figure 3 shows the structure of each dataset.

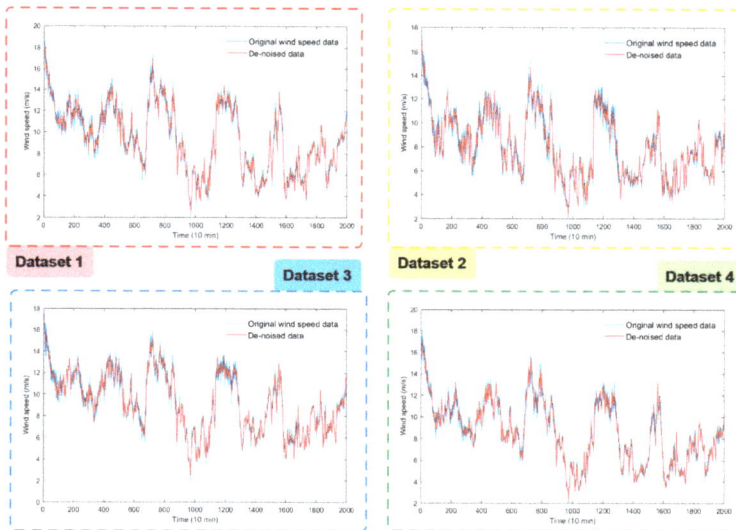

Figure 3. The original wind speed data and de-noised data from four datasets.

4.2. Forecasting Model Parameter Setting

Wind speed series from four datasets are chosen to test the forecasting accuracy of the proposed hybrid model. The results of the proposed model are also used to compare with other conventional methods; namely, BP neural networks, the ARIMA model, and the RBF model. This paper followed the standard of energy industry NB/T31046-2013 and the rules for measuring wind sources, which were published and made by the National Energy Administration in 2013.

(1) For BPNN, the newff function of the neural network toolbox is employed to build the network. The dimensions of the input, hidden, and output layers are 4, 5, and 1, respectively. The learning

rate is set to 0.1, the maximum number of iterations is set to 100, and the training precision is set to 0.00004.

(2) For ARIMA, the forecasting results are influenced by the moving average and the order of auto-regressive. The observed value's fitting effect is measured by the AIC criterion, and the AIC also calculates the most suitable number for the parameters. When the AIC reaches the lowest value, the ARIMA method can achieve the best order.

(3) For RBFNN, similar with BPNN, the newrb function of the neural network toolbox is employed to build the forecasting network. The same parameters as the BPNN are used in the RBFNN.

4.3. Experimental Results for Datasets

The original time series are first preprocessed through the EEMD. Figure 3 presents the preprocessed data obtained by the EEMD method for the four wind turbines. As indicated in Figure 3, for the #2, #3, and #4 wind turbines, 10 IMF sequences are obtained from the original training data of the time series. According to the principles of denoising, eliminating the high-frequency sequence from the IMF sequences can assist in obtaining a cleaner data sequence; that is, a data sequence with lower noise. For this paper, the first IMF sequence obtained by the EEMD method is eliminated from the original data sequence due to its high frequency, so that a stationary time series can be obtained to improve the accuracy of the prediction. Taking wind turbine #2's data as an example, the visualization of the de-noise preprocessing of the EEMD method is shown in Figure 4. The final result after the de-noise processing with the EEMD method is also presented in Figure 4. The preprocessing results of all four wind turbines are shown in Figure 3.

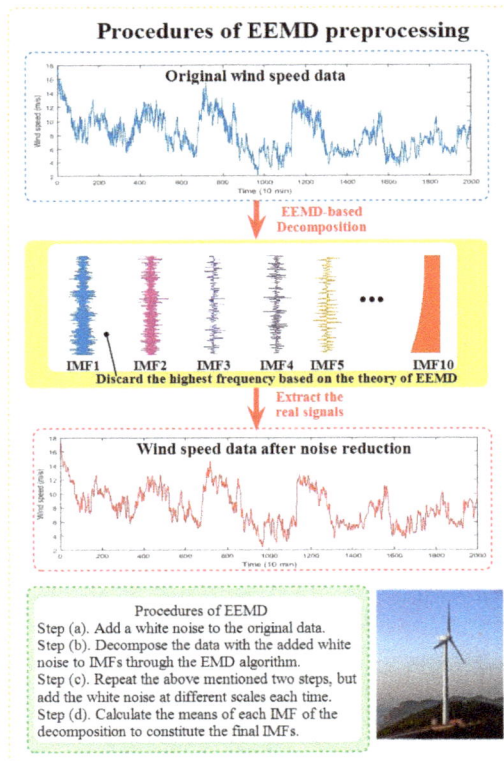

Figure 4. Wind speed data noise reduction using EEMD preprocessing process.

For the experiment, the proposed method is trained using the selected data from four datasets. Multistep forecasting is applied in the experiment, which performs the prediction through removing the old input data for each circulation. By applying the previous output values rather than the actual series, the multistep ahead method predicts the next wind speed value through this circulation [34]. The one-step forecasting result is calculated based on the observed values from x_1 to x_n, where n is the number of output value. Two-step forecasting result is calculated based on the observed value from x_2 to x_n and the one-step result. Additionally, three-step forecasting result is obtained based on the observed value from x_3 to x_n and the previous two-steps results. The validity of the proposed hybrid model was analyzed based on the results of multistep ahead forecasting. Tables 2–5 show the forecasting results of different models in four wind turbines, respectively. The results of multistep ahead forecasting are shown Table 6.

The performance of hybrid and traditional methods in one-step forecasting are compared. To testify the effectiveness of the CSO, Particle swarm optimization (PSO) is used for comparison. PSO, which is similar with CSO, optimizes the parameters based on the velocity of particle and the method of position updating. The MAPE values of the proposed hybrid are 4.79%, 3.07%, 2.69%, and 3.91% in four wind turbines, respectively. The values of Willmott's Index reach 0.9838, 0.9929, 0.9918, and 0.9839, respectively. The other values of forecasting error indexes also indicate that the hybrid approach performs better than traditional methods. The results are shown in Tables 2 and 3.

In the results of dataset 2, the MAPE values of the proposed model are 3.07%, 6.53%, and 10.70% in three-steps forecasting, separately, which shows that one step forecasting performs better. Multistep forecasting, which is based on the theory of removing old input data and adding new output data from the previous step, cannot achieve the same performance accuracy as single step forecasting.

Remark 1. *The hybrid approach includes more parameters so that better performance can be achieved. As for the multistep forecasting, its process is complex and leads to high error because less historical data is employed and the forecasted result of each step will be included as input in the next circulation. Therefore, with more steps of forecasting being conducted, more errors will appear, leading to poor performance.*

Table 2. Comparison of the value of RMSE, MAE, and MAPE between the proposed model and some related models.

Model	RMSE	MAE	MAPE (%)	RMSE	MAE	MAPE (%)
	Wind turbine 1			Wind turbine 3		
EEMDCSOSVM	0.3513	0.1815	4.79	0.2463	0.2120	2.69
EEMDPSOSVM	0.4652	0.2159	5.63	0.3027	0.2436	2.82
CSOSVM	0.8294	0.2975	10.67	0.5121	0.4251	5.08
PSOSVM	1.0135	0.7100	12.18	0.7928	0.6362	9.70
	Wind turbine 2			Wind turbine 4		
EEMDCSOSVM	0.2342	0.1841	3.07	0.2679	0.3028	3.91
EEMDPSOSVM	0.2744	0.2191	3.24	0.3147	0.1869	4.44
CSOSVM	1.1381	0.1927	7.63	0.9204	0.6137	9.27
PSOSVM	0.6680	0.5165	8.58	1.1478	0.8190	10.15

Table 3. Comparison of the value of WI, E_{NS}, and E_{LM} between the proposed model and some related models.

Model	WI	E_{NS}	E_{LM}	WI	E_{NS}	E_{LM}
	Wind turbine 1			Wind turbine 3		
EEMDCSOSVM	0.9838	0.9332	0.7600	0.9918	0.9670	0.8268
EEMDPSOSVM	0.9723	0.9251	0.7148	0.9744	0.9310	0.7201
CSOSVM	0.9152	0.6858	0.4460	0.9667	0.8720	0.6525
PSOSVM	0.8622	0.5367	0.3599	0.8239	0.4950	0.1197
	Wind turbine 2			Wind turbine 4		
EEMDCSOSVM	0.9929	0.9713	0.8388	0.9839	0.9350	0.7565
EEMDPSOSVM	0.9876	0.9424	0.7950	0.9786	0.9234	0.7159
CSOSVM	0.9206	0.7011	0.5134	0.8274	0.4885	0.2514
PSOSVM	0.9174	0.6852	0.4812	0.7490	0.3355	0.1253

5. Analysis of the Forecasting Results

This section analyzes the experimental results of the hybrid approach, and the effectiveness of the proposed method is verified. First, based on the results of single step forecasting, the hybrid approach is compared with some existing conventional models. Then, the results of multi-step forecasting are used for further analysis.

5.1. Single-Step Forecasting

This is divided into two parts to verify the effectiveness of the hybrid approach. First, the hybrid method is tested by comparing it with conventional methods. Then, the results of four different datasets are analyzed.

5.1.1. Analysis of the Proposed Method and Conventional Models

The forecasting results of the proposed model and conventional models are shown in Tables 4 and 5. The original time series and preprocessed data obtained by the EEMD technique is shown in Figure 3. By comparing the results illustrated above, the following conclusions are made:

The fluctuation and instability of the original time series are clearly shown in Figure 3. Additionally, in Figure 3, the reconstructed wind speed data in which the signals with high frequency have been removed obviously shows reduced noise of the raw data. By comparing the forecasting results of the new preprocessed data with the original data from Tables 2 and 3, the model achieves a high performance when using the de-noised wind speed data. The MAPE values of the CSOSVM model decrease by 5.88%, 5.56%, 2.39%, and 7.36%, respectively. Using the PSOSVM model for comparison, the MAPE values also decrease by 6.55%, 6.88%, 5.34%, and 5.71% in four datasets, respectively. Therefore, the EEMD technique has good validity.

The proposed hybrid approach achieves better performance than the conventional methods in wind speed prediction. From Table 4, the MAPE value of BPNN, RBFNN, ARIMA, and the proposed model in wind turbine 1 are 8.68%, 10.32%, 8.41%, and 4.79%, respectively. In the other three datasets the proposed method also achieves higher performance than other conventional methods. The values of the evaluation metrics shown in Tables 4 and 5 all suggest that the hybrid model achieves higher accuracy. The new hybrid approach can be regarded as more effective than traditional methods.

Remark 2. *The data preprocessing technique is a reliable method to reduce the fluctuation in the raw data, and the EEMD proves to be an effective method to achieve this purpose. Therefore, the proposed hybrid approach outperforms the conventional methods in forecasting.*

Table 4. Comparison of the value of RMSE, MAE, and MAPE between the proposed model and some traditional models (BP RBF ARIMA).

Model	RMSE	MAE	MAPE (%)	RMSE	MAE	MAPE (%)
	Wind turbine 1			Wind turbine 3		
BPNN	0.5939	0.4984	8.68	0.4794	0.3286	5.28
RBFNN	0.8867	0.5924	10.32	0.6587	0.4468	6.52
ARIMA	0.5784	0.4672	8.41	0.4472	0.2907	5.13
EEMDCSOSVM	0.3513	0.1815	4.79	0.2463	0.2120	2.69
	Wind turbine 2			Wind turbine 4		
BPNN	0.3998	0.3232	5.42	0.4585	0.3385	6.34
RBFNN	0.4847	0.3806	6.41	1.2374	0.5543	10.32
ARIMA	0.3627	0.2925	5.61	0.4122	0.3679	5.88
EEMDCSOSVM	0.2342	0.1841	3.07	0.2679	0.3028	3.91

Table 5. Comparison of the value of WI, E_{NS} and E_{LM} between the proposed model and some traditional models (BP RBF ARIMA).

Model	WI	E_{NS}	E_{LM}	WI	E_{NS}	E_{LM}
	Wind turbine 1			Wind turbine 3		
BPNN	0.9410	0.7822	0.5403	0.9708	0.8850	0.6806
RBFNN	0.8926	0.6017	0.4630	0.9429	0.7751	0.6128
ARIMA	0.9476	0.8137	0.6512	0.9755	0.9024	0.7148
EEMDCSOSVM	0.9838	0.9332	0.7600	0.9918	0.9670	0.8268
	Wind turbine 2			Wind turbine 4		
BPNN	0.9731	0.8889	0.6962	0.9412	0.7628	0.5877
RBFNN	0.9595	0.8320	0.6422	0.7035	0.3339	0.3248
ARIMA	0.9752	0.8933	0.7017	0.9503	0.7937	0.6128
EEMDCSOSVM	0.9929	0.9713	0.8388	0.9839	0.9350	0.7565

5.1.2. Analysis of the Four Different Wind Turbines

This paper employs four wind turbines to conduct the forecasting experiment of the wind speed. The original time series of the four datasets are presented in Figure 3. The results from the four different datasets using different approaches are shown in Tables 2–5.

From Figure 3, the general trend of the four time series is approximately the same, though there are still differences in some time points. The main reason for this difference is that in a general location, the wind speed is nearly the same. However, when narrowing down to a specific wind farm unit, the location and size of the wind farm varies, resulting in different wind speed data from one dataset to another.

When analyzing from the specific forecasting results, according to the data presented in Table 4 for instance, the MAPE values of the proposed model from the four wind turbines are 4.79%, 3.07%, 2.69%, and 3.91%. For the forecasting accuracy, dataset 3 achieves the highest performance.

Remark 3. *Because of the uncertainty of the magnitude and direction of the wind in different locations, the forecasting results of four different datasets vary. The wind speed also experiences enormous fluctuations in different time periods. As a result, the prediction precision also varies due to different locations and time periods.*

5.2. Multi-Step Forecasting

Multi-step forecasting is an effective way to test the accuracy of the forecasting method. Therefore, this paper employed multistep forecasting on the proposed hybrid model. The experimental results are used to verify the accuracy of the proposed approach. The results of the multi-step forecasting are presented in Table 6. Figures 5–7 show the output values of different methods using multi-step forecasting. In 1-step ahead forecasting, the MAPE values of the BP, ARIMA, RBF, and the proposed model are 5.46%, 5.22%, 4.83%, and 3.07%, respectively. In 2-step ahead forecasting, the MAPE values of these models are 6.70%, 6.42%, 5.97%, and 4.55%. Additionally, in 3-step ahead forecasting, the MAPE values of the four models are 7.83%, 7.15%, 6.92%, and 5.46%, which indicates that the proposed hybrid model achieves a better performance than other conventional models.

When comparing the results of three-step forecasting with each other, one-step forecasting clearly performs better than the others.

Remark 4. *These results can be summarized as follows: The proposed hybrid approach obtains higher accuracy compared with the traditional models used in our experiments in both single step and multi-step forecasting. Therefore, the proposed hybrid approach is valid.*

Table 6. Results of multi-step forecasting of the proposed model.

Dataset		MAE	MAPE (%)	RMSE	WI	E_{NS}	E_{LM}
Dataset 1	1-Step	0.1815	4.79	0.3513	0.9838	0.9332	0.7600
	2-Step	0.3622	6.19	0.4546	0.9720	0.8874	0.6699
	3-Step	0.5895	10.14	0.7354	0.9249	0.7115	0.4609
Dataset 2	1-Step	0.1841	3.07	0.2342	0.9929	0.9713	0.8388
	2-Step	0.3962	6.53	0.5034	0.9583	0.8355	0.6310
	3-Step	0.6579	10.70	0.8291	0.8887	0.6073	0.3925
Dataset 3	1-Step	0.2120	2.69	0.2463	0.9918	0.9670	0.8268
	2-Step	0.4814	5.97	05308	0.9529	0.8306	0.5820
	3-Step	0.8694	9.52	0.8329	0.8599	0.5705	0.3240
Dataset 4	1-Step	0.3028	3.91	0.2679	0.9839	0.9350	0.7565
	2-Step	0.4275	5.12	0.3652	0.9645	0.8604	0.6667
	3-Step	0.5310	8.03	0.5437	0.9189	0.6883	0.4797

6. Discussion

This section discusses the sample selection, experimental results, which consist of each element in the proposed hybrid approach, the error evaluation indexes, and the forecast performance based on the DM (Diebold-Mariano) test and the Forecasting Effectiveness test.

6.1. Sample Selection

Currently, there is no explicit theory about how to select the number of training and testing sample. Too small and the input sample cannot train the model well, while too large and the sample will cause over-fitting. In terms of wind speed forecasting, how to select the number of input set is still a difficult and challenging issue [35,36]. It can only get the optimal input set by experiments. Besides, we organized and list the simulation results based on different ratio between training and testing samples in Table 7. Taking the one-step forecasting result of dataset one as an example, the MAPE values of the proposed model are 5.40%, 4.92%, 4.79%, 4.95%, and 5.11% when the ratio of training and testing sample is set to 7:3, 8:2, 9:1, 14:1, and 29:1, respectively. The forecasting accuracy of other ratios are all lower than the ratio 9:1's accuracy. Table 1 below shows that the experiment we took before choosing a suitable ratio of training sample and testing sample. It can be indicated that when the ratio of training and testing sample is set to 9:1, the performance of wind speed forecasting outperforms the other ratios used in the experiment. Thus, based on the experiments and experience sample data are selected.

Table 7. MAPE (%) values of different ratios between training and testing sample of the proposed model.

Datase	Forecasting Steps	Raito Between the Training Sample and Testing Sample (MAPE (%))				
		7:3	8:2	9:1	14:1	29:1
Dataset 1	1-step	5.40	4.92	4.79	4.95	5.11
	2-step	9.57	8.15	6.19	8.82	8.97
	3-step	13.94	13.46	10.14	14.84	15.73
Dataset 2	1-step	4.16	3.91	3.07	3.54	3.76
	2-step	9.32	7.52	6.53	7.32	8.44
	3-step	12.69	11.24	10.70	11.15	11.80
Dataset 3	1-step	3.48	3.28	2.69	3.41	3.97
	2-step	6.62	6.12	5.97	6.15	6.72
	3-step	10.73	9.82	9.52	9.84	10.15
Dataset 4	1-step	4.88	4.09	3.91	4.16	4.51
	2-step	7.26	6.65	5.12	5.52	5.84
	3-step	11.01	10.39	8.03	8.85	8.99

6.2. Forecasting Error Analysis

The proposed hybrid approach outperforms the conventional methods. The forecasting results and precision of the proposed model in four different wind turbines in multi-step forecasting are shown in Table 6. Then, Figures 5–7 show the forecasting results of multi-step forecasting from the proposed approach and different traditional methods. The following conclusions are obtained from these results: (1) From the results of three-step forecasting, we can see that the hybrid method is more accurate than other conventional models because the error is the lowest many times. (2) The degree of fitting between the output series and the actual data from different models is shown in Figures 5–7. The proposed approach is superior to the corresponding traditional methods, with a higher precision.

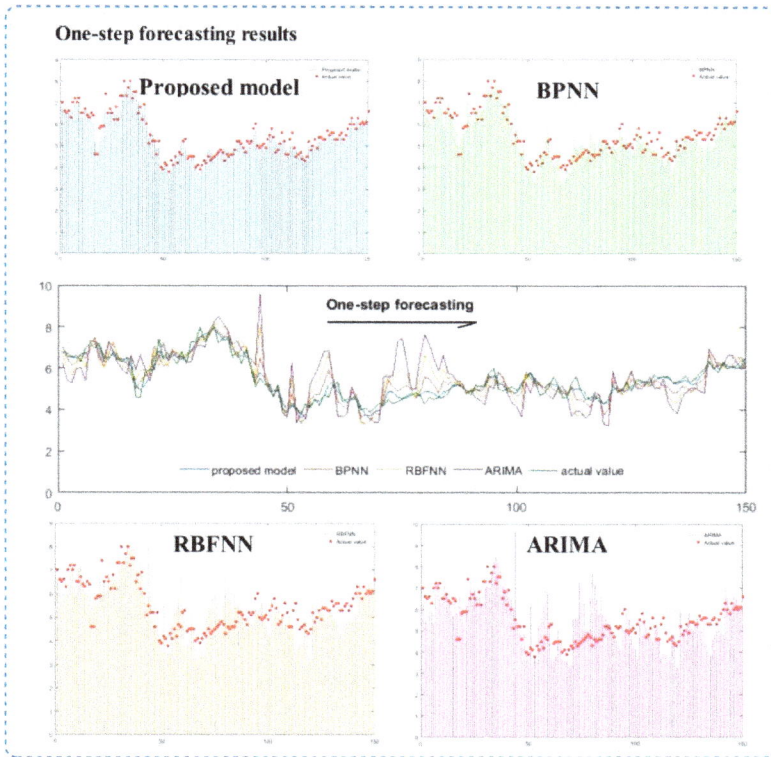

Figure 5. One-step forecasting results of the proposed model and other traditional models (BP RBF ARIMA).

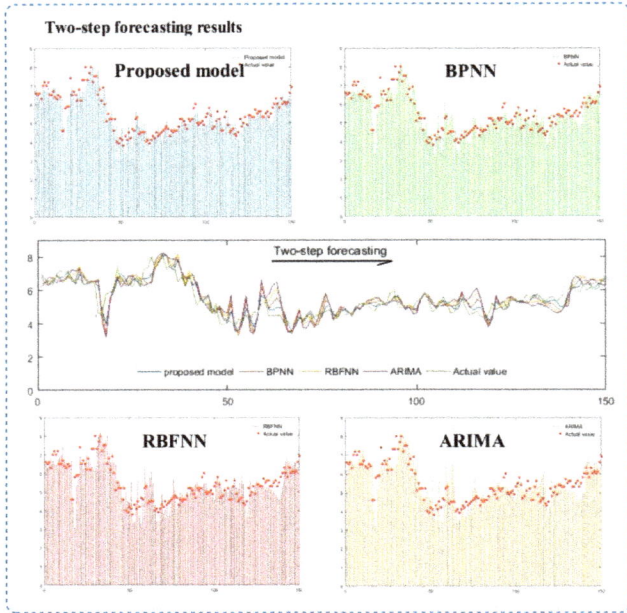

Figure 6. Two-step forecasting results of the proposed model and other traditional models (BP RBF ARIMA).

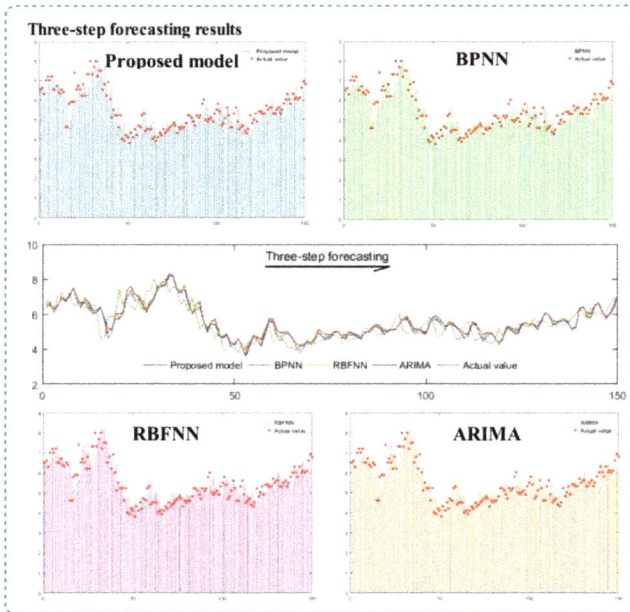

Figure 7. Three-step forecasting results of the proposed model and other traditional models (BP RBF ARIMA).

Multi-step forecasting is employed to further test the validity of the new hybrid approach. From Table 6, the results of multi-step forecasting of four wind turbines are shown. For wind turbine 1, the MAPE values of multi-step forecasting are 4.79%, 6.19%, and 10.14%, respectively. The other three wind turbines obtain similar results. As the results show, the new hybrid model is valid in multi-step forecasting.

Remark 5. *From the analysis, the performance of the hybrid method is better than those of the traditional models. The proposed model is able to adapt to the fluctuation of the input data so that it can obtain a more accurate result. Therefore, the new hybrid approach is more effective in wind speed prediction.*

6.3. Validity of the Data Preprocessing Technique

The irregularity of the raw wind speed series often makes the input data contain high noise and fluctuation. Therefore, removing the noise from the original data is important to obtain a better forecasting accuracy. To test the validity of this data preprocessing technique, the results of the proposed model for four datasets using original and preprocessed data are listed together for comparison. From Tables 2 and 3, the MAPE values of the proposed model decrease by 5.88%, 5.56%, 2.39%, and 7.36% in four wind turbines compared with the CSO-SVM model without the EEMD process. The MAE values also decreases by 0.1160, 0.0086, 0.2031, and 0.3109 in four wind turbines, respectively. The WI value increases by 0.0686, 0.0723, 0.0251, and 0.1565, respectively. The results indicate that the EEMD effectively improves the accuracy of the forecasting. As Tables 2 and 3 show, all six metrics changed positively, and the results demonstrate that after reducing the noise of the raw data, the performance of the hybrid model improved significantly.

6.4. Significance of the Error Evaluation Indexes

Based on the results of six error evaluation indexes, namely MAPE, MAE, RMSE, WI, E_{NS}, and E_{LM}, the significance of the proposed hybrid model is estimated. Taking the forecasting results of dataset 2 as an example, the MAPE values of the proposed model, BPNN, RBFNN, and ARIMA are 3.07%, 5.42%, 6.41%, and 5.61%, respectively. The WI values are 0.9929, 0.9731, 0.9595, and 0.9752, respectively. Combining the results of these together, the hybrid model is proven to be superior to the traditional methods. In conclusion, the forecasting ability of the proposed model can be better analyzed with the help of these evaluation indexes, and the comparison between the proposed model and the other forecasting methods is more comprehensive.

6.5. Results of DM Test and Forecasting Effectiveness

Besides the error evaluation indexes, which have been applied to evaluate the forecasting ability of the proposed hybrid approach, this paper also used the DM (Diebold-Mariano) test and Forecasting Effectiveness to further study the forecasting accuracy.

In this paper, the DM test is used to study the significant differences in forecasting ability between the new hybrid model and some conventional methods. The results of the DM test are shown in Table 8. The DM statistical values far exceed the critical value at the 1% significance level, at which the value of $|DM|$ is 2.1017. Thus, the proposed approach performs significantly different with the conventional methods at the 1% significance level. In conclusion, the proposed approach significantly outperforms the conventional methods.

Table 8. Results of DM test.

Datasets		BPNN	RBFNN	ARIMA
Dataset 1	1-step	7.2859	7.4392	8.0205
	2-step	6.8073	7.1682	7.5217
	3-step	6.5270	6.9463	7.2258
Dataset 2	1-step	6.9469	7.4245	6.2829
	2-step	6.7716	7.2839	6.5691
	3-step	6.8358	7.1475	6.2039
Dataset 3	1-step	7.1325	6.2161	6.2934
	2-step	7.0413	6.4755	6.7390
	3-step	6.9427	6.3586	6.4803
Dataset 4	1-step	6.8941	7.2790	7.1783
	2-step	7.2564	7.2387	7.0492
	3-step	6.4396	6.9257	6.5864

The results of forecasting effectiveness is shown in Table 9. The first-order forecasting effectiveness is based on the expected value of the forecasting accuracy sequence, while the second-order forecasting effectiveness is related to the difference between the standard deviation and expectation of the forecasting accuracy sequence. In Table 8, the forecasting effectiveness of the proposed model outperforms all the other models in both first order and second order.

Table 9. Forecasting Effectiveness of different models for four datasets.

Dataset		BPNN	RBFNN	ARIMA	EEMDCSOSVM
Dataset 1	First order	0.8951	0.8889	0.8874	0.9146
	Second order	0.8322	0.8257	0.8283	0.8638
Dataset 2	First order	0.9130	0.9029	0.9072	0.9367
	Second order	0.8384	0.8362	0.8436	0.8842
Dataset 3	First order	0.9283	0.9177	0.9220	0.9409
	Second order	0.8517	0.8582	0.8625	0.8975
Dataset 4	First order	0.8956	0.8934	0.9031	0.9269
	Second order	0.8344	0.8360	0.8492	0.8816

7. Conclusions

With a rapidly rising demand for clean energy obtained from renewable power resources, more attention and resources have been focused on the development of effectively utilizing these energy resources. Among those resources, wind power has the most promising future. In the area of the prediction, not only the accuracy, but also the stability, should be seen as key factors of the forecasting model. However, due to the uncertainty and intermittence of the raw wind speed data, the forecasting results obtained by conventional models cannot meet this goal. Additionally, with the poor results of forecasting, the power grids cannot make adjustments to the wind plan timely, causing low productivity and increasing the operation cost of the wind farms. Therefore, it is necessary to develop a short-term forecasting method that can achieve satisfactory accuracy and stability at the same time [37]. This paper suggests a new hybrid model that combines a parameter optimization algorithm (CSO) and a forecasting module (SVM). The key weight coefficients of the forecasting module are optimized by the proposed algorithm. To reduce the noise in the raw data, a data preprocessing technique (EEMD) is used to obtain a stationary time series. To verify the validity of the proposed model, multi-step ahead forecasting and several performance evaluation metrics are applied in this paper. Traditional forecasting methods are used for comparison. In one-step ahead forecasting, the MAPE values of the BP, ARIMA, RBF, and the proposed model are 5.46%, 5.22%, 4.83%, and 3.07%, respectively. In two-step ahead forecasting, the MAPE values of these models are 6.70%, 6.42%, 5.97%, and 4.55%, while in three-step ahead forecasting, the MAPE values of the four models are 7.83%, 7.15%, 6.92%, and 5.46%. Not only the figures of MAPE, but also other five variables and the two tests, show the excellent performance of the proposed approach. Based on the experimental results, the newly proposed approach achieved the highest performance in wind speed prediction in both one-step forecasting and multi-step forecasting, in comparison with the other three methods used in the experiment. In conclusion, based on the experimental results, the proposed approach achieved significant improvement in both forecasting accuracy and stability. The proposed a hybrid model, which effectively improved the accuracy and the stability of wind speed forecasting, can be a great tool in managing wind farms. Since this paper used the wind speed data from the power grids of China, the proposed model can adapt to the system smoothly, which can reduce the costs and risks of the power station. When employed in the dispatch of the power system, the proposed hybrid model can generate benefits economically. For instance, the wind station is able to make timely adjustments of the operating plan, and the accurate forecasting results can reserve the capacity of the wind farm, saving unnecessary costs, and avoiding generating waste. As for the focus of future studies, this proposed hybrid approach could be employed in other areas relating to trend forecasting, namely extreme weather prediction, profit forecasting, and traffic condition forecasting.

Author Contributions: Conceptualization, T.L.; Methodology, T.L.; Software, Y.G.; Validation, T.L. and J.H.; Formal Analysis, S.L.; Investigation, T.L.; Resources, S.L.; Data Curation, Y.G.; Writing-Original Draft Preparation, T.L.; Writing-Review & Editing, T.L.; Visualization, J.H.; Supervision, S.L.; Project Administration, S.L.

Conflicts of Interest: The authors declare no conflict of interest.

Abbreviation

Variables	Meaning
$s(t)$	Raw data
$c_i(t)$	Residuals in raw data
$r_n(t)$	IMFs of the raw data
ε	Amplitude of the added noise
ε_n	Standard deviation of the error
N_ε	Value of ensemble member
x_i^{t+1}	New solution of cuckoo search
x_i^t	Current solution of cuckoo search
α	Size of each step in cuckoo search
$\oplus$	Entry wise multiplications
$L(\lambda)$	Lévy distribution
$\phi(x)$	Nonlinear mapping
w	Weight vector of SVM
b	Scalar of SVM
C	Penalty factor of the error
$L(x_i, d_i)$	Loss function
$C\frac{n}{2}\sum_{i=1}^{n} L(x_i, d_i)$	Empirical error
$w^2/2$	Regularization term
l	Quantity of elements in the sample data series
α_i, α_i^*	Multipliers of Lagrange function
$K(x, x_i)$	Kernel function of SVM
γ	Parameter of the kernel function
N	Total output samples
y_i	Actual series
yp_i	Predicting results

References

1. Salcedo-Sanz, S.; Pérez-Bellido, Á.M.; Ortiz-García, E.G.; Portilla-Figueraa, A.; LuisPrieto, L.; Correosoc, F. Accurate short-term wind speed forecasting by exploiting diversity in input data using banks of artificial neural networks. *Neurocomputing* **2009**, *72*, 1336–1341. [CrossRef]
2. Soman, S.S.; Zareipour, H.; Malik, O.; Mandal, P. A review of wind power and wind speed forecasting methods with different time horizons. In Proceedings of the North American Power Symposium 2010, Arlington, TX, USA, 26–28 September 2010.
3. Wang, J.; Heng, J.; Xiao, L.; Wang, C. Research and application of a combined model based on multi-objective optimization for multi-step ahead wind speed forecasting. *Energy* **2017**, *125*, 591–613. [CrossRef]
4. Mohandes, M.A.; Halawani, T.O.; Rehman, S.; Hussaina, A.A. Support vector machines for wind speed prediction. *Renew. Energy* **2004**, *29*, 939–947. [CrossRef]
5. Liu, H.; Tian, H.-Q.; Chen, C.; Li, Y. A hybrid statistical method to predict wind speed and wind power. *Renew. Energy* **2010**, *35*, 1857–1861. [CrossRef]
6. Liu, H.; Tian, H.Q.; Pan, D.F.; Li, Y.F. Forecasting models for wind speed using wavelet, wavelet packet, time series and artificial neural networks. *Appl. Energy* **2013**, *107*, 191–208. [CrossRef]
7. Li, G.; Shi, J. On comparing three artificial neural networks for wind speed forecasting. *Appl. Energy* **2010**, *87*, 2313–2320. [CrossRef]
8. Barbounis, T.G.; Theocharis, J.B.; Alexiadis, M.C.; Dokopoulos, P.S. Long-term wind speed and power forecasting using local recurrent neural network models. *IEEE Trans. Energy Convers.* **2006**, *21*, 273–284. [CrossRef]
9. Cheng, C.H.; Wei, L.Y. A novel time-series model based on empirical mode decomposition for forecasting TAIEX. *Econ. Model.* **2014**, *36*, 136–141. [CrossRef]
10. Zhou, J.Y.; Shi, J.; Li, G. Fine tuning support vector machines for short-term wind speed forecasting. *Energy Convers. Manag.* **2011**, *52*, 1990–1998. [CrossRef]
11. Liu, D.; Niu, D.X.; Wang, H.; Fan, L. Short-term wind speed forecasting using wavelet transform and support vector machines optimized by genetic algorithm. *Renew. Energy* **2014**, *62*, 592–597. [CrossRef]
12. Hu, Q.H.; Zhang, S.G.; Xie, Z.X.; Mi, J.S.; Wan, J. Noise model based v-support vector regression with its application to short-term wind speed forecasting. *Neural Netw.* **2014**, *57*, 1–11. [CrossRef] [PubMed]
13. Guo, Z.H.; Wu, J.; Lu, H.Y.; Wang, J.Z. A case study on a hybrid wind speed forecasting method using BP neural network. *Knowl.-Based Syst.* **2011**, *24*, 1048–1056. [CrossRef]
14. Guo, Z.H.; Zhao, J.; Zhang, W.Y.; Wang, J.Z. A corrected hybrid approach for wind speed prediction in Hexi Corridor of China. *Energy* **2011**, *36*, 1668–1679. [CrossRef]
15. Pourmousavi Kani, S.A.; Aredhali, M.M. Very short-term wind speed prediction: A new artificial neural network-Markov chain model. *Energy Convers. Manag.* **2011**, *52*, 738–745. [CrossRef]
16. Pourmousavi Kani, S.A.; Riahy, G.D.; Mazhari, D. An innovative hybrid algorithm for very short-term wind speed prediction using linear prediction and markov chain approach. *Int. J. Green Energy* **2011**, *8*, 147–162. [CrossRef]
17. Liu, H.; Chen, C.; Tian, H.Q.; Li, Y.F. A hybrid model for wind speed prediction using empirical mode decomposition and artificial neural networks. *Renew. Energy* **2012**, *48*, 545–556. [CrossRef]
18. Mohammadi, A.; Dehghani, M.J.; Ghazizadeh, E. Game Theoretic Spectrum Allocation in Femtocell Networks for Smart Electric Distribution Grids. *Energies* **2018**, *11*, 1635. [CrossRef]
19. Kianoosh, G.; Boroojeni, M.; Hadi, A.; Bahrami, S.S.; Iyengar, A.I.; Sarwat, O.K. A novel multi-time-scale modeling for electric power demand forecasting: From short-term to medium-term horizon. *Electr. Power Syst. Res.* **2017**, *142*, 58–73.
20. Haque, A.U.; Mandal, P.; Kaye, M.E.; Meng, J.L.; Chang, L.C.; Senjyu, T. A new strategy for predicting short-term wind speed using soft computing models. *Renew. Sustain. Energy Rev.* **2012**, *16*, 4563–4573. [CrossRef]
21. Li, G.; Shi, J. On comparing three artificial neural networks for wind speed forecasting. *Appl. Energy* **2010**, *87*, 2313–2320.
22. Ortiz-García, E.G.; Salcedo-Sanz, S.; Perez-Bellido, A.M.; Gascon-Moreno, J.; Portilla-Figueras, J.A.; Prieto, L. Short-term wind speed prediction in wind farms based on banks of support vector machines. *Wind Energy* **2011**, *14*, 193–207.

23. Xiao, L.; Wang, J.; Hou, R.; Wu, J. A combined model based on data pre-analysis and weight coefficients. optimization for electrical load forecasting. *Energy* **2015**, *82*, 524–549. [CrossRef]
24. Vapnik, V. *Statistical Learning Theory*, 2nd ed.; Wiley: New York, NY, USA, 1998.
25. Vapnik, V. *The Nature of Statistical Learning Theory*; Springer: New York, NY, USA, 1999.
26. Wu, Z.; Huang, N.E. Ensemble empirical mode decomposition: A noise-assisted data analysis method. *Adv. Adapt. Data Anal.* **2009**, *1*, 1–41. [CrossRef]
27. Yang, X.S.; Deb, S. Engineering optimization by cuckoo search. *Int. J. Math. Model. Numer. Optim.* **2010**, *1*, 330–343.
28. Yang, W.; Wang, J.; Wang, R. Research and application of a novel hybrid model based on data selection and artificial intelligence algorithm for short term load forecasting. *Entropy* **2017**, *19*, 52. [CrossRef]
29. Yang, X.S.; Deb, S. Cuckoo search via Lévy flights. Presented at 2009 World Congress on Nature & Biologically Inspired Computing (NaBIC), Coimbatore, India, 9–11 December 2009.
30. Pai, P.; Lin, C. A hybrid ARIMA and support vector machines model in stock price forecast. *Omega* **2005**, *33*, 497–505. [CrossRef]
31. Huang, C.; Davis, L.; Townshend, J. An assessment of support vector machines for land cover classification. *Int. J. Remote Sens.* **2002**, *23*, 725–749. [CrossRef]
32. Sung, A.H.; Mukkamala, S. Identifying important features for intrusion detection using support vector machines and neural networks. In Proceedings of the 2003 Symposium on Applications and the Internet, Orlando, FL, USA, 27–31 January 2003.
33. Xiao, L.; Shao, W.; Wang, C.; Zhang, K.; Lu, H. Research and application of a hybrid model based on multi-objective optimization for electrical load forecasting. *Appl. Energy* **2016**, *180*, 213–233. [CrossRef]
34. Iversen, E.B.; Morales, J.M.; Møller, J.K.; Madsen, H. Short-term probabilistic forecasting of wind speed using stochastic differential equations. *Int. J. Forecast.* **2015**, *32*, 981–990. [CrossRef]
35. Chao, R.; Ning, A.; Wang, J.Z.; Li, L.; Hu, B.; Shang, D. Optimal parameters selection for BP neural network based on particle swarm optimization: A case study of wind speed forecasting. *Knowl.-Based Syst.* **2014**, *56*, 226–239.
36. Wang, J.; Zhu, S.; Zhao, W.; Zhu, W. Optimal parameters estimation and input subset for grey model based on chaotic particle swarm optimization algorithm. *Expert Syst. Appl.* **2011**, *38*, 8151–8158. [CrossRef]
37. Zhou, J.Z.; Sun, N.; Jia, B.J.; Peng, T. A Novel Decomposition-Optimization Model for Short-Term Wind Speed Forecasting. *Energies* **2018**, *11*, 1752. [CrossRef]

applied
sciences

MDPI

Article

Smart-Grid-Aware Load Regulation of Multiple Datacenters towards the Variable Generation of Renewable Energy

Peicong Luo [1], Xiaoying Wang [1,*], Hailong Jin [2], Yuling Li [1] and Xuejiao Yang [1]

[1] State Key Laboratory of Plateau Ecology and Agriculture, Department of Computer Technology and Applications, Qinghai University, Xining 810016, China; peicongluo@163.com (P.L.); lyl_feng@126.com (Y.L.); yxj_abc@126.com (X.Y.)
[2] Department of Computer Science and Technology, Tsinghua University, Beijing 100084, China; jinhl15@mails.tsinghua.edu.cn
*   Correspondence: wxy_cta@qhu.edu.cn; Tel.: +133-2767-3963

Received: 4 January 2019; Accepted: 29 January 2019; Published: 3 February 2019

Abstract: Recently, as renewable and distributed power sources boost, many such resources are integrated into the smart grid as a clean energy input. However, since the generation of renewable energy is intermittent and unstable, the smart grid needs to regulate the load to maintain stability after integrating the renewable energy source. At the same time, with the development of cloud computing, large-scale datacenters are becoming potentially controllable loads for the smart grid due to their high energy consumption. In this paper, we propose an appropriate approach to dynamically adjust the datacenter load to balance the unstable renewable energy input into the grid. This could meet the demand response requirements by taking advantage of the variable power consumption of datacenters. We have examined the scenarios of one or more datacenters being integrated into the grid and adopted a stochastic algorithm to solve the problem we established. The experimental results illustrated that the dynamic load management of multiple datacenters could help the smart grid to reduce losses and thus save operational costs. Besides, we also analyzed the impact of the flexibility and the delay of datacenter actions, which could be applied to more general scenarios in realistic environments. Furthermore, considering the impact of the action delay, we employed a forecasting method to predict renewable energy generation in advance to eliminate the extra losses brought by the delay as much as possible. By predicting solar power generation, the improved results showed that the proposed method was effective and feasible under both sunny and cloudy/rainy/snowy weather conditions.

Keywords: renewable energy; load regulation; datacenter; smart grid; demand response

1. Introduction

With the exploitation and usage of green energy, more and more attention has been paid to green energy generation and management issues of the smart grid. It is predicted that the domestic total solar power generation will reach 150 GW by the end of 2020 [1]. Although renewable energy generation is of great importance to environmental sustainability in the future, there are still various problems to widely penetrate renewable energy into the grid due to its instability and intermittency. With the large-scale integration of renewable energy generation, the power flow distribution of the grid will change and the power flow might be reversed. Meanwhile, problems such as voltage fluctuation or over-limit violations will occur, which will affect the safe and reliable operation of the grid. In order to maintain the reliability of the grid, traditional grids usually adopt a passive regulation mode where the power supply changes with power demand. For example, large-scale batteries as an energy storage element, combined with corresponding control strategies, could help achieve the stability of the grid [2,3]. An analysis of three potential future technical regulation systems has been presented [4] in which wind power and small and medium-scale CHP (combined heat and power) units are involved in balancing and grid stabilizing tasks. The results indicated that such systems could improve the ability to integrate renewable energy. Compared to traditional batteries, this system could reduce carbon dioxide emissions effectively. However, such passive methods might result in an increase in grid operations. Moreover, the movement of electricity over distances results in losses, leading to a great amount of economic expense, as reported in Reference [5]. According to statistics, the global electric power transmission and distribution losses account for nearly 8.3% of the output [6], which shows that the power transmission and distribution losses are worthy of consideration due to the possible capital costs they might bring. In other words, the operational costs of the regional grid could be remarkably reduced by a reduction in transmission losses. Hence, it is necessary to adopt a strategy to make full use of loads to realize the demand response to balance the power supply and demand, to improve the efficiency, and to maintain the reliability of the grid with renewable energy sources incorporated.

On the other hand, with the development and wide-spread usage of large-scale datacenters, the power consumption of one datacenter could reach 50 MW or even more. In China, the total power consumption of domestic datacenters has exceeded the annual power generation of the Three Gorges Dam since 2015. Besides, the power of the datacenter is still growing up by 10–20% every year, which is a similar rate to that of renewables [7]. Studies have shown that the power of the datacenters is flexible to adjust, which implies that the datacenter is a potentially controllable load to achieve a demand response. Meanwhile, as a large consumer of electricity, connecting the datacenters into the grid with renewable sources can also reduce energy expenses effectively.

In this paper, we attempted to explore dynamic load adjustment approaches for multiple datacenters, holistically in the smart grid, in order to balance the varying renewable energy input into the grid. By our approach, the total power losses of the whole grid could be reduced, which showed the value of datacenters participating in demand response programs for the balance of power generation and consumption. Furthermore, although the power of the datacenters was flexible to adjust, the actual adjustment actions would have to spend some time, which we called "the action delay" hereafter. By our investigation, this kind of delay would result in extra losses for the grid. Thus, we designed a forecasting method, based on the concept of a neural network, to predict renewable energy generation in advance so as to adjust the power of the datacenters as soon as possible and reduce the extra losses. Moreover, we also considered the impact of the possible adjustment range in practical environments, which we called "the flexibility". This paper is an extended version of our prior work [8]. The extensions include the results of extended experimental scenarios upon three datacenters, more experiments under both sunny and cloudy weather conditions, a suitable prediction method adopted to forecast renewable energy generation in advance, the analysis and comparison of different forecasting accuracy results on power losses, and discussion of the relationship between forecasting time and accuracy.

The rest of this paper is organized as follows. In Section 2, we present the background of this paper and some relevant work. In Section 3, we describe the system model and the establishment of the problem to be solved later. Section 4 elaborates on the optimization method used to solve the defined problem. Section 5 demonstrates the experimental results of our approach and the analysis of practical details. Concluding remarks and a discussion about the future work are given in Section 6.

2. Background and Related Work

2.1. Datacenters Power Consumption and Renewable Energy Generation

With the rapid development of information technology and the coming of the Internet era, especially the development of cloud computing around the world, the proportion of datacenters with more than 100 racks is increasing year by year. Hence, the problems of high energy consumption, high cost and high pollution are increasingly prominent. For example, for a datacenter with a construction scale of 2000 racks, the electricity consumption per hour will be around 6000 kWh and the annual electricity consumption will be about 52,560 MWh. Then, its total annual cost could reach up to $105 million including the electricity cost, the air conditioning, fresh air, lighting and other power consumption of the datacenter, with a PUE (power usage effectiveness) of 2 [3]. In addition, the impact of datacenters on the environment is attracting increasing attention from the public. According to the current development trends, the average annual electricity consumption of datacenters will account for 1% of the total global value in 2020 as predicted in Reference [9]. The IT industry emits about 35 million tons of carbon dioxide per year, accounting for 2% of global emissions. In order to reduce energy consumption and carbon emissions of datacenters, it is important to maintain sustainable datacenters and make full use of green energy [10,11]. Meanwhile, governments all over the world have also published laws and policies to encourage energy conservation and the reduction of emissions. The cost of deploying equipment to build new energy sources, such as photovoltaic panels and wind turbines, has dropped as manufacturing costs have decreased, along with massive investment and government incentives. More and more IT enterprises and organizations are gradually realizing full or partial new energy-driven datacenters, such as the wind power datacenters that the Green House Data built in Wyoming [12], and the solar datacenters that the Facebook built in Oregon [13]. Early in April 2012, eBay decided to use 30 Bloom Energy fuel cells to power its datacenter in Utah [14]. Apple will utilize 60% power from photovoltaic generation and battery station to drive their datacenters in Southern California [15]. On the other hand, datacenter energy efficiency is usually low, with huge energy waste. According to statistics of the Ministry of Industry and Information Technology, the average PUE value for datacenters in China is between 2.2 and 3.0, while the actual energy consumption may be much higher. For enterprises, electricity for datacenters has become a big expense, greatly eroding the operating profits of enterprises. According to data reported in Reference [3], the total electricity consumption of datacenters all over China was more than 110.8 billion kWh in 2016, and in 2017 it was 120–130 billion kWh, which is more than the total generating capacity of the Three Gorges dam in the whole year of 2017 (97.605 billion kWh) together with the Gezhouba dam power plant (19.05 billion kWh in 2017).

At the same time, with the rapid development of the world's economy, the demand for energy is enhancing day by day, while traditional energy sources are drying up gradually. People have begun to focus on new types of clean energy, hoping that it could change the current energy structure and realize more sustainable development. In recent years, photovoltaic power generation and wind power generation have developed quickly. By the end of 2015, the total installed capacity of solar cells had reached 20,000 MW all over the world [16]. In 2017, the cumulative installed capacity of photovoltaic generation in China reached 130 million kW with 69% year-on-year growth, accounting for 7.3% of the total power generation capacity of the whole country. Among them, the cumulative installed capacity of centralized photovoltaic is 100 million kW and the distributed photovoltaic is 29.66 million kW. The installed wind power generation capacity in China increased by 53.06 million kW, an increase of 54% year on year, accounting for 40% of the total installed power capacity in the whole country. At the same time, by the end of 2017, the cumulative grid-integration capacity of wind power in China had reached 163.67 million kW, growing 10.1% year on year, among which the cumulative grid-connected capacity of offshore wind power was 2.02 million kW, growing by 37% year on year [8]. On November 20th, 2018, Power Construction Corporation of China opened the tender for the centralized procurement project of 1 GW photovoltaic modules and inverter frames in 2019 [17].

2.2. Power Consumption Adjustment of Datacenters

Recently, the participation of datacenters towards demand response requirements has become increasingly important given their high and increasing energy consumption and their flexibility in demand management compared to conventional industrial facilities. The huge yet flexible power adjustments of datacenters make them promising resources for demand response, which requires a certain amount of power adjustment at a certain time. Datacenters could dynamically adjust the power consumption themselves by leveraging the IT computing knobs such as geographical load balancing [18,19], dynamic capacity provisioning [20], and workload shifting [21], as well as non-IT knobs including batteries and cooling systems [22,23]. One of the most comprehensive studies describing the potential of different hardware components in the datacenters and strategies providing a demand response was released by Lawrence Berkeley National Laboratories [24]. Some of the initial work in the area comes from Urgaonkar et al. [25], which proposed an approach for adjusting the power consumption of datacenters by using energy storage to shift peak demand away from high peak periods. While the design of workload planning algorithms for datacenters has received considerable attention over the recent years [20,26–35], all of the proposed flexible datacenter workload load planning methods could change the overall power consumption as needed. A more complex approach was presented in Reference [36]. It took advantage of two datacenter flexibility mechanisms—workload shifting and local generation (local diesel generators and local renewable energy). Using these mechanisms, algorithms were developed in order to avoid the coincidental peak and reduce the energy costs. They relied on the prediction of a coincident peak occurrence based on historical data to optimize the workload allocation and local generation and to minimize the expected cost. In Reference [37] a technique was proposed for balancing and keeping the peak power consumption of the datacenter under a given threshold according to the electricity pricing but at the same time allowing the datacenter to respond to the regulation control signals that may request an increase in power consumption. Dan et al in Reference [38], on the basis of real-time electricity price consideration, proposed to reduce the energy consumption cost of datacenters by dynamically adjusting the server capacity and performing workload transfers in each time slot.

Besides, some researches have focused on dynamically controlling the number of active servers based on the load and regard it as an effective means of power control [39–41]. In References [20,42,43], dynamic speed/voltage scaling (DVS) could change processor power consumption, which could also adjust the datacenter power consumption on demand by adjusting the frequency based on the instantaneous power demand. Furthermore, our research team also conducted relevant studies to dynamically adjust the power consumption amount according to the variation of the power supply. Zhang et al. [44] designed adaptive scheduling algorithms and deployed renewable energy in the datacenters, aiming at scheduling approximate applications, in order to meet the user demands as well as maximizing the utilization of renewable energy. The main purpose was to appropriately utilize the renewable energy and at the same time aimed at typical approximate applications based on the trade-off of performance and accuracy to schedule and manage these tasks. By running such applications, the power consumption of the computing nodes could follow a trend of changing energy input as much as possible, which showed the effect of adjusting the datacenter power consumption. Xiaoying et al. [45] also considered that the workloads of large-scale datacenters are variable, believing that a coordinated resource management and power management approach could help datacenters to use renewable energy more effectively and proposed a green-power-aware virtual machine migration strategy to manage resources and power in green datacenters powered by mixed supply of both grid and renewable energy. The results illustrated that the holistic power consumption of both IT devices and cooling devices in the datacenter could be controlled towards the variation of the mixed power supply. Based on the above relevant research and also our prior work, we regard datacenters as adjustable and controllable loads in the smart grid, which might help the grid to keep its stability and improve its efficiency.

2.3. Renewable Energy Generation Integrated in the Grid

As renewable energy power stations are being constructed all over the world, the problems resulting from a grid penetrated with renewables have emerged. The photovoltaic power generation will be greatly affected by the change of solar radiation, while the wind generation will be affected by the wind speed variation. This makes renewable power generation intermittent and unstable.

To improve the utilization efficiency of interconnected devices, improve the power quality of the renewable energy interconnection grid system, maintain the normal operation of the grid and get good economic profit, Ahamed et al. [46] proposed a hybrid power generation system that connected the photovoltaic power station and batteries in parallel, which absorbed and supplied unbalanced power from the battery and flattened grid-connected power. Zhijiang et al. and Shang et al. [47,48] established a hybrid energy storage system composed of batteries and supercapacitors and optimized the power distribution of the storage equipment through different filtering algorithm controls, which improved the economy of the mixed energy storage equipment. Feng et al. [49] proposed the IP-IQ current detection method of two kinds of control schemes of the photovoltaic grid inverter based on the instantaneous reactive power theory in order to realize the dynamic reactive power compensation and flexible combination of an active filter, while at the same time improving the power quality. In Reference [50], a control strategy of the unified power quality conditioner was proposed to stabilize the power system and increase the utilization ratio of the equipment. Although battery storage systems can effectively increase grid stability after integrating renewable energy, electricity storage is not the optimum solution to integrate the large inflows of fluctuating renewable energy since more efficient and cheaper options can be found by integrating the electricity sector with other parts of the energy system and by creating a smart energy system. Lund et al. in Reference [51] investigated the most efficient and lowest cost storage options as a part of a smart energy systems approach, as defined in Reference [52]. By using this approach, it was explained why the best storage solutions could be found by integrating the individual sub-sectors of the energy system. One of the main reasons why a cross-sector approach could identify more economically viable solutions was that the cheaper and more efficient storage technologies that existed in the thermal and transport sectors, compared to the electricity sector. Lund et al. [53] also made a state-of-the-art description of different single-sector approaches for the transformation towards future sustainable energy solutions within the electricity, gas, building and industrial sectors. They discussed the smart energy systems concept with regard to the issues of the definition of the term, identification of renewable systems design, the integration of holistic storage solutions and the modelling of national energy systems. In addition, some researchers have also studied various kinds of controllable loads. Short et al. [54] put forward the dynamically controllable load-refrigerators. The refrigerator could change its temperature automatically as the grid frequency changed, which meant changing the power of the load. In the case of an increase (or decrease) in power generation, the load power can be increased (or decreased) by effectively controlling the controllable dynamic load. However, a controllable load changing with the frequency of power grid is still relatively rare at present, so it will take some time for large-scale application. Kondoh et al. [55] considered that each customer's load and each generator's active and reactive power should be controlled in order to stabilize and optimize the grid, comparing independent and cooperative control techniques as applied to load regulation using electric water heaters. However, both refrigerators and electric water heaters are low-power household appliances and the power control required a large number of such appliances. Besides, on account of the intermittent and unstable power supply, the controllable load should be adjusted frequently, which would affect the normal life of users and the service life of the electrical appliances. Although the strategies mentioned in the above papers could achieve stability after renewable energy is integrated into the grid, they all required some additional devices or even the devices of the user terminal, leading to additional costs.

In view of solving the instability and intermittency of the grid integrating with renewable energy, this paper proposed to add datacenters as electric loads into the grid so as to consume surplus power generated from renewable energy stations. This strategy could help achieve a win-win effect since on the one hand it could help maintain the normal operation of the grid integrating renewable energy through the adjustment of the datacenters and, on the other hand, the datacenter itself, as a big consumer of electricity, could leverage the renewable energy more effectively to further reduce the operational costs.

3. System Model and Problem Statement

Since the power of a datacenter could be dynamically adjusted, it can meet the demand response requirements in a smart grid integrated with renewable energy generation by adjusting the power itself according to the power of renewable energy generation within one day. However, considering the large-scale renewable energy generation integrated into the smart grid, the number, location and the real-time power consumption of datacenters will influence the normal operation and the power losses of the whole grid, which could comprehensively reflect the rationality and efficiency of the grid planning and design, production operation, and management level. This paper studies the smart grid operation with both renewable energy generation and multiple datacenters integrated proposes an optimal load allocation strategy in several different cases and mainly focuses on the total power losses. Meanwhile, considering the delay, an appropriate forecasting method has been adopted to predict the renewable energy generation power, so as to reduce the extra losses caused by the action delay as much as possible. At the same time, we also pay attention to the bus voltage and branch power flow in case they exceed the limitation, which might lead to the instability of the smart grid.

3.1. Power Grid System

In this paper, we focus on the interaction of datacenters and renewable power plants in the transmission network of the smart grid. Take the IEEE 30 Bus system [56] as an example of the transmission network and the topology of case_ieee30 is shown in Figure 1. The case_ieee30 dataset was converted from an IEEE common data format. The data of renewable energy generation is the photovoltaic generation from the Green Power Network [57], which is a third-party monitoring data management service platform for photovoltaic power stations. It realizes intelligent housekeeping services of photovoltaic power stations with high-quality products and technologies.

1. Load model: To model the load of this system, we used constant power according to the specifications in Reference [49]. Loads are connected to 21 buses respectively as shown by the arrows in Figure 1.
2. Generator model: There are six generators, which are connected to buses 1, 2, 5, 8, 11 and 13 respectively. Specifically, Gen1 is a balancing bus node.
3. Branch model: There are 41 branches each connecting two buses in the system. Each branch has its own line capacity limit and the voltage limitation of all branches is 0.95 pu to 1.1 pu. We will check and try to avoid the possible violations based on these limits. Moreover, when power is transmitted along the branches, there will be some losses on the line. We will use the summarized losses as the main metric in the experiments in later sections.

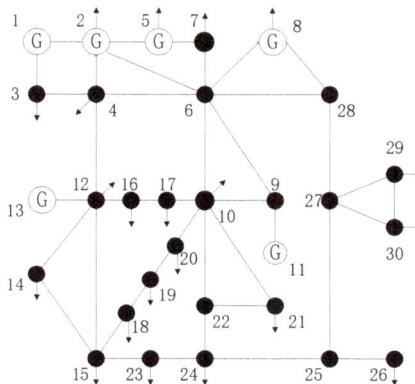

Figure 1. Model of the IEEE 30-bus test system.

3.2. The Model of Renewable Energy Generation Station

The power generation of the solar station can be obtained as a function, which can be described by Equation (1).

$$P = M \cdot A_p \cdot \varphi \cdot \eta_p \cdot \eta_{DC} \tag{1}$$

where M is the number of solar photovoltaic panels; A_p is the superficial area of each PV panel; φ is the solar radiant quantity; η_p denotes the efficiency of the PV panel; η_{DC} denotes the efficiency of the maximum power tracker of the solar PV panel array. In these factors, φ is essentially random, which will be impacted by seasons, solar radiation, temperature and pressure.

When it is decided to construct a renewable energy station at a certain location, we connect it to the bus corresponding to the region.

3.3. The Datacenter Model

In this paper, we regard the datacenter as a large and non-neglectable load for the utility grid system. The capacity of a datacenter is usually described by the power consumption amount when it is running at peak load and thus fully utilized. When we place a datacenter at a certain location, it is connected to the bus corresponding to that area where the datacenter is located. Then the load of this bus should be added to the current load of the datacenter, as follows:

$$\begin{cases} P_i = P_i + L_{DC}^P \\ Q_i = Q_i + L_{DC}^Q \end{cases} , i = 1, 2, \ldots, N \tag{2}$$

where P_i and Q_i denote the active and the reactive power load at bus i respectively; N is the number of buses in the grid and L_{DC}^P and L_{DC}^Q represents the active load and reactive load of the datacenter respectively.

3.4. Problem Formulation

In the grid system described above, we assume that a large renewable energy generation station is to be connected, which might incur oscillating power input to the grid. In order to consume the extra power generated from renewable sources, multiple datacenters can be established at several different locations and connected to the grid system as controllable loads. The problem here is to determine how much power each datacenter should consume in the case of multiple datacenters and make adjustment accordingly in a real-time manner.

Assume there are n datacenters deployed in the grid. Denote the solar power generated as P_S, the load of the i-th datacenter as L_i. Then, the total power losses will be impacted by them, denoted as a function Loss $(P_S, L_1, L_2, \ldots, L_n)$. Then, the problem we need to address can be defined as:

$$\text{minimize Loss } (P_S, L_1, L_2, \ldots, L_n) \tag{3}$$

$$P_S = \sum_{i=1}^{n} L_i \tag{4}$$

$$U_M(1 - \varepsilon_1) \le U_i \le U_M(1 + \varepsilon_2), \ 1 \le i \le n \tag{5}$$

$$|P_{ij}| \le P_{ij}^{MAX}, 1 \le i, j \le n \tag{6}$$

wherein U_M is the system nominal voltage; ε_1 and ε_2 are the allowable deviation rates specified internationally; P_{ij} is the branch power flowing from bus node i to bus node j, and P_{ij}^{MAX} is the maximum power allowed to pass through the branch (i.e., the branch capacity limit value). As shown, Equations (5) and (6) are both constraints to ensure the stability of the grid by avoiding any over-limit violations across all of the buses and branches. Especially, Equation (5) is the constraint for the bus voltage of each node and Equation (6) gives the limitation of the power flow on each branch.

4. Dynamic Load Management of Datacenters Based on Forecasting

4.1. Dynamic Load Adjustment of Datacenters

An imbalance between load and generation might lead to the failure of the normal operation of the power grid. Therefore, facing the special situation of renewable energy integrated into the grid, we selected datacenters as dynamic loads to maintain the basic stability of the smart grid. However, while maintaining stability, we also need to focus on the efficiency of the grid, which can be reflected by the power losses. In the scenario of multiple datacenters integrated into the smart grid, it will be difficult to solve the problem defined in Section 3. Furthermore, the Loss, as described in Section 3.4, is not an explicit function, which means that we cannot use linear or non-linear programming directly to solve this problem. Thus, a stochastic algorithm is proposed to implement an optimal load allocation strategy in order to obtain the minimum loss value.

In computer science and operations research, the genetic algorithm (GA) is a metaheuristic inspired by the process of natural selection that belongs to the larger class of evolutionary algorithms (EA). Genetic algorithms are commonly used to generate high-quality solutions for optimization and search problems by relying on bio-inspired operators such as mutation, crossover and selection [58]. As one of the stochastic algorithms, a genetic algorithm is good at solving global optimization problems and is always used to solve practical problems. In this experiment, we adopt the main concept of the genetic algorithm and use the toolbox '*Deap*', a novel evolutionary computation framework for rapid prototyping and testing of ideas to get an optimal load allocation strategy, to implement the entire algorithm. The steps of the dynamic load management algorithm can be presented as follows:

1. randomly initialize the population
2. determine the fitness of the current population
3. repeat.

- Select parents from the current population.

In this paper, the tournament selection strategy is adopted, i.e., to select a certain number of individuals from the population each time. In our experiment, we selected three individuals and then choose the best one among the three to enter the child population each time.

- Perform crossover operations on parents creating the population.

In this paper, we adopted the two-point crossover strategy, which was helpful to create more new individuals. Two-point crossover refers to the random setting of two crossover points in the individual coding string and then the partial genes between the two crossover points will be exchanged. An example of a two-point crossover operation is shown as Figure 2, wherein the dotted lines represent the two crossover points.

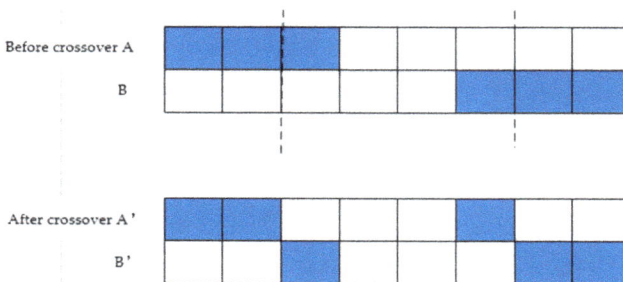

Figure 2. Two-point crossover operation. (**A**) and (**B**) are two randomly selected individuals, and A′ and B′ are the resultant individuals after the crossover operation.

- Perform mutation operations of the current population.

Here we adopt the gauss mutation strategy, wherein the mean is 0, the standard deviation is 1, and the independent probability for each attribute to be mutated is 0.1 to enter the child population.

- Determine the fitness of the population until the best individual is good enough.

4. output the best individual of the final population as the result.

When the iteration times of the evolution reach the maximum value, the individual with the maximum fitness obtained in the evolutionary process will be the output as the optimal solution. In this algorithm, we used vectors to represent the individual and the load capacity of each datacenter at every time interval was denoted as a gene. Then, the load capacity values of all datacenters in the grid at each time interval composed the individual $(x_1, x_2 \ldots)$ and the dimension of the vector could be determined according to the number of added datacenters. When an individual completes coding, a population of a specified size will be created and then we can call the 'power' module to calculate the fitness to assess the relative merits of each individual and determine their genetic opportunity.

4.2. Power Generation Forecasting

Renewable energy resource simulation and power prediction technology have always been a hot research direction. In recent years, the influence on large-scale renewable energy integrated into the grid appeared gradually. Renewable energy resource simulation and power prediction technology are developing in the direction of detailed simulation and customized prediction of resources, multi-space-time scale power prediction, probability prediction and event prediction [59].

Under the consideration of datacenter action delay, we adopted the neural network and called the framework "Keras" to forecast renewable energy generation power in advance. Keras is a high-level neural network API (application program interface), written in Python. It was developed with a focus on enabling fast experimentation [60]. In this paper, we selected N-days of historical renewable generation power data as a training set to forecast the renewable generation power of the next day. In the training set, the power data of n time points were used as a group of historical data to obtain the power data of the (n + 1)-th time point for training. We used a sequential model to stack the network layers. Then, the compile () method was used to compile the model. After that, we trained the network with the function fit () under a certain number of iterations, in which the optimization and loss function used were "Adam" and "mean_squared_error" respectively. Adam is an algorithm for the first-order gradient-based optimization of stochastic objective functions, based on adaptive estimates of lower-order moments, which was straightforward to implement and computationally efficient. It has little memory requirements, is invariant to the diagonal rescaling of the gradients, and is well suited for problems that are large in terms of data and/or parameters [61].

5. Experiment Results and Analysis

5.1. Testbed Setup and Parameter Settings

Here, we used MATpower [60] to simulate the operation of the grid in the following experiments. MATpower is a package of MATLAB M-files for solving power and optimizing power flow problems, designed to give the best performance possible while keeping the code simple to understand and modify, which is a simulation tool for researchers and educators.

In our experiments, we chose the monitoring power data from the photovoltaic station in Xuhui district (which belongs to the city of Shanghai) government from two typical days on 14 July 2018, and 19 December 2018, corresponding to sunny and cloudy/snowy weather conditions respectively. The monitoring time interval was 10 min and thus 144 data records were collected for one-day simulation.

Here, we simulated three different scenarios of MATpower in the following experiments:

(1) only one datacenter deployed. The datacenter load varied along with the input solar generation power;

(2) multiple datacenters deployed. Here we assume there are multiple distributed datacenters put into the grid and the loads are evenly allocated to them;

(3) multiple datacenters deployed and running with dynamically allocated loads.

We selected bus 9, 15 and 25 to integrate the solar station and datacenters respectively in the tested grid system. As parameter settings of GA, we define both the size of the population and the times of iterations as 100 and the fitness function consists of two parts, as follows:

$$loss1 + \alpha * loss2 \tag{7}$$

where loss1 is to guarantee the loss value is small enough, loss2 is to ensure the residual error between the allocated load and the input solar is small enough, and α is used to control the two variables. Here, we set $\alpha = 0.25$. Figure 3 shows the input solar generation power on the two days, where Figure 3a,b correspond to sunny and cloudy/snowy weather conditions respectively.

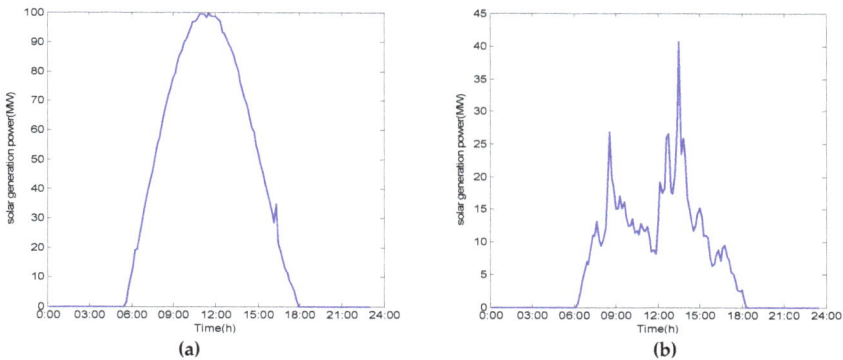

Figure 3. Input solar generation power. (a) Sunny day; (b) cloudy/snowy day.

5.2. Results of Power Losses under Accurate Responses

The experimental results under three scenarios are shown in Figure 4 as a holistic view.

In Figure 4a, the upper curve illustrates the first scenario mentioned in Section 5.1, which means only one datacenter was integrated into the case_ieee30 at bus 15. The power of the datacenter varied along with the input solar generation. As shown in the figure, the peak loss is 25.6 MW. The curve in the middle illustrates the second scenario, in which two datacenters were integrated into case_ieee30 at bus 15 and 25. In this case, the two datacenters consumed an equal amount of power and the sum of the two datacenter's power was consistent with the input solar generation. Besides, the solar station was still at bus 9. We can find the peak loss value was smaller than in scenario 1, which was 25.25 MW. The lowest curve demonstrates the third scenario mentioned in Section 5.1. In this case, two datacenters were integrated into to case_ieee30 at bus 15 and 25 respectively and the power of the two datacenters was dynamically adjusted according to changes in solar power generation, while the total power consumption of the two datacenters was kept constant. We found that the loss value was smaller than in scenarios 1 and 2, which was 23.93MW at the peak. In Figure 4b, which illustrates that for the results under cloudy/snowy weather conditions, the peak loss values in the same three scenarios were 20.99 MW, 19.82 MW, 18.08 MW respectively and the average value in the third scenario was still the lowest.

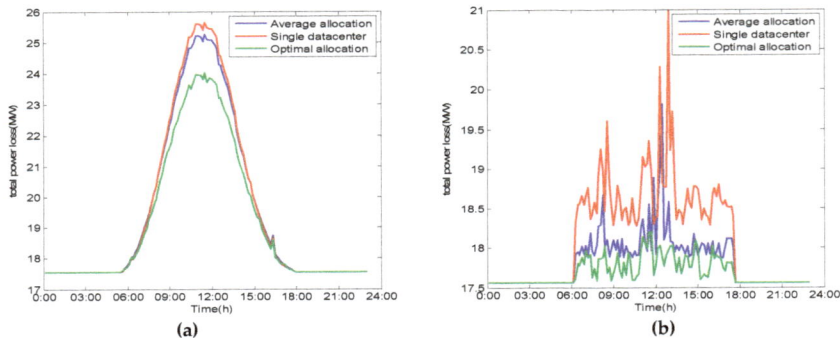

Figure 4. Comparison of loss values over three scenarios. (**a**) sunny day; (**b**) cloudy/snowy day.

From Figure 4 we can see that the curve of the optimal load-allocation strategy was almost always below the other two curves, which means that the dynamic load management methods can help to reduce the power losses of the grid system and thus provide savings in the operational costs. To examine the values more clearly, the corresponding statistical results are listed explicitly in Table 1. We also examined the scenario of putting three datacenters into the grid and the results are shown in Table 2. The datacenters were put at bus 15, 20, and 25 respectively. Here the two tables only give the results on a sunny day since it can be seen from Figure 4 that the other condition exhibited the same comparative trend.

Table 1. Statistical results with two datacenters.

Scenario	(1)	(2)	(3)
Average Total loss (MW)	19.548	19.492	19.19
percentage of loss/generation	8.52%	8.39%	8%

Table 2. Statistical results with three datacenters.

Scenario	(1)	(2)	(3)
Average Total loss (MW)	19.548	19.212	18.889
percentage of loss/generation	8.52%	8.02%	7.49%

5.3. Considering Practical Factors Including Flexibility and Action Delays

The above experiments are based on the ideal scenario that the datacenter can act as soon as possible to adjust the power consumption of itself to the target value exactly. In a practical environment, the power adjustment range of a datacenter will be constrained, which means that the datacenter cannot vary the power consumption as much as possible to reach the target value. We defined "flexibility" to describe such characteristics of the datacenter. Furthermore, the datacenter has to spend some time (which we defined as "action delay") to dynamically change its power consumption through the combination of multiple methods, such as load shedding or cooling temperature adjustment. Hereafter, we also conducted a series of experiments to study the impact of practical factors. Since the results of sunny or cloudy scenarios are similar, we have only shown the results of data from the sunny day in this subsection for clarity.

To examine the impact of the flexibility of the datacenter, the adjustable range of the datacenter can be expressed as follows:

$$[(1-e) \times p(t), (1+e) \times p(t)] \tag{8}$$

wherein e represents the flexibility of the datacenter, and the $p(t)$ means the power of the datacenter in the last time interval. We specify that if the power of the datacenter in the next interval is out of the allowed range, the upper or lower limits of the adjustable range will be taken as the power of datacenter in the next interval. On the other hand, to evaluate the impact of the action delay, we simulated different scenarios with a delay of 5, 10, 15 and 20 min respectively.

Figure 5 shows the impact of different flexibility values, where the abscissa represents different flexibility values and the ordinate represents the mean power losses (MW). We focused on the average loss under different flexibility values, where $e = 0$ means the datacenter kept a constant load, and in such scenario, the loss value was remarkably large. This shows that the adjustments made to the datacenter loads can help to effectively reduce the total power losses.

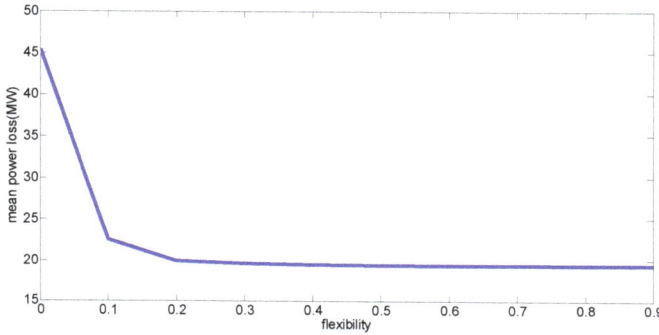

Figure 5. The impact of different flexibility values.

The impact of different action delay times on the total power losses of the grid can be seen in Figure 6. As mentioned, we tested different delay time settings from 0 to 20 min and the average loss values were recorded under each condition. From the figure, we can see that the loss value increases proportionally with the delay, which has a great impact on the action delay on the power loss of the entire grid. Besides, comparing the results with Figure 5, we can see that when the flexibility value is equal to or greater than 0.3, the loss value does not change substantially anymore and the value is close to the result in Figure 6 when there is no delay. This implies that even an adjustable range of ±30% of the datacenter power consumption can lead to fairly satisfactory results.

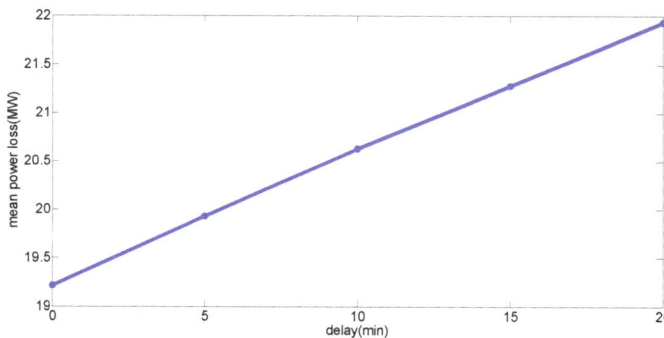

Figure 6. The impact of datacenter action delays.

5.4. Improvement Based on Forecasting Methods

Since we found that the action delay had a great impact on the power losses, in order to eliminate such impact as much as possible, we tried to adopt some forecasting method to help predict the future data as accurately as possible.

We used the neural network to forecast the solar generation power on July 14th and December 19th. Considering that July 14th was a sunny day, in order to improve the accuracy of the prediction, we chose data obtained from June 1st, 14th, 15th, 26th 2018, which were four sunny days, as the training set and the power data of 5 time points were used as a group of historical data to obtain the power data of the next time point for training. Considering that December 19th was a cloudy day, we chose data obtained from Dec 2th, 5th, 7th, 8th, 9th, 10th, 11th, 15th 2018 which were also cloudy/rainy/snowy days as a group of historical data to obtain the power data for the next time point for training. In addition, by adjusting the number of iterations, we obtained the forecasting results of different errors and obtained the loss values of the grid according to different settings of prediction accuracy.

In the following experiments, we adopted the framework "Keras" based on neural networks to forecast the generation of power. Figure 7 shows that forecasting results under sunny weather conditions, in which Figure 7a shows the comparison of the predicted values and the actual values, and Figure 7b shows the average power losses under three different accuracy settings. It can be observed that leveraging forecasting techniques are helpful to eliminate the negative impact of action delay time.

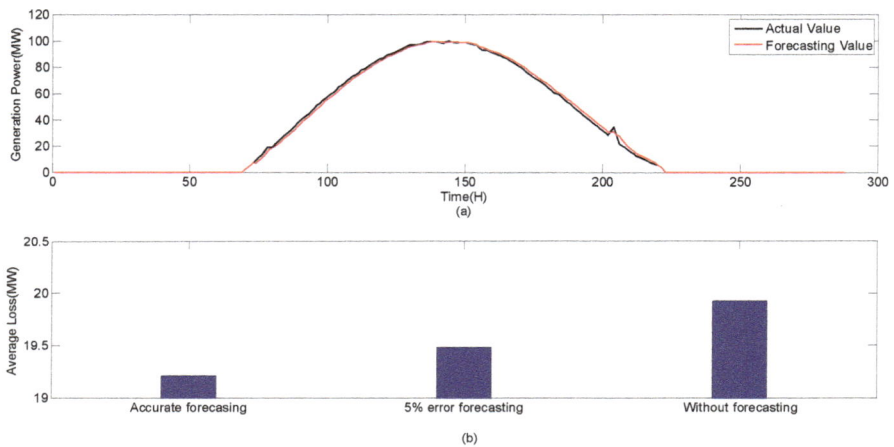

Figure 7. Forecasting results under 5-min action delay on a sunny day. (**a**) Comparison of the actual generation power and predicted values after 150 iteration times of one day; (**b**) comparison of the average power losses under different forecasting accuracy.

Figure 8 shows the forecasting results under a 10-min action delay of the datacenters, in which the forecasting average error in Figure 8a,c was 5% and 10% respectively. Figure 8b,d show the average power losses under different accuracy settings, including 5%/10% forecasting error, accurate forecasting and without forecasting. From the results in Figures 7 and 8, we can see that improving the forecasting accuracy can effectively reduce the extra power losses.

In order to verify the effectiveness of our prediction method, we also carried out the same experiments under cloudy/snowy weather conditions and the results are shown in Figure 9. The forecasting average error in Figure 9a,c is 10% and 15% respectively. Figure 9b,d shows the average power losses under different accuracy settings, including 10%/15% forecasting error, accurate forecasting and without forecasting. Figure 9 also illustrates that although on cloudy/snowy days the data fluctuations make the forecasting difficult and less accurate, we can still see the advantages of leveraging the forecasting techniques to help reduce extra power losses. To sum up, a reasonable forecasting method can eliminate extra losses from the action delay of the datacenters as much as possible, so it is meaningful to adopt an appropriate forecasting method to improve the impact of datacenter reaction on the grid.

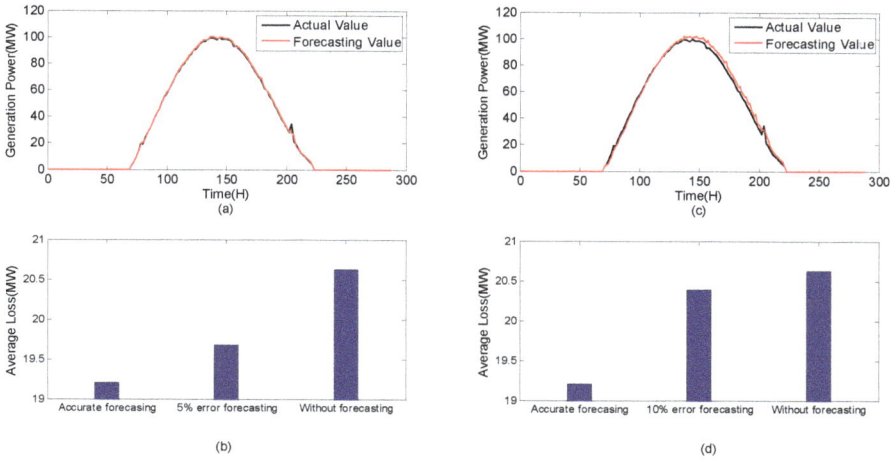

Figure 8. Forecasting results under a 10-min action delay on a sunny day. (**a**,**c**) show the forecasting results by different iteration time, where the iteration time is 150 s in (**a**) and 125 s in (**c**). (**b**,**d**) show the average power losses under different settings.

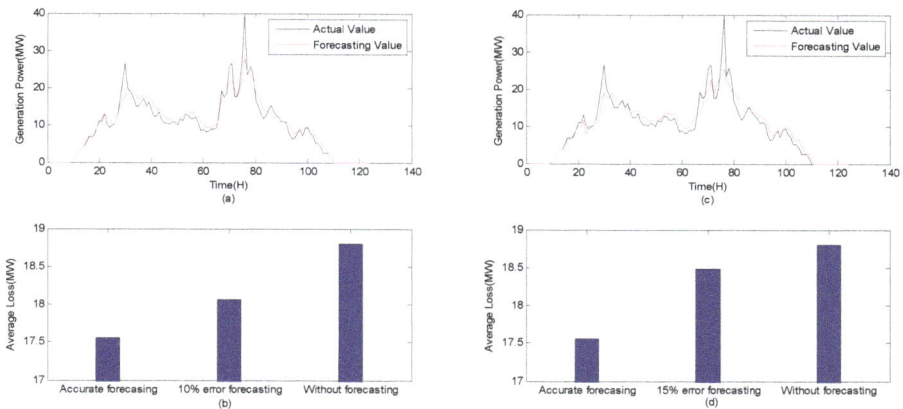

Figure 9. Forecasting results under a 10-min action delay. (**a**,**c**) show the different iteration time forecasting results, where the iteration time is 150 s in (**a**), and 120 s in (**c**). (**b**,**d**) show the average power losses under different settings.

In addition, we also examined the relationship between forecasting accuracy and the computation time. In general, in order to get more accurate prediction results, it has to spend more time on the computation process. Figure 10 illustrates the evaluation results for both the sunny and cloudy/rainy/snowy weather conditions under a 10-min action delay. From Figure 10a we can see that nearly-accurate forecasting takes more than twice the time compared to achieving 95% accuracy, while in contrast, to achieve 95% accuracy only takes 3 s more than to achieve 90% accuracy. Similarly, from Figure 10b we can see that accurate forecasting takes more than twice the time compared to achieving 85% accuracy, while to achieve 90% accuracy only takes 4 s more than to achieve 85% accuracy. This implies it could be practical and helpful to spend a little more time to get better forecasting results. Besides, comparing Figure 10a,b, it can be observed the forecasting time cost of the cloudy/rainy/snowy day is much higher than the sunny day. For example, in Figure 10a it only takes about 17 s to achieve 90% accuracy but in Figure 10b it takes nearly 26 s.

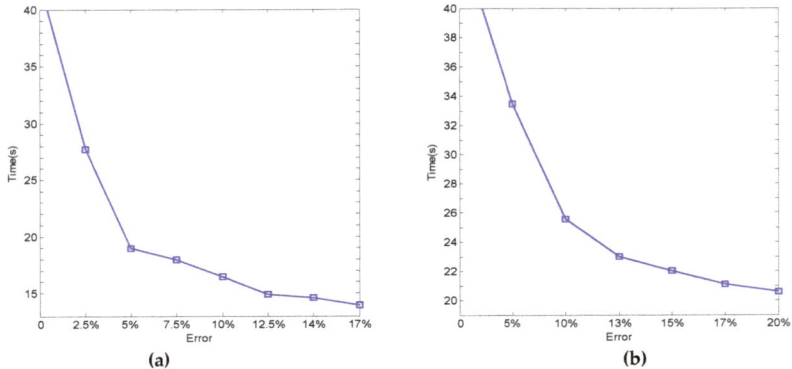

Figure 10. The relationship between forecasting error and forecasting time. (a) sunny day; (b) cloudy/rainy/snowy day.

6. Conclusions and Future Work

In this paper, we proposed the idea that putting the datacenters into the grid integrated with renewables as dynamic loads to maintain the normal operation of the grid. In addition, we designed an optimal load allocation strategy based on stochastic searching ideas, which could achieve a reasonable distribution of the load to get the minimum power losses when deploying multiple datacenters in the grid. Our results show that multiple datacenters in the smart grid as dynamic loads can effectively reduce the power losses and especially when adopting the optimal load-allocation scheme obtained by the GA-based method to adjust the power usage of datacenters; the advantages would more obvious. Stability issues of the grid were also considered in the problem we defined and the experiments we conducted, including bus voltage variations and the overloading of transmission lines. Besides, we also incorporated practical factors into the experiments such as the flexibility and action delay during the demand response process of datacenters. Moreover, in order to eliminate the inescapable influence of the action delay when adjusting the power of the datacenters as much as possible, we adopted forecasting techniques to predict the renewable energy generation amount looking ahead, under both sunny and cloudy/snowy weather conditions. Comparing the results under different action delay settings, it can be seen that adopting the forecasting method could reduce the extra power losses effectively. During our experiments, the relationship of computational time and the achieved accuracy was also investigated, which illustrated the feasibility of the proposed method based on forecasting.

From the study about forecasting results and forecasting time, we also found that the time cost of the forecasting method could further be optimized. As the next step, we will focus on the combination of multiple kinds of demand response strategies and approaches in the datacenter towards the requirements of the electricity providers. Besides, we will also pay more attention to the power management and other kinds of reliability issues of the smart grid after integrating renewable energy sources and take frequency and voltage excursion as the knobs to facilitate better demand response in the future work.

Author Contributions: Conceptualization, P.L. and X.W.; methodology, P.L.; software, H.J.; validation, Y.L., X.Y.; formal analysis, P.L.; investigation, H.J.; resources, P.L.; data curation, Y.L.; writing—original draft preparation, P.L.; writing—review and editing, X.W.; supervision, X.W.; project administration, X.W.; funding acquisition, Y.W.

Funding: This paper is partially supported by The National Natural Science Foundation of China (No. 61762074, No. 91847302, No. 61563044 and No. 61862053) and National Natural Science Foundation of Qinghai Province (No. 2019-ZJ-7034, No. 2015-ZJ-725, No. 2017-ZJ-902).

Conflicts of Interest: The authors declare no conflict of interest.

References

1. Singh, G.K. Solar power generation by PV (photovoltaic) technology: A review. *Energy* **2013**, *2*, 1–13. [CrossRef]
2. Hill, C.A.; Such, M.C.; Chen, D.; Gonzalez, J.; Grady, W.M. Battery energy storage for enabling integration of distributed solar power generation. *IEEE Trans. Smart Grid* **2012**, *2*, 850–857. [CrossRef]
3. Freitag, M.; Teuscher, J.; Saygili, Y.; Zhang, X.; Giordano, F.; Liska, P.; Hua, J.; Zakeeruddin, S.M.; Moser, J.; Grätzel, M.; et al. Dye-sensitized solar cells for efficient power generation under ambient lighting. *Nat. Photonics* **2017**, *2*, 372. [CrossRef]
4. Lund, H. Electric grid stability and the design of sustainable energy systems. *Int. J. Sustain. Energy* **2005**, *2*, 45–54. [CrossRef]
5. PJM. *A Survey of Transmission Cost Allocation Issues, Methods and Practices*; Valley Forge: New York, NY, USA, 2010.
6. Koomey, J. *Growth in Data Center Electricity Use 2005 to 2010*; Analytics Press: Oakland, CA, USA, 2011.
7. The Worldbank Data. Available online: https://data.worldbank.org (accessed on 25 January 2018).
8. Luo, P.; Wang, X.; Jin, H.; Li, Y.; Yang, X. Load Management for Multiple Datacenters towards Demand Response in the Smart Grid Integrating Renewable Energy. In Proceedings of the 2018 2nd International Conference on Computer Science and Artificial Intelligence (CSAI 2018), Shenzhen, China, 8–10 December 2018.
9. Chec. Available online: http://www.chec.com.cn (accessed on 26 December 2018).
10. Alashkar, A.; Gadalla, M. Thermo-economic analysis of an integrated solar power generation system using nanofluids. *Appl. Energy* **2017**, *2*, 469–491. [CrossRef]
11. Wang, Q.; Xia, C.; Tang, Z. Discussion on Feasibility for Distributed Energy in Internet Data Center. *Power Syst. Clean Energy* **2013**, 87–91. (In Chinese)
12. Gao, P.X.; Curtis, A.P.; Wong, B.; Keshav, S. It's not easy being green. In Proceedings of the ACM SIGCOM 2012, Helsinki, Finland, 13–17 August 2012; pp. 211–222. [CrossRef]
13. Ren, C.-G.; Wang, D.; Urgaonkar, B.; Sivasubramaniam, A. Carbon- aware energy capacity planning for datacenters. In Proceedings of the IEEE International Symposium on Modeling, Analysis and Simulation of Computer and Tele-Communication Systems (MASCOTS2012), Washington, DC, USA, 7–9 August 2012; pp. 391–400. [CrossRef]
14. Greenpeace. Available online: http://www.greenpeace.org (accessed on 26 December 2018).
15. Cook, G.; Horn, J.V. *How Dirty Is Your Data? A Look at the Energy Choices That Power Cloud Computing*; Greenpeace International Technical Report; Greenpeace International: Amsterdam, The Netherlands, 2011.
16. Ben, K. The country plans to add 6000MW of distributed photovoltaic power generation capacity next year. *East China Electric Power* **2013**, *10*.
17. POWERCHINA. Available online: http://www.powerchina.cn/col/col382/index.html (accessed on 26 December 2018).
18. Qureshi, A.; Weber, R.; Balakrishnan, H.; Guttag, J.; Maggs, B. Cutting the electric bill for internet-scale systems. In Proceedings of the ACM SIGCOMM, Barcelona, Spain, 17–21 August 2009.
19. Liu, Z.; Lin, M.; Wierman, A.; Low, S.H.; Andrew, L.L. Greening geographical load balancing. In Proceedings of the ACM SIGMETRICS, San Jose, CA, USA, 7–11 June 2011.
20. Lin, M.; Wierman, A.; Andrew, L.L.; Thereska, E. Dynamic right-sizing for power-proportional data centers. *IEEE/ACM Trans. Network. (TON)* **2013**, *21*, 1378–1391.
21. Liu, Z.; Chen, Y.; Bash, C.; Wierman, A.; Gmach, D.; Wang, Z.; Marwah, M.; Hyser, C. Renewable and cooling aware workload management for sustainable data centers. In Proceedings of the ACM SIGMETRICS, London, UK, 11–15 June 2012.
22. Guo, Y.; Fang, Y. Electricity cost saving strategy in data centers by using energy storage. *IEEE Trans. Parallel Distrib. Syst.* **2013**, *24*, 1149–1160. [CrossRef]
23. Guo, Y.; Gong, Y.; Fang, Y.; Khargonekar, P.P.; Geng, X. Energy and network aware workload management for sustainable data centers with thermal storage. *IEEE Trans. Parallel Distrib. Syst.* **2014**, *25*, 2030–2042. [CrossRef]

24. Ghatikar, G.; Ganti, V.; Matson, N.; Piette, M.A. Demand Response Opportunities and Enabling Technologies for Data Centers: Findings from Field Studies. 2012. Available online: http://eetd.lbl.gov/sites/all/files/LBNL-5763E.pdf (accessed on 14 September 2018).

25. Urgaonkar, R.; Urgaonkar, B.; Neely, M.; Sivasubramaniam, A. Optimal Power Cost Management Using Stored Energy in Data Centers. In Proceedings of the ACM SIGMETRICS, San Jose, CA, USA, 7–11 June 2011.

26. Gandhi, A.; Chen, Y.; Gmach, D.; Arlitt, M.; Marwah, M. Minimizing data center sla violations and power consumption via hybrid resource provisioning. In Proceedings of the IGCC, Orlando, FL, USA, 25–28 July 2011.

27. Chen, Y.; Gmach, D.; Hyser, C.; Wang, Z.; Bash, C.; Hoover, C.; Singhal, S. Integrated management of application performance, power and cooling in data centers. In Proceedings of the NOMS, Osaka, Japan, 19–23 April 2010.

28. Govindan, S.; Choi, J.; Urgaonkar, B.; Sivasubramaniam, A.; Baldini, A. Statistical profiling-based techniques for effective power provisioning in data centers. In Proceedings of the 4th ACM European Conference on Computer Systems, Nuremberg, Germany, 1–3 April 2009.

29. Choi, J.; Govindan, S.; Urgaonkar, B.; Sivasubramaniam, A. Profiling, prediction, and capping of power consumption in consolidated environments. In Proceedings of the MASCOTS, Baltimore, MD, USA, 8–10 September 2008.

30. Heo, J.; Jayachandran, P.; Shin, I.; Wang, D.; Abdelzaher, T.; Liu, X. Optituner: On performance composition and server farm energy minimization application. *IEEE Trans. Parallel Distrib. Syst.* **2011**, *22*, 1871–1878. [CrossRef]

31. Verma, A.; Dasgupta, G.; Nayak, T.; De, P.; Kothari, R. Server workload analysis for power minimization using consolidation. In Proceedings of the USENIX ATC, San Diego, CA, USA, 14–19 June 2009.

32. Meisner, D.; Sadler, C.; Barroso, L.; Weber, W.; Wenisch, T. Power management of online data-intensive services. In Proceedings of the ISCA, San Jose, CA, USA, 4–8 June 2011.

33. Zhang, Q.; Zhani, M.; Zhu, Q.; Zhang, S.; Boutaba, R.; Hellerstein, J. Dynamic energy-aware capacity provisioning for cloud computing environments. In Proceedings of the 9th ACM International Conference on Autonomic Computing, San Jose, CA, USA, 18–20 September 2012.

34. Xu, H.; Li, B. Cost efficient datacenter selection for cloud services. In Proceedings of the IEEE International Conference on Communications, Beijing, China, 15–17 August 2012.

35. Yao, Y.; Huang, L.; Sharma, A.; Golubchik, L.; Neely, M. Data centers power reduction: A two time scale approach for delay tolerant workloads. In Proceedings of the INFOCOM, Orlando, FL, USA, 25–30 March 2012; pp. 1431–1439.

36. Liu, Z.; Wierman, A.; Chen, Y.; Razon, B.; Chen, N. Data center demand response: Avoiding the coincident peak via workload shifting and local generation. In *Proceedings of the ACM SIGMETRICS/International Conference on Measurement and Modeling of Computer Systems*; ACM: New York, NY, USA, 2013; pp. 341–342.

37. Aksanli, B.; Rosing, T. Providing regulation services and managing data center peak power budgets. In Proceedings of the Design, Automation and Test in Europe Conference and Exhibition (DATE), Dresden, Germany, 24–28 March 2014; pp. 1–4.

38. Xu, D.; Liu, X.; Fan, B. Efficient server provisioning and offloading policies for internet datacenters with dynamic load demand. *IEEE Trans. Comput.* **2015**, *2*, 682–697. [CrossRef]

39. Chen, Y.Y.; Das, A.; Qin, W.; Sivasubramaniam, A.; Wang, Q.; Gautam, N. Managing Server Energy and Operational Costs in Hosting Centers. In Proceedings of the ACM SIGMETRICS International Conference on Measurement and Modeling of Computer Systems, Banff, AB, Canada, 6–10 June 2005.

40. Chen, G.; He, W.; Liu, J.; Nath, S.; Rigas, L.; Xiao, L.; Zhao, F. Energy-Aware Server Provisioning and Load Dispatching for Connection-Intensive Internet Services. In Proceedings of the Fifth USENIX Symposium Networked Systems Design and Implementation (NSDI), San Francisco, CA, USA, 16–18 April 2008.

41. Weiser, M.; Welch, B.; Demers, A.; Shenker, S. Scheduling for Reduced CPU Energy. In Proceedings of the First USENIX Conference Operating Systems Design and Implementation (OSDI), Monterey, CA, USA, 14–17 November 1994.

42. Yao, F.F.; Demers, A.; Shenker, S. A Scheduling Model for Reduced CPU Energy. In Proceedings of the 36th Annual Symposium Foundations of Computer Science (FOCS), Milwaukee, WI, USA, 23–25 October 1995.

43. Lorch, J.R.; Smith, A.J. Improving Dynamic Voltage Scaling Algorithms with PACE. In Proceedings of the ACM SIGMETRICS Int'l Conf. Measurement and Modeling of Computer Systems, Cambridge, MA, USA, 16–20 June 2001.

44. Wang, X.; Zhang, G.; Yang, M.; Zhang, L.A. Green-aware Virtual Machine Migration Strategy for Sustainable Datacenter Powered by Renewable Energy. *Simul. Model. Pract. Theory* **2015**, *58*, 3–14. [CrossRef]

45. Gungor, V.C.; Sahin, D.; Kocak, T.; Ergut, S.; Buccella, C.; Cecati, C.; Hancke, G.P. A Survey on Smart Grid Potential Applications and Communication Requirements. *IEEE Trans. Ind. Inform.* **2013**, *2*, 28–42. [CrossRef]

46. Ahamed, M.H.F.; Dissanayake, U.; De Silva, H.M.P.; Pradeep, H.R.C.G.P.; Lidula, N.W.A. Modelling and simulation of a solar PV and battery based DC microgrid system. In Proceedings of the IEEE International Conference on Electrical, Electronics, and Optimization Techniques (ICEEOT), Chennai, India, 3–5 March 2016; pp. 1706–1711. [CrossRef]

47. Cheng, Z.; Li, Y.; Xie, Y. Control strategy for hybrid energy storage of photovoltaic generation microgrid system with super capacitor. *Power Syst. Technol.* **2015**, *2*, 2739–2745. (In Chinese)

48. Shang, G.; Xin-Chun, Q.I.; Tao, X.; Zheng, X. Common DC bus based PV-hybrid energy storage power system and optimal control using double filters. *Power System Prot. Control* **2014**, *2*, 92–97. (In Chinese)

49. Zheng, S.C.; Wang, P.Z.; Ge, L.S. Study on Pwm Control Strategy of Photovoltaic Grid-connected Generation System. In Proceedings of the 2006 CES/IEEE 5th International Power Electronics and Motion Control Conference, Shanghai, China, 14–16 August 2006.

50. Zhang, G.; Zhang, T.; Ding, M.; Su, J.; Wang, H.; Lv, S.; Chen, J. Simulation research on unified power quality conditioner with PV grid connected generation. *Proc. CSEE* **2007**, *2*, 82–86. (In Chinese)

51. Lund, H.; Østergaard, P.A.; Connolly, D.; Mathiesen, B.V. Smart Energy and Smart Energy Systems. *Energy* **2017**, *137*, 556–565. [CrossRef]

52. Lund, H. Renewable Energy Systems—A Smart Energy Systems Approach to the Choice and Modeling of 100% Renewable Solutions. *Chem. Eng. Trans.* **2014**, *39*, 1–6.

53. Lund, H.; Østergaard, P.A.; Connolly, D.; Ridjan, I. Energy Storage and Smart Energy Systems. *Int. J. Sustain. Energy Plan. Manag.* **2016**, *11*, 3–14.

54. Short, J.A.; Infield, D.G.; Freris, L.L. Stabilization of Grid Frequency Through Dynamic Demand Control. *IEEE Trans. Power Syst.* **2007**, *2*, 1284–1293. [CrossRef]

55. Kondoh, J.; Aki, H.; Yamaguchi, H.; Murata, A.; Ishii, I. Consumed power control of time deferrable loads for frequency regulation. In Proceedings of the IEEE Power Systems Conference and Exposition, New York, NY, USA, 10–13 October 2004.

56. Zimmerman, R.D.; Murillo-Sánchez, C.E.; Thomas, R.J. MATPOWER: Steady-State Operations, Planning and Analysis Tools for Power Systems Research and Education. *IEEE Trans. Power Syst.* **2011**, *26*, 12–19. [CrossRef]

57. SmartPV. Available online: http://www.lvsedianli.com (accessed on 26 December 2018).

58. Mitchell, M. *An Introduction to Genetic Algorithms*; MIT Press: Cambridge, MA, USA, 1998.

59. Yang, C.; Du, Z.; Shi, T.; Fang, L. Development status and trend of large-scale new energy and renewable energy power friendly access technology. *China Acad. J.* **2008**, *8*, 14–17. (In Chinese)

60. Keras Documentation. Available online: https://keras.io (accessed on 26 December 2018).

61. Chow, J.H.; Frederick, D.K.; Chbat, N.W. *Discrete-Time Control Problems Using MATLAB and the Control System Toolbox*; Thomson-Brooks/Cole: London, UK, 2003; 269p, ISBN 05343847.

applied
sciences

MDPI

Article

Meteorological Variables' Influence on Electric Power Generation for Photovoltaic Systems Located at Different Geographical Zones in Mexico

Jose A. Ruz-Hernandez [1], Yasuhiro Matsumoto [2], Fernando Arellano-Valmaña [1],
Nun Pitalúa-Díaz [3,*], Rafael Enrique Cabanillas-López [3], José Humberto Abril-García [3],
Enrique J. Herrera-López [4] and Enrique Fernando Velázquez-Contreras [3]

[1]   Facultad de Ingeniería, Universidad Autónoma del Carmen, Calle 56 No. 4 Esq. Avenida Concordia Col.
      Benito Juárez, Cd. Del Carmen C.P. 24180, Campeche, Mexico; jruz@mail.unacar.com (J.A.R.-H.);
      ff.arellano@hotmail.com (F.A.-V.)
[2]   Departamento de Ingeniería Eléctrica, Centro de Investigación y de Estudios Avanzados del IPN, Av.
      Instituto Politécnico Nacional 2508, San Pedro Zacatenco, Ciudad de México C.P. 07360, CDMX, Mexico;
      ymatsumo@cinvestav.mx
[3]   Departamento de Ingeniería Industrial, Departamento de Ingeniería Química y Metalurgia, Departamento
      de Investigación en Polímeros y Materiales, Universidad de Sonora, Blvd. Luis Encinas y Rosales S/N,
      Col. Centro, Hermosillo C.P. 83000, Sonora, Mexico; rcabani@iq.uson.mx (R.E.C.-L.);
      jose.abril@unison.mx (J.H.A.-G.); evlzqz@guaymas.uson.mx (E.F.V.-C.)
[4]   Biotecnología Industrial, Sublínea Bioelectrónica, Centro de Investigación y Asistencia en Tecnología y
      Diseño del Estado de Jalisco A.C., Camino Arenero 1227, Col. El Bajío del Arenal,
      Zapopan C.P. 45019, Jalisco, Mexico; eherrera@ciatej.mx
*   Correspondence: nun.pitalua@unison.mx; Tel.: +52-1662-114-0682

Received: 1 March 2019; Accepted: 10 April 2019; Published: 20 April 2019

Abstract: In this study, the relation among different meteorological variables and the electrical power from photovoltaic systems located at different selected places in Mexico were presented. The data was collected from on-site real-time measurements from Mexico City and the State of Sonora. The statistical estimation by the gradient descent method demonstrated that solar radiation, outdoor temperature, wind speed, and daylight hour influenced the electric power generation when it was compared with the real power of each photovoltaic system. According to our results, 97.63% of the estimation results matched the real data for Sonora and 99.66% the results matched for Mexico City, achieving overall errors less than 7% and 2%, respectively. The results showed an acceptable performance since a satisfactory estimation error was achieved for the estimation of photovoltaic power with a high determination coefficient R^2.

Keywords: photovoltaic systems; meteorological variables; electric power; gradient descent; sustainable development

---

1. Introduction

Renewables energies represent a potential alternative in the transition towards a low-carbon society, where photovoltaic sources play a key role; however, consumers are also investors and a project is implemented only if economic conditions are verified [1].

Solar technologies are characterized depending on the way they capture, convert, and distribute sunlight such as photovoltaic (PV) systems and their corresponding requirements for an energy storage arrangement [2,3]. Consequently, these technologies feed power to the electric grid by using solar panels as generators [4–6]. In addition, concentrating solar power plants (CSP) use mirrors to focus the energy from the sun to drive traditional steam turbines or engines to create electricity; solar heating and cooling (SHC) systems which collect the thermal energy from the sun use this heat to provide hot water, space heating, and cooling for residential, commercial, and industrial applications. These technologies displace the need to use electricity or natural gas [7–9].

A photovoltaic system is composed of several components: the solar panel and the inverter for grid-connected systems and additionally energy storage for stand-alone systems. The fabrication of the solar panel involves diverse stages. The first step is to define the type of panel, where the most known is monocrystalline or polycrystalline and where the solar cells are manufactured using wafers made of silicon [10]. Nowadays, the PV systems are widely used to generate electricity due its accessible cost [5,6,11,12]. Some studies aimed at reducing this cost even more, e.g., in Reference [13], a design optimization model for the residential PV systems in South Korea was proposed, where the objective function to be minimized consisted of three costs, such as the monthly electric bill, the PV-related construction costs, and the PV-related maintenance cost.

The solar radiation (photons) is responsible for the photovoltaic effect; nonetheless, some weather factors have an effect on the amount of energy generated even with the optimum radiation. A cloudy day generates a shadow on the solar panel by the time it has to capture the photons so that the incidence of these particles will be less, achieving a minor electric power. It is well-known that clouds are water steam concentrated in the air; in other words, humidity and temperature work together, and such a relation is an example of how the meteorological variables impact the generation of electric power [14,15].

Solar energy, besides wind energy, is currently the most resourceful renewable source worldwide [16,17]. Its obtainment, unlike many others currently used, does not mean any harm to the environment, and its resources overcome, by far, everyone else [12,18–20]. Some countries such as Germany, Italy, Spain, United States of America, and China are ahead on solar energy research; meanwhile in Mexico, being a country rich in solar radiation, with a great territorial extension, and having some solar energy studies, solar energy still does not have the necessary research compared to countries leading solar technology [4–6,11,12,21].

Figure 1 depicts the behavior of the horizontal solar radiation on the world. Similarly, if Mexico is compared with Europe, it can be seen that the only country with a notorious radiation incidence is Spain, achieving a maximum value between 4.8 and 5.4 kWh/m^2 [22]. Mexico exceeds Spain both in incidence territory and in radiation intensity, with an average between 5.6 and 6.2 kWh/m^2, as can be seen in Figure 2, exceeding even China, which mostly contemplates values of 4.6 kWh/m^2.

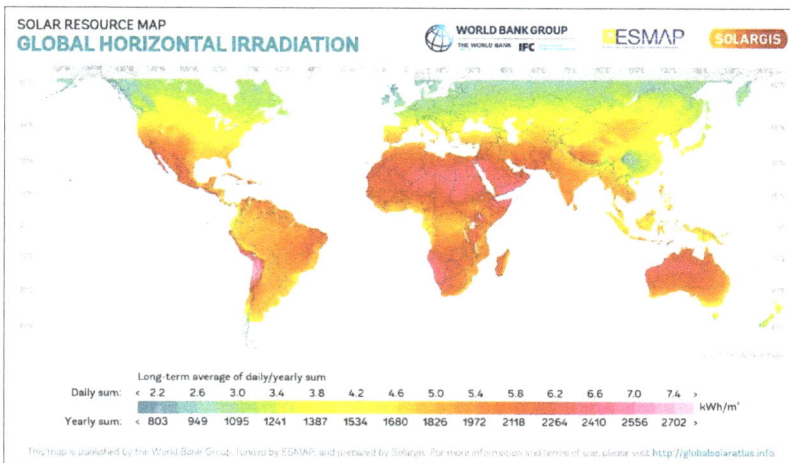

Figure 1. The worldwide horizontal solar radiation [23].

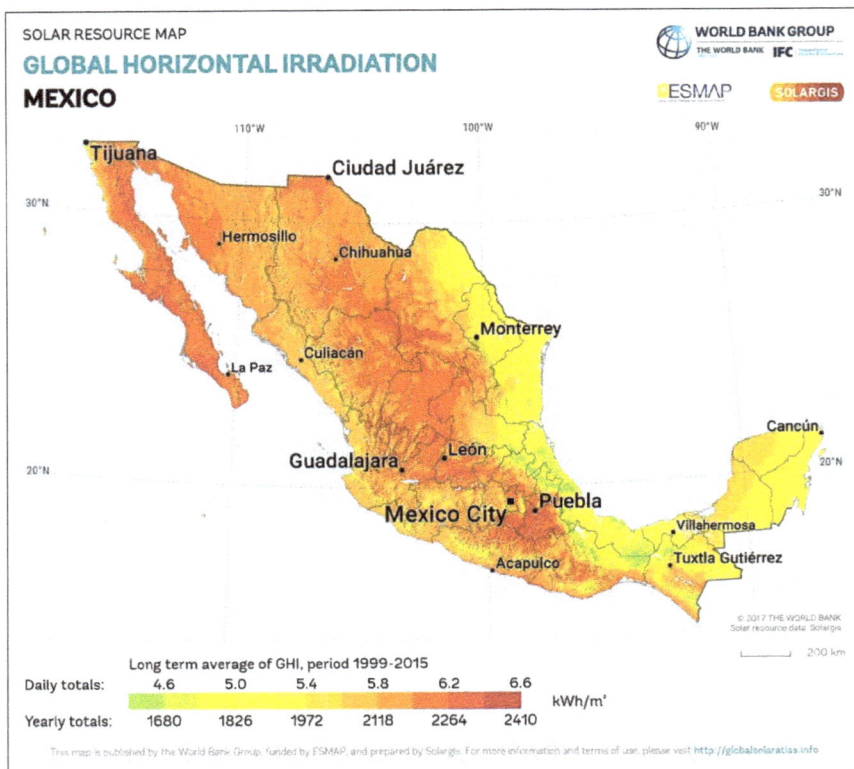

Figure 2. The Mexican solar radiation atlas [24].

Table 1 shows the data compilation from 2010 describing the global-horizontal solar radiation for some locations in Mexico. The total irradiation provided by solar energy in the year 2018 was summarized in Reference [25]. Both reports agree that the Northwest of Mexico reached the maximum values, the states of Sonora, Baja California, Coahuila, and Chihuahua being the main producers/receivers.

Recently, diverse reports have studied the relations between some meteorological variables and the electric power generated in a photovoltaic system. In Reference [26], an analysis of the climatic factors and solar data from the Andes site was performed; nonetheless, all the data gathered for the study was averaged every 10 min and the power measurements were estimated. In [27], the effect of diverse meteorological variables (outdoor temperature, air pressure, humidity, wind speed, and solar radiation) on the generated energy are mentioned, although a representative model of the plant was not obtained and all the data was experimentally generated. In Reference [28], a statistical method was applied to forecast the energy generated by a solar plant, though a database of 30 plants from different locations is necessary and although it does not have real power measurements from the site and the computational load is heavy. In Reference [29], an artificial neural network (ANN) is used to obtain a model to forecast the photovoltaic energy; however, the solar radiation was the only meteorological variable analyzed.

Higher Education Institutions (HEI) currently play a major role in the generation of human capital and the associated impact on societal development; HEIs are ideal locations to focus the resources in terms of the deployment and experimentation of decarbonization technologies to demonstrate the best practice for a further replication within wider society [30].

Careful planning is required to manage the future electricity demand of PV systems due to its increasing potential demand in Mexico [31,32]. Therefore, it is vital to understand the influence of meteorological variables on energy consumption in which a better understanding of it can contribute to a more useful strategy in meeting the energy efficiency goal for the country.

According to the above and considering References [1,13,30], the aim of this work is to present a statistical analysis based on the gradient descent method that is easy to implement and has a low computational load to estimate the electric power generation from the meteorological data such as the solar radiation, outdoor temperature, wind speed, and daylight time collected from PV systems located in Mexico City and Sonora [33]. This is important because most of the current photovoltaic system deployments do not monitor these factors or employ them in adaptation and prediction tasks [34–36].

The proposed methodology achieves a satisfactory estimation of the PV power with a high determination coefficient and a fair error percentage value.

Table 1. The solar radiation on select places in Mexico (data in kWh/m^2 per day) [37].

State	City	Jan	Feb	Mar	Apr	May	Jun	Jul	Aug	Sep	Oct	Nov	Dec	Avg
Sonora	Hermosillo	4.0	4.6	5.4	6.6	8.3	8.5	6.9	6.6	6.7	6.0	4.7	3.9	6.0
Sonora	Guaymas	4.5	5.7	6.5	7.2	7.3	6.8	5.9	5.8	6.3	5.9	5.0	5.6	5.9
Chihuahua	Chihuahua	4.1	4.9	6.0	7.4	8.2	8.1	6.8	6.2	5.7	5.2	4.6	3.8	5.9
SLP	SLP	4.3	5.3	5.8	6.4	6.3	6.1	6.4	6.0	5.5	4.7	4.2	3.7	5.4
Zacatecas	Zacatecas	4.9	5.7	6.6	7.5	7.8	6.2	6.2	5.9	5.4	4.8	4.8	4.1	5.8
Guanajuato	Guanajuato	4.4	5.1	6.1	6.3	6.6	6.0	6.0	5.9	5.8	5.2	4.8	4.6	5.6
Aguascalientes	Aguascalientes	4.5	5.2	5.9	6.6	7.2	6.3	6.1	5.9	5.7	5.1	4.8	4.0	5.6
Oaxaca	Salina Cruz	5.4	6.3	6.6	6.4	6.1	5.0	5.6	5.9	5.2	5.9	5.7	5.2	5.8
Oaxaca	Oaxaca	4.9	5.7	5.8	5.5	6.0	5.4	5.9	5.6	5.0	4.9	4.8	4.4	5.3
Jalisco	Colotlán	4.6	5.7	6.5	7.5	8.2	6.6	5.8	5.6	5.8	5.3	4.9	4.1	5.9
Jalisco	Guadalajara	4.6	5.5	6.3	7.4	7.7	5.9	5.3	5.3	5.2	4.9	4.8	4.0	5.6
Durango	Durango	4.4	5.4	6.5	7.0	7.5	6.8	6	5.6	5.7	5.1	4.8	3.9	5.7
Baja California	La Paz	4.4	5.5	6.0	6.6	6.5	6.6	6.3	6.2	5.9	5.8	4.9	4.2	5.7
Baja California	San Javier	4.2	4.6	5.3	6.2	6.5	7.1	6.4	6.3	6.4	5.1	4.7	3.7	5.5
Baja California	Mexicali	4.1	4.4	5.0	5.6	6.6	7.3	7.0	6.1	6.1	5.5	4.5	3.9	5.5
Querétaro	Querétaro	5.0	5.7	6.4	6.8	6.9	6.4	6.4	6.4	6.3	5.4	5.0	4.4	5.9
Puebla	Puebla	4.9	5.5	6.2	6.4	6.1	5.7	5.8	5.8	5.2	5.0	4.7	4.4	5.5
Hidalgo	Pachuca	4.6	5.1	5.6	6.8	6.0	5.7	5.9	5.8	5.3	4.9	4.6	4.2	5.4

2. Methodology

2.1. Statistical Method

The amount of data collected and the correlation among them are quite important to accomplish an estimation. If the total data is scarce or the correlation between any input and the output is low, the estimation will not be satisfactory.

The correlation analysis is one of the most used and reported statistical methods on scientist and medical researches; its visual representation is known as a dispersion graphic. It is used to prove or reject the existence of a relation between two different variables based on the Pearson correlation coefficient described by Equation (1).

$$\rho\,(a,b) = \frac{E\,(ab)}{\sigma_a \sigma_b} \tag{1}$$

where $E\,(ab)$ is the crossed correlation between a and b and where $\sigma_a^2 = E\,(a^2)$ and $\sigma_b^2 = E\,(b^2)$ are the variations of the variables a and b, respectively [38].

If "a" and "b" are the two considerate variables, then a dispersion graphic shows the location of each ordered pair in a coordinated system. If most points appear to be close to a straight line, this correlation is known as linear. If most points appear to be close to a curve, the correlation is known as nonlinear. On the other hand, if a clear pattern among the ordered pairs is inexistent, then there is no relation between both variables [39]. In order to represent the curve (curve or line regression) that best fits the behavior of ordered pairs, Equation (2) is used.

$$y = a_0 + a_1 x + a_2 x^2 + a_3 x^3 + \cdots + a_m x^m \tag{2}$$

The correlation coefficient or R requires that both a magnitude and a direction be either positive (from 0 to 1) or negative (from -1 to 0). If R gets close to ± 1, the correlation will be stronger. The correlation does not depend on the direction or the sign: A correlation of 0.57 is equal to a -0.57 one.

It can be also stated that the greater the absolute value of R, the greater the correlation will be. The determination coefficient R^2 is defined as the percentage of the variation of the dependent variable values that can be explained as variations of the independent variable. In other words, a determination coefficient $R^2 = 0.23$ symbolizes that 23% of the dependent variables is attached to the changes of the independent variable. Therefore, if a correlation factor of $R = 0.20$ was found between two variables, the determination coefficient would be $R^2 = 0.04$ so that only 4% of the dependent variable is affected by the variations of the independent variable [40].

In order to create a statistical representative model of the generated power by solar energy, the concept of gradient descent optimization (GDO) was considered due its ability to minimize the model error by the LSR of a linear regression model, which is often used on estimation studies, but it is not as complex to implement as an intelligent technique [29,41,42].

This optimization method has multiple applications. In Reference [43], this method was used to optimize the geometry of the strut-and-tie truss to minimize the difference in the share of resisting actions with respect to the prediction of the multi-action shear model. In Reference [44], a model based on gradient descent is proposed that integrates several important parameters for ranking channels in order to select the best communication channel from a radio spectrum for transmission, enhanced by cognitive radio as an intelligent wireless solution.

The GDO is based on the linear regression method known as the relation among all the input variables and the output one. Depending on the number of input variables, the regression can be simple or multiple [45–47]. To organize all the gathered data, every input variable is considered as a column in a matrix called "X" and the output parameter is considered as a vector "y", as seen in Equation (3).

$$X = \begin{bmatrix} x_{11} & x_{12} & x_{13} & \cdots & x_{1k} \\ x_{21} & x_{22} & x_{23} & \cdots & x_{2k} \\ x_{31} & x_{32} & x_{33} & \cdots & x_{3k} \\ \vdots & \vdots & \vdots & & \vdots \\ x_{n1} & x_{n2} & x_{n3} & \cdots & x_{nk} \end{bmatrix} \qquad y = \begin{bmatrix} y_1 \\ y_2 \\ \vdots \\ y_n \end{bmatrix} \tag{3}$$

The purpose of GDO is to find an estimation of the real output through an equation involving all the collected data as shown in Equation (4).

$$h_\theta(x) = \theta_0 + \theta_1 x_1 + \theta_2 x_2 + \ldots + \theta_k x_k + \varepsilon \tag{4}$$

where $h_\theta(x)$ is the estimated output, x_k is the kth input variable, θ_k is the characteristic coefficient of every variable, and ε is the error between the model and the real data [48].

As can be seen from Equation (4), θ_0 is not a value of influence for any input; however, it does locate the resulting estimation graphic on the Y-axis. If θ_0 is greater than expected, the plot will have a wide gap against the real data in the lower values; nonetheless, if θ_0 is smaller than necessary, the gap between the real and estimated model will be large on the upper zone. Therefore, both cases have a greater error. In order to achieve the minimum error value and instead of a linear regression which uses the least squares method, the GDO finds the right θ_0 by recursive partial derivatives of the cost function described by Equation (5)

$$J[\theta] = \frac{1}{2m} \sum_{i=1}^{m} \left(h_\theta\left(x^{(i)}\right) - y^{(i)} \right)^2 \tag{5}$$

where m is the total amount of rows in the matrix, $x\,(i)$ represents the ith row of "X", and $y\,(i)$ is the value of the ith row of "y". The gradient descent, denoted by Equation (6), aims to converge to the cost function minimum through its partial derivative. The quickness of the convergence is given by α.

$$\theta_k := \theta_k - \alpha \frac{\partial}{\partial \theta_k} J\,[\theta] \tag{6}$$

Equation (7) shows the substitution of Equation (5) into (6) which has to be repeated itself n-times until the convergence is done.

$$\theta_k := \theta_k - \alpha \frac{1}{m} \sum_{i=1}^{m} \left(h_\theta \left(x^{(i)} \right) - y^{(i)} \right) \cdot x_k^{(i)} \tag{7}$$

2.2. Solar Data

The data gathered from the State of Sonora, specifically the city of Hermosillo, were obtained with the support of the University of Sonora (UNISON), while the ones from Mexico City were issued by the Centro de Investigación y de Estudios Avanzados (CINVESTAV) campus Zacatenco, with the solar radiation (global-horizontal), outside temperature, wind speed, and daylight hour (time) as the input meteorological variables and the electric power as the output for both PV systems. Considering that both the solar plants are from different locations and do not have the same arrangement, their respective output power magnitudes vary among them; Hermosillo site has an electric power maximum around 2500 W, while Mexico City site reaches values around 45 kW.

The matrix "X" and the vector "y" from Equation (3) are shown in Equation (8).

$$X = \begin{bmatrix} Solar\,radiation & Temperature & Wind\,speed & Time \end{bmatrix}$$
$$y = [Electric\,Power] \tag{8}$$

By using the same variables from both locations, the aim is to achieve an acceptable statistical estimation of the electric power even if those variables are gathered from distant geographic areas.

3. Results and Discussion

3.1. Hermosillo Site

Figure 3 shows the geographical zone where Hermosillo Site (HS) is located. This site covers a period data of 6 months. Figures 4–7 depict the dispersion graphic and fitted curve for each input variable against the electric power. In Tables 2 and 3, the monthly and total results of the characteristic equations for each fitted curve and their determination coefficients are described, respectively.

Equation (7) is applied in order to calculate the values of θ_k for Hermosillo by using the Matlab® software. The optimal values found for θ are mentioned in Table 4 and substituted into Equation (4), thereby obtaining its mathematical representation given by Equation (9). Figure 8 is obtained by implementing six months of gathered data into Equation (9). It describes the behavior of the statistical estimation (red) against the real data (green) obtained by the PV system in Hermosillo. For a better appreciation of these results, Figure 9 shows a close-up from 24 November to 24 December 2018. A tracing practically coincident with the rise and fall times for each day as a satisfactory reach of the maximum and minimum values is observed.

$$h_\theta = 0.0044 + 2.2387x_1 + 0.1156x_2 + 0.0154x_3 + 0.0022x_4 \tag{9}$$

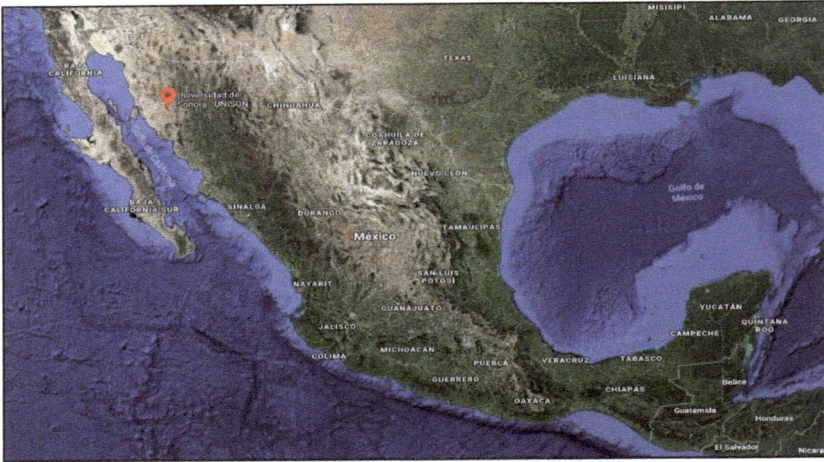

Figure 3. The geographical location of the Hermosillo Site (HS) [49] (29.0843293 N, −110.9583558 W).

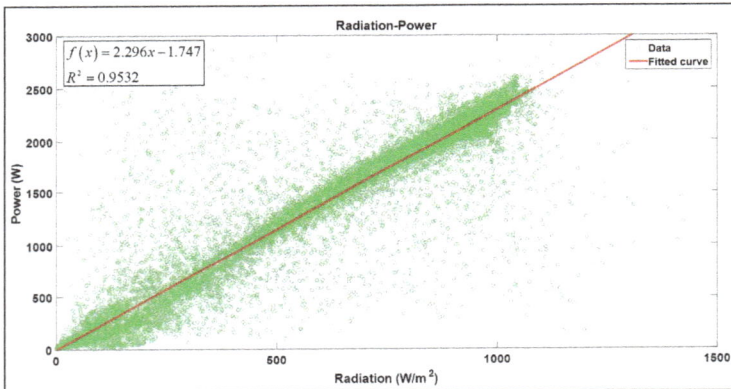

Figure 4. The Power vs. Solar radiation of HS.

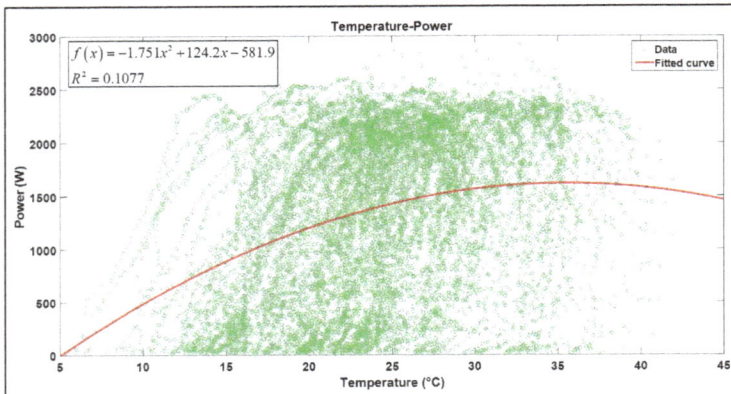

Figure 5. The Power vs. Temperature of HS.

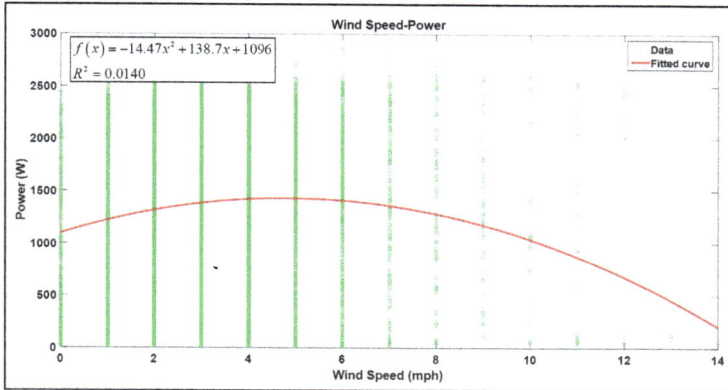

Figure 6. The Power vs. Wind speed of HS.

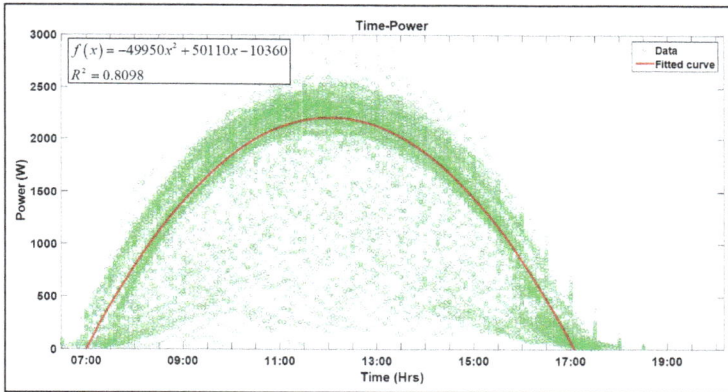

Figure 7. The Power vs. Time of HS.

Table 2. The fitted curve characteristic equation of every input from HS.

Month	Input Variables	
	Solar radiation	Temperature
Aug	$f(x) = 2.245x + 92.59$	$f(x) = -17.98x^2 + 1274x - 20{,}920$
Sep	$f(x) = 2.308x + 28.45$	$f(x) = -11.43x^2 + 796.7x - 12{,}210$
Oct	$f(x) = 2.276x + 50.32$	$f(x) = -8.751x^2 + 556.5x - 7217$
Nov	$f(x) = 2.326x - 29.48$	$f(x) = -6.074x^2 + 354.8x - 3505$
Dec	$f(x) = 2.25x - 36.17$	$f(x) = 0.08391x^2 + 45.57x + 219.1$
Jan	$f(x) = 2.398x + 8.88$	$f(x) = -2.016x^2 + 144.8x - 752.7$
TOTAL	$f(x) = 2.296x - 1.747$	$f(x) = -1.751x^2 + 124.2x - 581.9$
	Wind speed	Hour
Aug	$f(x) = -60.94x^2 + 438.2x + 753.3$	$f(x) = -43{,}250x^2 + 44{,}070x - 8893$
Sep	$f(x) = -21.89x^2 + 111.6x + 1352$	$f(x) = -42{,}970x^2 + 43{,}290x - 8593$
Oct	$f(x) = -20.63x^2 + 118.4x + 1292$	$f(x) = -46{,}940x^2 + 46{,}740x - 9314$
Nov	$f(x) = 4.208x^2 + 12.31x + 1348$	$f(x) = -50{,}610x^2 + 50{,}160x - 10{,}180$
Dec	$f(x) = -11.68x^2 + 120.2x + 1005$	$f(x) = -51{,}600x^2 + 51{,}860x - 10{,}900$
Jan	$f(x) = -14.18x^2 + 206.8x + 886.7$	$f(x) = -55{,}500x^2 + 56{,}360x - 11{,}960$
TOTAL	$f(x) = -14.47x^2 + 138.7x + 1096$	$f(x) = -49{,}950x^2 + 50{,}110x - 10{,}360$

Table 3. The determination coefficients (R^2) of every input from HS.

Month	Solar Radiation	Temperature	Wind Speed	Hour
Aug	0.9917	0.1679	0.0541	0.9825
Sep	0.9981	0.1502	0.0242	0.9968
Oct	0.9523	0.1664	0.0380	0.9850
Nov	0.9708	0.1955	0.0135	0.9962
Dec	0.9851	0.0965	0.0136	0.8865
Jan	0.9609	0.1390	0.0565	0.9760
TOTAL	0.9532	0.1077	0.0140	0.8098

Table 4. The values of θ_k from HS by gradient descent.

θ_k	HS
θ_0	0.0044
θ_1	2.2387
θ_2	0.1156
θ_3	0.0154
θ_4	0.0022

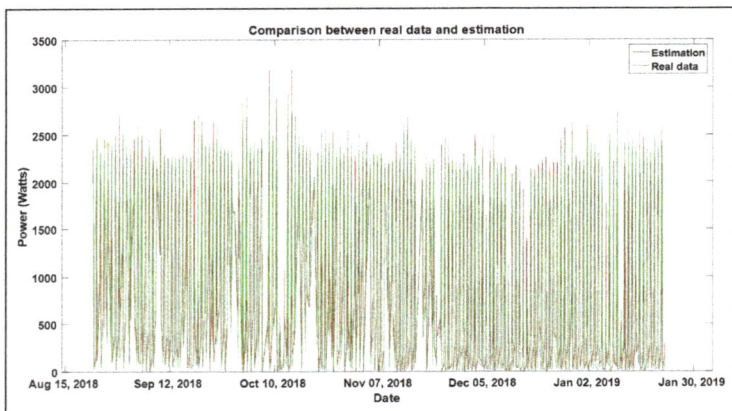

Figure 8. The results obtained from HS by gradient descent.

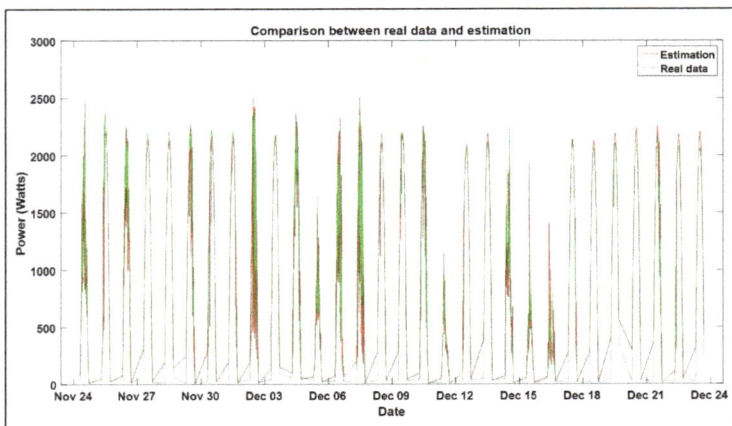

Figure 9. A close-up of the results from HS.

3.2. Data from Mexico City

The geographical zone of the Mexico City Site (MCS) is presented in Figure 10. The data covers a period of 6 months. From Figures 11–14 the dispersion graphic and fitted curve for each input variable against electric power are shown. In Tables 5 and 6, the monthly and total results of the characteristic equations for each fitted curve and their determination coefficients are described, respectively.

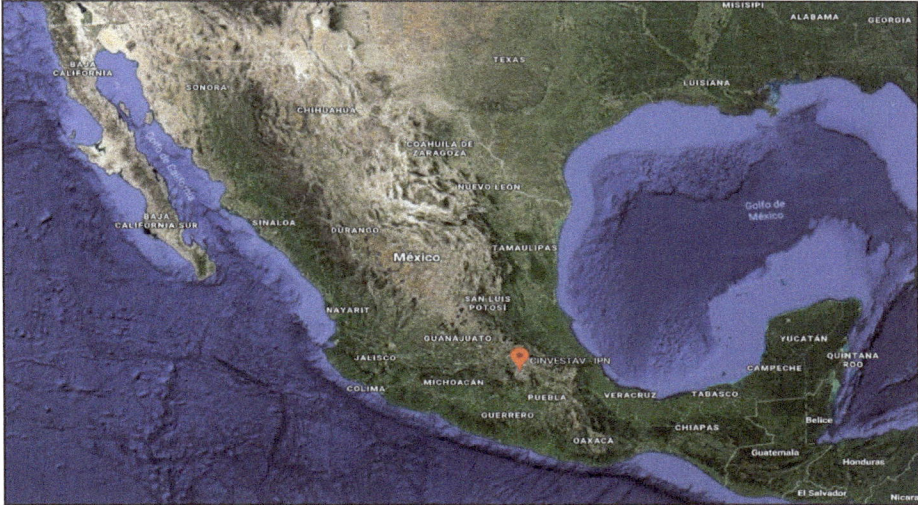

Figure 10. The geographical location of Mexico City Site (MCS) [50] (19.5099425 N, −99.1317477 W).

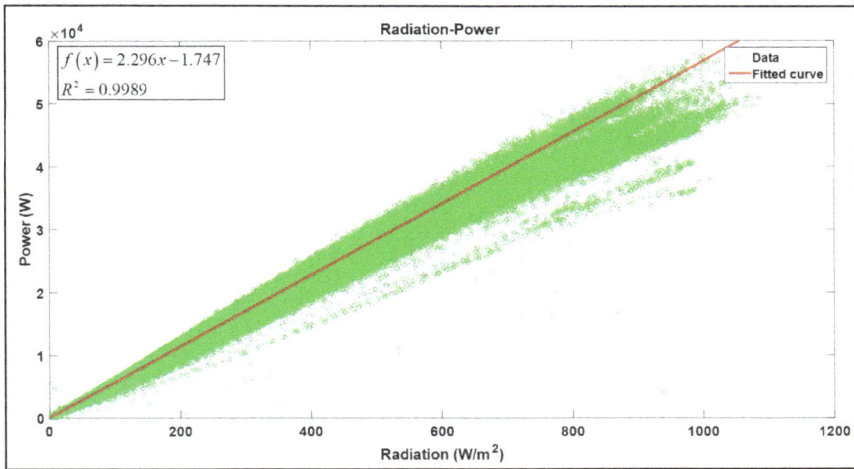

Figure 11. The Power vs. Solar radiation of MCS.

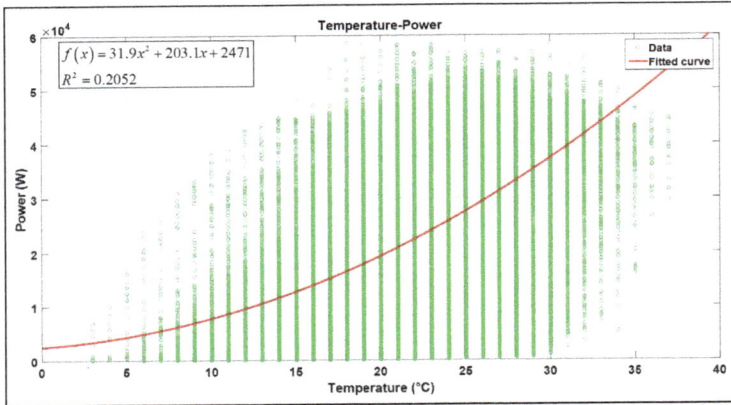

Figure 12. The Power vs. Temperature of MCS.

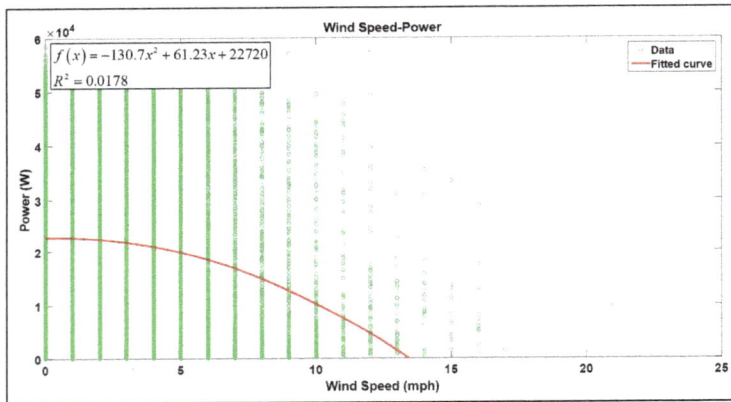

Figure 13. The Power vs. Wind speed of MCS.

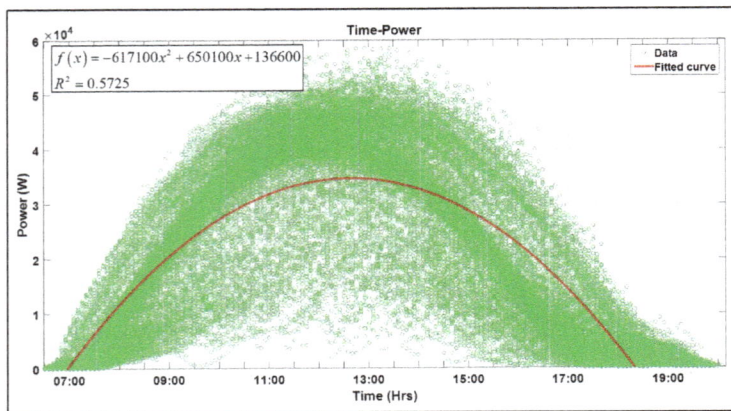

Figure 14. The Power vs. Time of MCS.

Table 5. The fitted curve characteristic equation of every input from MCS.

Month	Input Variables	
	Solar radiation	Temperature
Jul	$f(x) = 53.58x + 135$	$f(x) = -20.07x^2 + 3392x - 44{,}530$
Aug	$f(x) = 56.62x + 241.2$	$f(x) = -43.78x^2 + 4162x - 51{,}030$
Sep	$f(x) = 59.48x + 245.7$	$f(x) = 9.184x^2 + 1875x - 25{,}960$
Oct	$f(x) = 57.58x + 189.8$	$f(x) = -6.143x^2 + 2115x - 18{,}990$
Nov	$f(x) = 57.64x + 1.454$	$f(x) = -132x^2 + 5978x - 36{,}790$
Dec	$f(x) = 53.52x - 7.362$	$f(x) = -104.9x^2 + 4312x - 18{,}320$
TOTAL	$f(x) = 56.75x + 65.85$	$f(x) = 31.9x^2 + 203.1x + 2471$
	Wind speed	Hour
Jul	$f(x) = -173.8x^2 + 531.6x + 20{,}420$	$f(x) = -584{,}100x^2 + 632{,}300x - 137{,}600$
Aug	$f(x) = -157.7x^2 + 325.2x + 21{,}850$	$f(x) = -599{,}500x^2 + 649{,}400x - 141{,}400$
Sep	$f(x) = -156.4x^2 + 459.8x + 19{,}960$	$f(x) = -632{,}500x^2 + 686{,}200x - 153{,}700$
Oct	$f(x) = -113x^2 + 47.06x + 23{,}490$	$f(x) = -693{,}400x^2 + 735{,}600x - 159{,}300$
Nov	$f(x) = 66.96x^2 - 1894x + 28{,}680$	$f(x) = -918{,}400x^2 + 908{,}400x - 181{,}500$
Dec	$f(x) = 44.08x^2 - 706.9x + 22{,}910$	$f(x) = -837{,}800x^2 + 843{,}100x - 175{,}100$
TOTAL	$f(x) = -130.7x^2 + 61.23x + 22{,}720$	$f(x) = -617{,}100x^2 + 650{,}100x + 136{,}600$

Table 6. The determination coefficients (R^2) of every input from MCS.

Month	Solar Radiation	Temperature	Wind Speed	Hour
Jul	0.9997	0.4952	0.0295	0.6032
Aug	0.9990	0.3261	0.0214	0.5652
Sep	0.9950	0.3684	0.0093	0.5014
Oct	0.9923	0.2148	0.0177	0.5803
Nov	0.9995	0.1674	0.0320	0.8684
Dec	0.9995	0.1440	0.0015	0.8865
TOTAL	0.9989	0.2052	0.0178	0.5725

The determination coefficient values for Mexico City are found applying the same procedure used in the case of Hermosillo, as shown in Table 6. Analogous to the case from Hermosillo, the coefficients θ for Mexico City data and its mathematical representation are displayed in Table 7 and Equation (10), respectively. Figure 15 shows a six months estimation result (red) against the real data behavior (green) from Mexico City. In order to a better appreciation, Figure 16 shows a close-up of Figure 15 covering the same period as that in the case of Hermosillo. The estimation result matches with the rise and fall times of the real power.

$$h_\theta = 0.1005 + 55.9815x_1 + 2.2910x_2 + 0.1290x_3 + 0.0533x_4 \tag{10}$$

Table 7. The values of θ_k from MCS by gradient descent.

θ_k	MCS
θ_0	0.1005
θ_1	55.9815
θ_2	2.2910
θ_3	0.1290
θ_4	0.0533

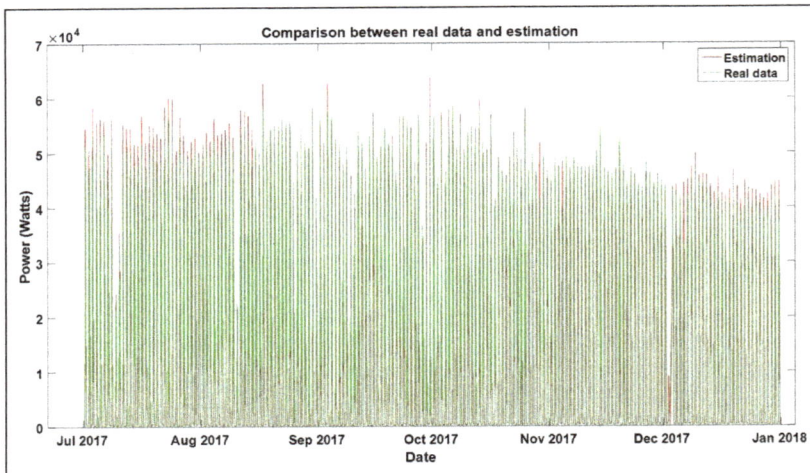

Figure 15. The results obtained from MCS by gradient descent.

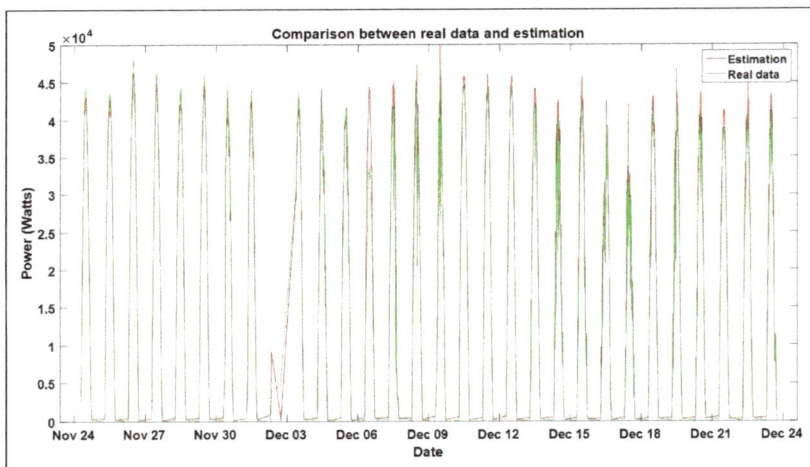

Figure 16. A close-up of the results from MCS.

3.3. Error Analysis On-Site

As seen in Section 1 and according to Reference [51], the output power of a PV array greatly depends, among other parameters, on solar radiation; however, this variable has an intermittent nature and suffers from rapid fluctuations. In Reference [41,51], the above is considered and some parameters besides the solar radiation such as clear sky or weather data are added; nonetheless, for these cases, the solar radiation is either the estimated output or an estimated input, contrary to this paper where this variable is measured on-site.

Each coefficient θ_k from Equations (9) and (10) represents the characteristic magnitude of its respective input variable x_k. Its quantity values prove the influence in the mathematical model approximation, establishing an existing relationship between the meteorological variables involved and (responsible for solar fluctuations) the power output.

Comparing Figures 9 and 16, an adequate behavior is observed between both results, achieving a valid statistical estimation. Regardless of the amount of stored data, similar outcomes were obtained from both locations.

This statement is mathematically proved through the correlation and error analysis.

3.3.1. Correlation Analysis

According to Section 2.1, the smaller an error between the estimation and real data, the greater the determination coefficient will be. Figures 17 and 18 show the dispersion graphic of the estimation against the real electric power for Hermosillo and Mexico City respectively. Moreover, in Table 8, the monthly and total fitted curve characteristic equation and determination coefficients for both cases are displayed.

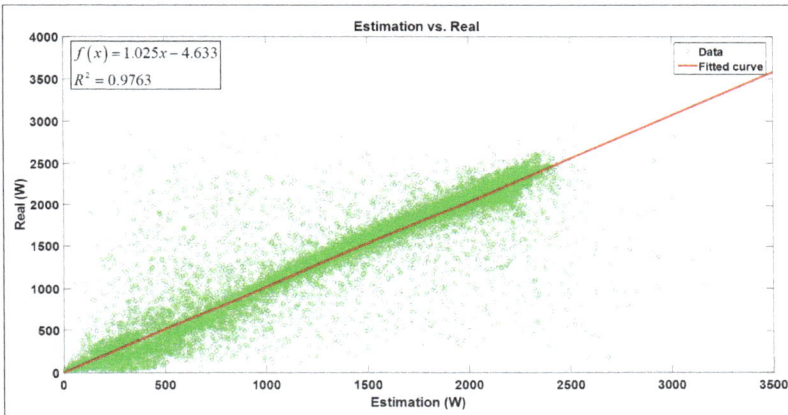

Figure 17. The estimated vs. real photovoltaic power of HS.

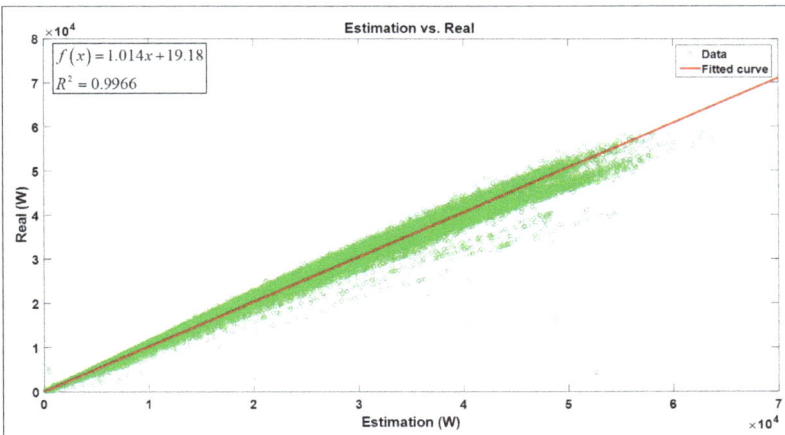

Figure 18. The estimated vs. real photovoltaic power of MCS.

Table 8. The monthly and total fitted curve characteristic equation and determination coefficient values of estimation for both locations.

Location	Month	Fitted Curve	R^2
HS	Aug	$f(x) = 1.005x + 84.92$	0.9916
	Sep	$f(x) = 1.032x + 20.92$	0.9981
	Oct	$f(x) = 1.031x + 44.45$	0.9877
	Nov	$f(x) = 1.022x - 29.74$	0.9926
	Dec	$f(x) = 1.05x - 37.57$	0.9849
	Jan	$f(x) = 1.009x + 6.707$	0.9982
	TOTAL	$f(x) = 1.025x - 4.633$	0.9763
MCS	Jul	$f(x) = 1.003x + 91.51$	0.9989
	Aug	$f(x) = 1.01x + 184.3$	0.9961
	Sep	$f(x) = 0.9843x + 525.7$	0.9946
	Oct	$f(x) = 1.003x + 140.6$	0.9997
	Nov	$f(x) = 1.007x - 40.05$	0.9982
	Dec	$f(x) = 1.004x - 44.28$	0.9995
	TOTAL	$f(x) = 1.014x + 19.18$	0.9966

Both statistical models estimate favorably the real electric power for each tested day as we can observe through the values in Table 8. A great closeness can be confirmed between the estimation and the real data for each location, which shows that the implementation of a gradient descent optimization achieves a satisfactory result.

3.3.2. Error Analysis

The less the difference between the estimated and the real value, the better the estimation. Two kinds of errors were applied, where "P_m" was the measured power, "P_e" was the estimated power, "s" was the sample in consideration, and "N" was the total amount of samples. The first one was called Mean Absolute Error (MAE) defined by Equation (11) and was used as a standard statistical metric to measure the model performance in meteorology, air quality, and climate research studies [42,51,52]. The second one was the percentage of the MAE known as Mean Absolute Percentage Error (MAPE) defined by Equation (12).

Given that the results shown in Figures 9 and 16 resemble a sinusoidal behavior, $(P_m - P_e)/P_m$ does not present the real error between both signals. Considering Figure 19a, if the measured value is on a high level, then the rate between the $P_m - P_e$ and P_m values will be low; however, if the measured value is on a low level, then the same rate will be high or close to 1, as seen in Figure 19b. According to this, the rate is changed regarding the full range of the measured signal (max-min) instead in order to obtain a reliable value. Equation (13) represents the modified MAPE.

$$MAE = \frac{\sum_{s=1}^{N} |P_m - P_e|}{N} \tag{11}$$

$$MAPE_\% = \frac{\sum_{s=1}^{N} \left| \frac{P_m - P_e}{P_m} \right|}{N} \cdot 100 \tag{12}$$

$$MAPE_{(range)\%} = \frac{\sum_{s=1}^{N} \left| \frac{P_m - P_e}{\max(P_m) - \min(P_m)} \right|}{N} \cdot 100 \tag{13}$$

With respect to the data from HS and MCS, Table 9 shows the values of the errors mentioned in Equations (11) and (13).

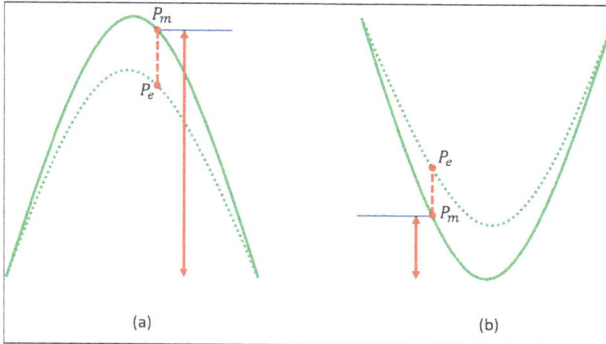

Figure 19. The difference errors between the high and low values: (**a**) Error value when measured data is high; (**b**) Error value when measured data is low.

Table 9. The error values between the estimated and real data.

Location	Month	MAE	MAPE (Range)
HS	Aug	299.7928 W	11.9436%
	Sep	203.7386 W	7.5643%
	Oct	267.9841 W	9.2382%
	Nov	138.1913 W	5.1418%
	Dec	178.9316 W	7.0381%
	Jan	184.5001 W	6.7978%
	TOTAL	200.0924 W	6.8859%
MCS	Jul	590.6460 W	1.0480%
	Aug	1192.3 W	2.0392%
	Sep	726.2990 W	1.2646%
	Oct	741.8355 W	1.2660%
	Nov	741.4300 W	1.3534%
	Dec	812.8113 W	1.7237%
	TOTAL	1130.4 W	1.9291%

Table 9 shows the errors for both locations, where the MAE result values for HS have a range from 138.1913 W to 299.7928 W and an overall of 200.6071 W. Considering MCS, its MAE goes from 590.6460 W to 1192.3 W, with an overall of 1130.4 W. From these results we can highlight that the values are low according to the total value range for each location, being up to 2500 W and 45 kW for HS and MCS, respectively. This statement is proven by column 4 from Table 9, achieving a MAPE of 6.9036% in general for HS; meanwhile, for MCS, its respective results is a MAPE of 1.9291%. The above demonstrates the estimation robustness of the statistical model.

4. Conclusions

A gradient descent method was used to calculate, through a characteristic equation, the relationship between several variables which can be easily computed using quantitative data.

The results showed that solar radiation and daylight time are relevant in estimating the electric power, with solar radiation being the one with the most influence over the photovoltaic power generation; nonetheless, temperature, wind speed, and daylight hour affect this process, with their inclusion being fundamental in the analysis to achieve a proper electric power estimation.

According to Figures 9 and 16, an acceptable approximation between the statistical simulation results and real-time power generation has achieved. This is proven both by Table 8, where 97.63% of the estimation results matches with the real data for HS and 99.66% match for MCS, and by Table 9, achieving overall error values no greater than 7% and 2% for HS and MCS, respectively. An observable relationship between the determination coefficient values and error results is clear, concluding that the percentage error will be lower while R^2 is higher.

Several causes could perturb the correlation, resulting in a wider dispersion, with the weather conditions as the most important for this study. Contrary to Mexico City, Hermosillo has abrupt changes in climate throughout the year, reaching temperatures around 50 °C and 0 °C in summer and winter, respectively, as well as sudden weather changes from a sunny day to a cloudy or even a rainy day within hours.

However, according to the above and even if the results have a satisfactory behavior, by definition, a statistical estimation will always have an error value. Nevertheless, these error values can be reduced, increasing the gathered input data or by other alternative methods such as intelligent systems which have proven to be an efficient methodology [53–56]. The above will be discussed elsewhere.

Author Contributions: Research ideas and global design: N.P.-D., J.A.R.-H., and F.A.-V.; analysis of the results: Y.M., E.F.V.-C., and E.J.H.-L.; revision of the document: R.E.C.-L. and J.H.A.-G.

Funding: This research received no external funding.

Acknowledgments: The authors would like to thank the University of Sonora (UNISON), Centro de Investigación y de Estudios Avanzados (CINVESTAV) campus Zacatenco, and Consejo Nacional de Ciencia y Tecnología (CONACYT), with the mobility scholarship 291249.

Conflicts of Interest: The authors declare no conflict of interest.

Abbreviations

The following abbreviations are used in this manuscript:

ANN	Artificial Neural Network
CSP	Concentrated Solar Power Plants
GDO	Gradient Descent Optimization
HEI	Higher Education Institutions
HS	Hermosillo Site
kWh/m^2	Quantity of kilowatts during an hour striking a squared meter area
LSR	Least Square Regression
MAE	Mean Absolute Error
MAPE	Mean Absolute Percentage Error
MCS	Mexico City Site
PV	Photovoltaic
SHC	Solar Heating and Cooling
W	Watt. It is a unit of power defined as a derived unit of 1 joule per second (J/s)

References

1. D'Adamo, I. The Profitability of Residential Photovoltaic Systems. A New Scheme of Subsidies Based on the Price of CO_2 in a Developed PV Market. *Soc. Sci.* **2018**, *7*, 148. [CrossRef]. [CrossRef]
2. SEIA. Solar Energy Industries Association. 2018. Available online: https://www.seia.org/initiative-topics/solar-technologies (accessed on 12 April 2019).
3. Järvelä, M.; Valkealahti, S. Ideal operation of a photovoltaic power plant equipped with an energy storage system on electricity market. *Appl. Sci.* **2017**, *7*, 749. [CrossRef]. [CrossRef]
4. Awan, A.B.; Zubair, M.; Abokhalil, A.G. Solar energy resource analysis and evaluation of photovoltaic system performance in various regions of Saudi Arabia. *Sustainability* **2018**, *10*, 1129. [CrossRef]. [CrossRef]
5. Zahedi, A. Solar photovoltaic (PV) energy; latest developments in the building integrated and hybrid PV systems. *Renew. Energy* **2006**, *31*, 711–718. [CrossRef]. [CrossRef]
6. Perea-Moreno, A.J.; Hernandez-Escobedo, Q.; Garrido, J.; Verdugo-Diaz, J.D. Stand-Alone Photovoltaic System Assessment in Warmer Urban Areas in Mexico. *Energies* **2018**, *11*, 284. [CrossRef]. [CrossRef]

7. Trujillo-Camacho, M.; García-Gómez, C.; Hinojosa-Palafox, J.; Castillón-Barraza, F. Evaluación de compositos TiO$_2$/clinoptilolita en la fotodegradación del tinte MV-2B en un reactor-concentrador solar cpc. *Revista Mexicana de Ingeniería química* **2010**, *9*, 139–149.

8. López-Sosa, L.; Hernández-Ramírez, L.; González-Avilés, M.; Servín-Campuzano, H.; Zárate-Medina, J. Desarrollo de un recubrimiento absorbente solar de bajo costo basado en hollín de biomasa forestal: Caracterización térmica y aplicación en un sistema de cocción solar. *Revista Mexicana de Ingeniería Química* **2018**, *17*, 651–668. [CrossRef]

9. González-Avilés, M.; López-Sosa, L.; Servín-Campuzano, H.; González-Pérez, D. Adopción tecnológica sustentable de cocinas solares en comunidades indígenas y rurales de Michoacán. *Revista Mexicana de Ingeniería Química* **2017**, *16*, 273–282.

10. Saga, T. Crystalline and Polycrystalline Silicon PV Technology. *NPG Asia Mater* **2010**, *2*, 96–102. [CrossRef]

11. Kazem, H.A.; Yousif, J.H.; Chaichan, M.T. Modeling of daily solar energy system prediction using support vector machine for Oman. *Int. J. Appl. Eng. Res.* **2016**, *11*, 10166–10172.

12. Zhang, W.-T.; Wang, S.; Du, X.-H. Research of power prediction about photovoltaic power system: Based on BP neural network. *J. Environ. Prot. Ecol.* **2017**, *18*, 1614–1623.

13. Shin, H.; Geem, Z.W. Optimal Design of a Residential Photovoltaic Renewable System in South Korea. *Appl. Sci.* **2019**, *9*, 1138. [CrossRef] [CrossRef]

14. Hogg, D.C.; Decker, M.; Guiraud, F.; Earnshaw, K.; Merritt, D.; Moran, K.; Sweezy, W.; Strauch, R.; Westwater, E.; Little, C. An automatic profiler of the temperature, wind and humidity in the troposphere. *J. Clim. Appl. Meteorol.* **1983**, *22*, 807–831. [CrossRef] [CrossRef]

15. Lawrence, M.G. The relationship between relative humidity and the dewpoint temperature in moist air: A simple conversion and applications. *Bul. Am. Meteorol. Soc.* **2005**, *86*, 225–234. [CrossRef] [CrossRef]

16. Dudley, B. BP Energy Outlook 2040. *BP plc* 2019. Available online: http://oilproduction.net/files/OilProduction-bp-energy-outlook-2019.pdf (accessed on 12 April 2019).

17. IRENA. *Renewable Power Generation Costs in 2017*; International Renewable Energy Agency: Abu Dhabi, UAE, 2018. Available online: https://www.irena.org/-/media/Files/IRENA/Agency/Publication/2018/Jan/IRENA_2017_Power_Costs_2018.pdf (accessed on 12 April 2019).

18. Rasero, C.M. Energía Solar Fotovoltaica. Energía Solar Fotovoltaica, Situación Actual 2011. Available online: https://static.eoi.es/savia/documents/componente75553.pdf (accessed on 12 April 2019).

19. Philibert, C. *Solar Energy Perspectives*; OECD: Paris, France, 2011.

20. Elshurafa, A.M.; Albardi, S.R.; Bigerna, S.; Bollino, C.A. Estimating the learning curve of solar PV balance–of–system for over 20 countries: Implications and policy recommendations. *J. Clean. Prod.* **2018**. [CrossRef] [CrossRef]

21. Nemet, G.F.; O'Shaughnessy, E.; Wiser, R.; Darghouth, N.; Barbose, G.; Gillingham, K.; Rai, V. Characteristics of low-priced solar PV systems in the US. *Appl. Energy* **2017**, *187*, 501–513. [CrossRef] [CrossRef]

22. Acceso a Datos de Radiación Solar en España. 2018. Available online: http://www.adrase.com (accessed on 12 April 2019).

23. The World Bank, Solar Resource Data: Solargis, World. 2017. Available online: https://globalsolaratlas.info/downloads/world (accessed on 12 April 2019).

24. The World Bank, Solar Resource Data: Solargis, Mexico. 2017. Available online: https://globalsolaratlas.info/downloads/mexico (accessed on 12 April 2019).

25. Yaneva, M.; Tisheva, P.; Tsanova, T. Mirecweek—The Big Mexico Renewable Energy Report. 2018. Available online: http://www.awex-export.be/files/library/Infos-sectorielles/Ameriques/2017/MEXIQUE/Mirec-Report-2018-The-BIG-Mexico-renewable-energy-report-ENG.pdf (accessed on 12 April 2019).

26. Della Ceca, L.; Micheletti, M.; Freire, M.; Garcia, B.; Mancilla, A.; Salum, G.; Crinó, E.; Piacentini, R. Solar and Climatic High Performance Factors for the Placement of Solar Power Plants in Argentina Andes Sites—Comparison With African and Asian Sites. *J. Sol. Energy Eng.* **2019**, *141*, 041004. [CrossRef] [CrossRef]

27. Sağlam, Ş. Meteorological parameters effects on solar energy power generation. *WSEAS Trans. Circuits Syst.* **2010**, *9*, 637–649.

28. Nam, S.; Hur, J. Probabilistic Forecasting Model of Solar Power Outputs Based on the Naïve Bayes Classifier and Kriging Models. *Energies* **2018**, *11*, 2982. [CrossRef] [CrossRef]

29. Ding, M.; Wang, L.; Bi, R. An ANN-based approach for forecasting the power output of photovoltaic system. *Procedia Environ. Sci.* **2011**, *11*, 1308–1315. [CrossRef] [CrossRef]

30. Horan, W.; Shawe, R.; O'Regan, B. Ireland's Transition towards a Low Carbon Society: The Leadership Role of Higher Education Institutions in Solar Photovoltaic Niche Development. *Sustainability* **2019**, *11*, 558. [CrossRef]. [CrossRef]

31. Viscidi, L. Mexico's Renewable Energy Future. 2018. Available online: https://www.wilsoncenter.org/sites/default/files/mexico_renewable_energy_future_0.pdf (accessed on 12 April 2019).

32. SENER. Prospectivas de Energías Renovables 2012–2026. 2012. Available online: https://www.gob.mx/cms/uploads/attachment/file/62954/Prospectiva_de_Energ_as_Renovables_2012-2026.pdf (accessed on 12 April 2019).

33. Gordon, G.; Tibshirani, R. Gradient descent revisited. *Optimization* **2012**, *10*, 725.

34. Hernández, L.; Baladrón, C.; Aguiar, J.M.; Calavia, L.; Carro, B.; Sánchez-Esguevillas, A.; Cook, D.J.; Chinarro, D.; Gómez, J. A study of the relationship between weather variables and electric power demand inside a smart grid/smart world framework. *Sensors* **2012**, *12*, 11571–11591. [CrossRef]. [CrossRef]

35. Ooi, S.C.; Mardiana, A.; Yusup, Y. Analysing Meteorological Variables, Energy Consumption and Occupant Behaviour in an Office Building in Hot-Humid Climate Zone. *Int. J. Sci. Res. IJSR* **2015**, *4*, 88–93.

36. Hor, C.L.; Watson, S.J.; Majithia, S. Analyzing the impact of weather variables on monthly electricity demand. *IEEE Trans. Power Syst.* **2005**, *20*, 2078–2085. [CrossRef]. [CrossRef]

37. Mendoza, C. Viabilidad Técnica-Económica de una Central Solar Termoeléctrica de Colectores Cilíndricos Parabólicos para su Implementación de México. UNAM, Marzo. 2012. Available online: http://132.248.52.100:8080/xmlui/handle/132.248.52.100/277 (accessed on 12 April 2019).

38. Benesty, J.; Chen, J.; Huang, Y.; Cohen, I. Pearson correlation coefficient. In *Noise Reduction in Speech Processing*; Springer: Berlin, Germany, 2009; pp. 1–4. [CrossRef].

39. Spiegel, M.; Stephens, L. *Estadística*, 4th ed.; Graw Hill: Contadero, Mexico, 2009; ISBN 978-6-071-51188-1. Available online: https://www.google.com/url?sa=t&rct=j&q=&esrc=s&source=web&cd=1&ved=2ahUKEwj12s7rtLDhAhWB7Z8KHQIiDwAQFjAAegQIAxAC&url=http%3A%2F%2Fensfep.edu.mx%2Fenlinea%2Fpluginfile.php%2F1531%2Fmod_folder%2Fcontent%2F0%2FEstad%25C3%25ADstica.%2520Serie%2520Schaum-%2520 (accessed on 12 April 2019).

40. Taylor, R. Interpretation of the correlation coefficient: A basic review. *J. Diagn. Med. Sonogr.* **1990**, *6*, 35–39. [CrossRef]. [CrossRef]

41. Clack, C.T. Modeling solar irradiance and solar PV power output to create a resource assessment using linear multiple multivariate regression. *J. Appl. Meteorol. Climatol.* **2017**, *56*, 109–125. [CrossRef]. [CrossRef]

42. Aggarwal, S.; Saini, L. Solar energy prediction using linear and non-linear regularization models: A study on AMS (American Meteorological Society) 2013–14 Solar Energy Prediction Contest. *Energy* **2014**, *78*, 247–256. [CrossRef]. [CrossRef]

43. Bairán, J.M.; Marí, A.; Cladera, A. Analysis of shear resisting actions by means of optimization of strut and tie models taking into account crack patterns. *Hormigón y acero* **2018**, *69*, 197–206. [CrossRef]. [CrossRef]

44. Arjoune, Y.; Mrabet, Z.; Kaabouch, N. Multi-Attributes, Utility-Based, Channel Quality Ranking Mechanism for Cognitive Radio Networks. *Appl. Sci.* **2018**, *8*, 628. [CrossRef]. [CrossRef]

45. Montgomery, D.C. *Design and Analysis of Experiments*; John Wiley & Sons: Hoboken, NJ, USA, 2017; ISBN 978-1-119-11347-8. Available online: https://www.wiley.com/en-mx/Design+and+Analysis+of+Experiments%2C+9th+Edition-p-9781119113478 (accessed on 12 April 2019).

46. Gardiner, S.K.; Crabb, D.P. Examination of different pointwise linear regression methods for determining visual field progression. *Investig. Ophthalmol. Vis. Sci.* **2002**, *43*, 1400–1407.

47. Verma, S.; Bartosova, A.; Markus, M.; Cooke, R.; Um, M.J.; Park, D. Quantifying the Role of Large Floods in Riverine Nutrient Loadings Using Linear Regression and Analysis of Covariance. *Sustainability* **2018**, *10*, 2876. [CrossRef]. [CrossRef]

48. Ng, A. CS229 Lecture Notes. 2000, Volume 1, pp. 1–3. Available online: http://backspaces.net/temp/ML/CS229.pdf (accessed on 12 April 2019).

49. Google Maps, UNISON. 2019. Available online: https://www.google.com/maps/place/Universidad+de+Sonora+-+UNISON/@23.9391074,-102.172878,2504267m/data=!3m1!1e3!4m5!3m4!1s0x86ce8447973925f7:0xcf527709b7555a3!8m2!3d29.0834761!4d-110.9603621 (accessed on 12 April 2019).

50. Google Maps, CINVESTAV. 2019. Available online: https://www.google.com/maps/place/CINVESTAV+-+IPN/@23.9391074,-102.172878,2504267m/data=!3m1!1e3!4m5!3m4!1s0x85d1f77a076bd911:0x63af0ad86ca91f65!8m2!3d19.5099425!4d-99.129559 (accessed on 12 April 2019).

51. Dev, S.; AlSkaif, T.; Hossari, M.; Godina, R.; Louwen, A.; van Sark, W. Solar Irradiance Forecasting Using Triple Exponential Smoothing. In Proceedings of the 2018 International Conference on Smart Energy Systems and Technologies (SEST), Sevilla, Spain, 10–12 September 2018; pp. 1–6. [CrossRef].

52. Chai, T.; Draxler, R.R. Root mean square error (RMSE) or mean absolute error (MAE)?—Arguments against avoiding RMSE in the literature. *Geosci. Model Dev.* **2014**, *7*, 1247–1250. [CrossRef]. [CrossRef]

53. Peterson, M.G. Intelligent medical systems and the interface with statistics. In Proceedings of the 11th IEEE Symposium on Computer-Based Medical Systems, Lubbock, TX, USA, 12–14 June 1998; p. 300. [CrossRef].

54. Chen, S.H.; Jakeman, A.J.; Norton, J.P. Artificial intelligence techniques: An introduction to their use for modelling environmental systems. *Math. Comput. Simul.* **2008**, *78*, 379–400. [CrossRef]. [CrossRef]

55. Mosavi, A.; Ozturk, P.; Chau, K.w. Flood prediction using machine learning models: Literature review. *Water* **2018**, *10*, 1536. [CrossRef]. [CrossRef]

56. Inman, R.H.; Pedro, H.T.; Coimbra, C.F. Solar forecasting methods for renewable energy integration. *Prog. Energy Combust. Sci.* **2013**, *39*, 535–576. [CrossRef]. [CrossRef]

applied
sciences

MDPI

Article

Optimal Strategy to Select Load Identification Features by Using a Particle Resampling Algorithm

Hengjing He [1], Xiaohong Lin [2], Yong Xiao [1,*], Bin Qian [1] and Hong Zhou [2]

[1]   Electric Power Research Institute, CSG, Guangzhou 510663, China
[2]   Department of Artificial Intelligence and Automation, Wuhan University, Wuhan 430072, China
*   Correspondence: xiaoyong@csg.cn; Tel.: +86-180-2880-9966

Received: 20 May 2019; Accepted: 26 June 2019; Published: 28 June 2019

Abstract: This paper proposes a robust strategy to select the load identification features, which is based on particle resampling to promote the performance for the successive load identification. Firstly, the sliding window incorporated with the bilateral cumulative sum control chart (CUSUM) method is utilized to obtain the load event. Then, the minimum inner-class variance, using the time-serial data, is introduced to judge the happened time as precise as possible, thus marking the changing point of the state of load for the following feature extraction. Due to the fluctuating data of current and voltage sampled by the monitoring device, the particle resampling method, containing the importance principle, is applied to find the steady and effectiveness point, ensuring that the obtained features have the desired fit with its actual features. The fitness measurement is then carried out by using the 2-D fuzzy theory. Finally, the proposed method was tested on the real household measurements in the labs. The result demonstrates an improvement in obtaining the desired load features when applied to the real household for the following load identification.

Keywords: resampling; feature extraction; non-intrusive load monitoring

1. Introduction

Recently, non-intrusive load monitoring (NILM) has gained major attention in the research field of smart grid [1], which aims to separate household energy consumption data collected from a single point of measurement into appliance-level consumption data. Since this technology was invented by George W. Hart et al. [2] in the early 1980s, non-intrusive load monitoring has been considered as a low-cost alternative to attaching individual monitors on each appliance, in contrast to the intrusive load monitoring method. In addition, it can provide information, such as energy usage, the state of the device, and so on. Up to now, the technology of NILM is becoming the state-of-the-art in the field of smart grid, which is built on signal processing, pattern recognition, and such deep learning algorithms for recognizing the power consumed in a household.

In general, the existing NILM methods can be divided into two main categories: event-based and non-event based [3]. The event-based approaches attempt to detect the changes of the state according to the significant change in power and then the features are obtained according to the changes in states. The recognition of appliances is then carried out by using a method, like the matching method and the clustering method, which is modeled by the real appliance recorded in the database. Non-event based NILM approaches are related to the machine learning from the big data collected by the single appliance or mixture of appliances during their working. However, the fundamental principle of load identification is dependent on the extraction of a stable and reliable load feature.

Load feature, also named load signature, is often represented during the appliance's work, including the start, running, and the stop state. Generally speaking, the feature of the start state and the stop state is the most remarkable during the extraction of the load feature. In the literature, the feature extraction stage can judge the algorithm type of the appliance, such as the transient state features-based algorithms and the steady state features-based algorithms [4–6]. Chang and Lian [7] adopted the wavelet transform coefficients (WTCs) to get the turn-on/off transient signal identification of load events. Similarly, a multi-resolution S-transform-based transient feature extraction scheme was proposed and presented in [8]. However, transient state features were obtained only if the sampling frequency exceeded 1000 Hz [9]. Taking into account the implementation, the steady state features-based algorithms seem to be more economical than the transient state features-based algorithms. Dinesh and Perera proposed a feature extraction method in [10] by using a modified mean shift algorithm at a low sampling rate. In [11], Leen's team shared a modified version of the chi-square goodness-of-fit test for event detection and getting the load features. Moreover, there are also many other methods based on stable load characteristics [12].

Those works above, however, almost all focused on the local information of the load during its turning on or off. The features extracted by the existing methods might not reach satisfactory predictive performance for load identification. Hence, in this paper, a particle resampling algorithm is proposed to select the desired load identification features, and then using the fuzzy theory, the method is tested with real data. The results showed that the proposed method that employed the steady state of load signatures, such as active power and reactive power, can identify the load accurately during load disaggregation.

The rest of the paper is organized as follows: Section 2 provides the proposed method, Section 3 describes the NILM platform in this work and the experimental results, and, finally, Section 4 concludes the paper.

2. Materials and Methods

This section mainly describes the materials and methods used for selecting the desired features and their evaluation. Generally speaking, the load identification method structure mainly consisted of the following three parts: (1) event detection, (2) feature extraction, and (3) fuzzy evaluation and identification. In this work, we mainly focused on feature extraction, and the whole framework is illustrated in Figure 1. In the event detection, the bilateral cumulative cum control chart (CUSUM) algorithm, combined with minimum inner-class variance rule method, was used to ensure that the load events and the change points were detected accurately. According to these change points, the resampling method was introduced to avoid the influence by the fluctuations of the voltage and current. Besides, the extracted features were validated by using the fuzzy evaluation, and, thus, can be applied to load identification. And then, the work is described as follows.

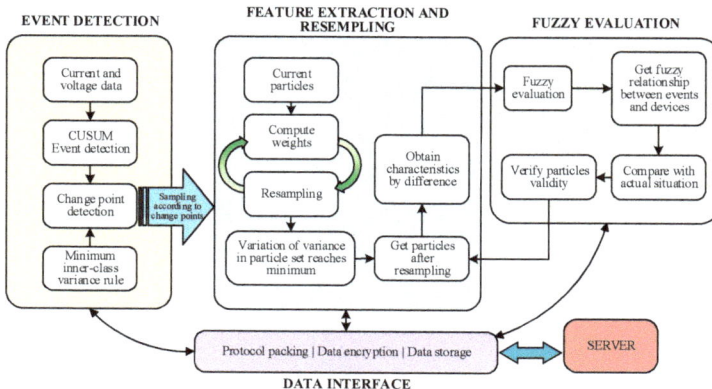

Figure 1. The method structure.

2.1. Event Detection Algorithm

Load event, which is defined as changes in load characteristics caused by switching on/off or state changes of individual devices [13], is the first and significant step in the load identification. In practical applications, the reliability and accuracy of event detection could be affected by the unpredicted switching and the interference of voltage and current fluctuations. In this paper, a non-parametric cumulative sum control chart (CUSUM) event detection algorithm was used for load detection. This method accumulates the sample data as well as the small deviation of the process. Since the accumulated value is significantly higher, a load event occurs. In addition, the method can be extended to the algorithm bilateral CUSUM due to the fact that the load events about turning on and turning off usually happened in pairs.

Let the time series of extracting load data be $X = \{x\,(k)\}$, $k = 1, 2, \ldots$ The statistic function in nonparametric bilateral CUSUM algorithm is defined as:

$$\begin{cases} f_0^+ = 0 \\ f_k^+ = \max\left(0, f_{k-1}^+ + x_k - (\mu_0 + \theta)\right) \\ f_0^- = 0 \\ f_k^- = \max\left(0, f_{k-1}^+ - x_k + (\mu_0 - \theta)\right) \end{cases}, \tag{1}$$

where μ_0 is the average value before the occurrence of the load event, θ is random noise introduced from outside, and f_k^+ and f_k^- are the random variable with 0 being the mean (i.e., random fluctuations around zero). When the load is turning on, x_k will increase and f_k^+ will have an increasing trend. On the contrary, the load when turned off will make f_k^- decrease. So the load event can be detected when the change exceeds the threshold h. Usually, the threshold h is set according to the lowest power value of the load.

To make the Equation (1) more understandable, Figure 2 gives a detailed description of the CUSUM. The load is a continuous change on the time axis, i.e., the mean μ_0 in the Equation (1) is changed with time. The sliding window model was then constructed to constrain the accumulated sum to ensure the load event was acquired accurately. So the W1 window and W2 window were modeled in this paper. The W1 window was used to calculate the mean value μ_0 of the sampling sequence. The W2 window was considered as the basis for judging whether a load event has occurred. From Equation (1), the value of f_k^+ in window W2 gradually accumulates when the value of x_i increases. So when f_k^+ exceeds a certain threshold h, the load occurs. On the contrary, the value of f_k^+ in window W2 fluctuates within a small range if no load event is detected. In this case, the W1 and W2 windows slide to the new sampling point and continue to detect.

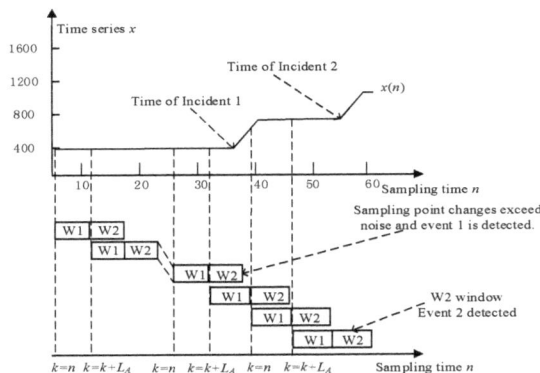

Figure 2. Two-sided cumulative sum control chart (CUSUM) algorithm sliding window model.

Considering that the threshold h is a global parameter, it is usually determined by the minimum load characteristic value. Therefore, to reduce the influence of the manually set threshold h, the minimum inner-class variance rule can be taken as the change point detection method. In the load event detection window, the active power data samples are classified into two categories: class C_0 $\{x_1, x_2, \ldots, x_k\}$ and class C_1 $\{x_{k+1}, x_{k+2}, \ldots, x_V\}$, where V is the sample length in the window, let:

$$m(C_0) = \frac{1}{k} \sum_{i=1}^{k} x_i \text{ and} \tag{2}$$

$$m(C_1) = \frac{1}{V-k} \sum_{i=k+1}^{V} x_i. \tag{3}$$

When the objective function

$$\min_k \sum_{i=1}^{k} (x_i - m(C_0))^2 + \sum_{i=k+1}^{V} (x_i - m(C_1))^2, \tag{4}$$

reaches the minimum and $|m(C_1) - m(C_0)|$ is greater than the set active power change value, the time of the change point could be found.

2.2. Identification Feature Extraction and Resampling

Since the change point was found by the above method, the feature of the load could be obtained by using the load characteristics of the changes [14]. Usually, the load event is determined based on physical changes in current, voltage, and other power information. Therefore, the load characteristics of these changes can be considered as the characteristics of the switching of the electrical device. For example, Figure 3 illustrates three types of load, named resistive load, capacitive load, and inductive load, which have different current phases for their capacitive reactance and opposite impedance performances. We can figure out the active power P by the voltage U, current I, and their phase difference φ that:

$$P = UI\cos\varphi. \tag{5}$$

Similarly, the reactive power Q can be described as:

$$Q = UI\sin\varphi. \tag{6}$$

Active power and reactive power can be calculated by Equations (5) and (6), and they can distinguish between the different types of loads according to their values. Moreover, active power and reactive power can be captured by low-frequency meters. In this paper, active power P and reactive power Q were adopted as the features.

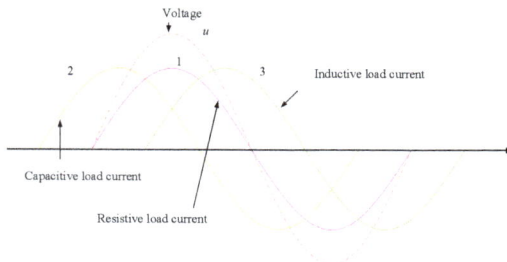

Figure 3. Current of the linear load.

In order to clearly illustrate the extraction of the load features, including the active and reactive characteristics, let $P(t)$ denote the active power variation with time t for an example. Usually, it is statistically stable. However, once the load changes the status, the $P(t)$ may undergo large changes at that time. So, the difference of P can represent the change of the status of the device, thus the value of $P(t)$ can be disaggregated. Here, we denote the $\Delta P = P(t + \Delta t) - P(t)$ as the difference of P, and the $P(t)$ satisfies the condition as follows:

$$\left. \begin{array}{l} \min \left| P(t) - \sum\limits_{i=1}^{m} a_i(t)P_i \right| \\ \min\limits_{\Delta t=T} \left| P(t + \Delta t) - \sum\limits_{i=1}^{m} a_i(t + \Delta t)P_i \right| \end{array} \right. , \tag{7}$$

where $P(t)$ is the active power at time t; m is the total number of load in the database; a_i is the mark of the state of load, where $a_i = 1$ indicates the running state and $a_i = 0$ means turned off; and T is the time interval.

Similarly, the reactive characteristic (or called the difference of Q) $\Delta Q = Q(t + \Delta t) - Q(t)$ at time t satisfies the condition as follows:

$$\left. \begin{array}{l} \min \left| Q(t) - \sum\limits_{i=1}^{m} a_i(t)Q_i \right| \\ \min\limits_{\Delta t=T} \left| Q(t + \Delta t) - \sum\limits_{i=1}^{m} a_i(t + \Delta t)Q_i \right| \end{array} \right. . \tag{8}$$

From Equations (7) and (8), it can be observed that the extraction of load characteristics from power load switching is related to the time of change point, i.e., the time interval T.

Although the change point is found according to the rules of the minimum inner-class variance, the voltage and current fluctuations make it difficult to determine this time interval T. Usually, the different time T can obtain the different P and Q features. In some papers [15–17], the time T is selected as the point after the change point. For some situations, the changes in P and Q may mismatch the load during load identification. So, in this paper, the importance resampling method was proposed to avoid the uncertainty of time interval T and the influence of power load information fluctuation.

The resampling algorithm is often used to solve the sequential importance sampling algorithm [18]. At present, there are many kinds of resampling algorithms [19,20]. In this paper, the importance resampling algorithm is adopted.

To further explain, Figure 4 describes the process of resampling. It can be seen that this method regards the characteristics at each time as a particle and resamples the importance of each particle according to the distribution of particles before and after the change point. The specific process is as follows:

Step 0: Assuming that the load is put into operation, the change point time t is obtained. Let $k = 0$, which randomly gets N particles after the current time t and before the next change point, and initializes each particle x_i with equal weight $\widetilde{\omega}_k(x_i), i = 1, \dots , N$.

Step 1: Importance sampling is used to distribute the weight of each particle. For each particle $i = 1, \dots , N$, estimate the weight of importance ω_k according to the degree of center deviation:

$$\omega_k(x_i) = 1 / \sum_{y \in \Omega} \|x_i - y\|^2 , \tag{9}$$

where Ω is the particle set. And we normalize Ω to get new weight $\widetilde{\omega}_k(x_i)$.

Step 2: We discard those particles with smaller weights and substitute sampling near the particles with larger weights.

Step 3: Set $k \to k+1$ and repeat the process of Step 1 and Step 2 to minimize the variation of variance in particle set Ω.

Step 4: We differentiate the current load characteristics in the particle set from the previously recorded load characteristics to extract the load variation characteristics.

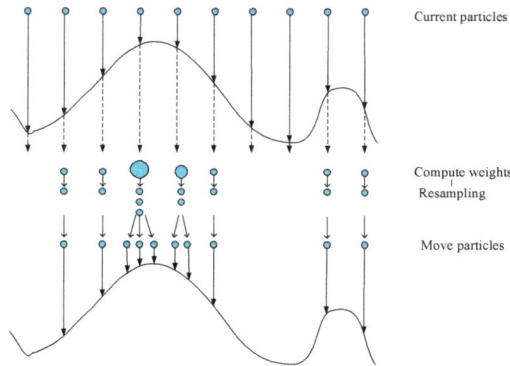

Figure 4. Process of resampling.

2.3. Fuzzy Evaluation Method

It is necessary to propose an evaluation method based on fuzzy membership after the identification features obtained by using the resampling method. The concept of the fuzzy set was first introduced in [21]. Fuzzy theory is a kind of transaction that copes with the concept of uncertainty through membership degree [22]. So, it can evaluate the relationship between load identification features and real load characteristics in the database. Considering the use of P and Q features as identification features, the 2-dimension fuzzy set are used in this paper.

Suppose that there are n loads, $A_1, A_2, A_3, \dots, A_n$, with two evaluation factors, active power (f_1) and reactive power (f_2). Consider m linguistic hedges Ψ. Note that it is possible to consider an objective application between the finite chain L and the ordinal scale Ψ, which keeps the order. Thus, each normal convex fuzzy subset defined on the ordinal scale Ψ can be considered as a discrete fuzzy number with the support L, $L = \{1, 2, \dots, m\}$. Then the following data can be set for the load A_i ($i = 1, 2, \dots, n$):

$$A_i = \begin{pmatrix} A_{i1} \\ A_{i2} \end{pmatrix} = \begin{pmatrix} x_{i11} & x_{i12} & \cdots & x_{i1m} \\ x_{i21} & x_{i22} & \cdots & x_{i2m} \end{pmatrix}, \tag{10}$$

where A_{i1} and A_{i2} are two discrete fuzzy numbers of the metric feature, m is the evaluation coefficient level, and x_{ijk} is the evaluation factor of the object A_i ($i = 1, 2, \dots, n$).

Then, the mean value can be worked out as $\mu(A_{ij}) = \sum_{k=1}^{m} x_{ijk} \cdot k / \sum_{k=1}^{m} x_{ijk}$. Let elements in K be the number (or numbers) that is (or are) closest to the mean value $\mu(A_{ij})$ in $L = \{1, 2, \dots, m\}$, i.e., $K = \{k \in L : |k - \mu(A_{ij})| \le 0.5\}$. It is obvious that the number of the elements of K can only be one or two. Then the following method can be given to construct one-dimensional discrete fuzzy number $u_{A_{ij}}: R \to [0, 1]$ for any $i = 1, 2, \dots, n$ and any $j = 1, 2$.

As K only has one element (denoted by k_0), $u_{A_{ij}}$ can be defined as:

$$u_{A_{ij}}(x) = \begin{cases} 1 - \frac{1 - u_{A_{ij}}(l_{ij})}{k_0 - l_{ij}}(k_0 - x), & x \in \left[l_{ij}, k_0\right] \cap L \\ 1 + \frac{1 - u_{A_{ij}}(\bar{l}_{ij})}{k_0 - \bar{l}_{ij}}(x - k_0), & x \in \left(k_0, \bar{l}_{ij}\right] \cap L \\ 0 & \text{otherwise} \end{cases} \tag{11}$$

As K has two elements (denoted by k_0 and $k_0 + 1$), $u_{A_{ij}}$ can be defined as:

$$u_{A_{ij}}(x) = \begin{cases} 1 - \dfrac{1 - u_{A_{ij}}\left(\underline{l}_{ij}\right)}{k_0 - \underline{l}_{ij}}(k_0 - x), & x \in \left[\underline{l}_{ij}, k_0\right] \cap L \\ 1, & x = k_0, k_0 + 1 \\ 1 + \dfrac{1 - u_{A_{ij}}\left(\bar{l}_{ij}\right)}{k_0 + 1 - \bar{l}_{ij}}(x - k_0 - 1), & x \in \left(k_0 + 1, \bar{l}_{ij}\right] \cap L \\ 0, & \text{otherwise} \end{cases} \tag{12}$$

where $\underline{l}_{ij} = \min\left\{k \middle| x_{ijk} \neq 0, k = 1, 2, \cdots m\right\}$, $\bar{l}_{ij} = \max\left\{k \middle| x_{ijk} \neq 0, k = 1, 2, \cdots m\right\}$, $u_{A_{ij}}(\underline{l}_{ij}) = x_{ij\underline{l}_{ij}} / \sum\limits_{k=1}^{m} x_{ijk}$, $u_{A_{ij}}(\bar{l}_{ij}) = x_{ij\bar{l}_{ij}} / \sum\limits_{k=1}^{m} x_{ijk}$, and $i = 1, 2, \dots, n, j = 1, 2$. We stipulate that $\left(1 - u_{A_{ij}}\left(\underline{l}_{ij}\right)\right) / \left(k_0 - \underline{l}_{ij}\right) = 0$ as $k_0 = \underline{l}_{ij}$, $\left(1 - u_{A_{ij}}\left(\bar{l}_{ij}\right)\right) / \left(k_0 - \bar{l}_{ij}\right) = 0$ as $k_0 = \bar{l}_{ij}$, and $\left(1 - u_{A_{ij}}\left(\bar{l}_{ij}\right)\right) / \left(k_0 + 1 - \bar{l}_{ij}\right) = 0$ as $k_0 = \bar{l}_{ij} - 1$ in Equations (10) and (11).

Then, we can construct the two-dimensional unite discrete fuzzy number $u_{A_{ij}}$ of $u_{A_{i1}}$ and $u_{A_{i2}}$ to express device A_i according to $u_{A_i}(X) = u_{A_i}(x_1, x_2) = \min\left\{u_{A_{i1}}(x_1), u_{A_{i2}}(x_2)\right\}$ for any $X = (x_1, x_2) \in R^2$ $(i = 1, 2, \dots, n)$. Next, the centroid can be calculated based on the resulting matrix:

$$C = (c_1, c_2) = \sum_{i=1}^{m} u(X_i) X_i / \sum_{i=1}^{m} u(X_i). \tag{13}$$

In order to obtain the final evaluation value, it is necessary to combine the ratios of the two criteria of the centroid $p = (p_1, p_2)$, where p_1 and p_2 describe the importance of the features of the centroid counterpart. Considering that the combination of the centroid and weight is more conducive to the comprehensive evaluation of the possibility of the category, the metric can be established as follows:

$$v = p_1 c_1 + p_2 c_2. \tag{14}$$

Finally, through comparing the v values of different objects, the actual object, which has the highest evaluation value, is found. Therefore, if the obtained load identification features had the largest evaluation value, the actual object was determined.

3. Experiments and Results

This section describes the experimental procedure and discusses the obtained results for demonstrating the efficiency of our method. Electrical characteristics, such as current and voltage, of the home were obtained through the monitoring device that was installed at the power inlet of the experimental resident user.

3.1. Laboratory Validation

The experiments used five devices, including a rice cooker, microwave oven, induction cooker, air conditioner, and kettle. Firstly, the individual device underwent the state of turning on and turning off several times, as seen in Figure 5 for an example. Then, the active power and reactive power were obtained through data statistical analysis, as shown in Table 1. Finally, these load features were recorded in the MySQL database. Meanwhile, the threshold was $h = 350$, the minimum change load active power was 100 watts, and the event detection window length was 20 sample points for the load event detection in this example.

Table 1. Household load power information. P: the active power; Q: the reactive power.

Devices	P/W	Q/Var	Mean of P/W	Mean of Q/Var
Induction cooker	1556.8–1622.1	54.5–63.5	1586.1	54.8
Kettle	1397.6–1429.4	−20.4−−2	1405.2	−8.6
Rice cooker	584.3–600.3	−8.5–1.9	594.6	−2.8
Air conditioner	435.9–516.9	−192.8−−145.1	467.3	−173.3
Microwave oven	1170.5–1223.6	−272.9−−196.2	1207.9	−221.6

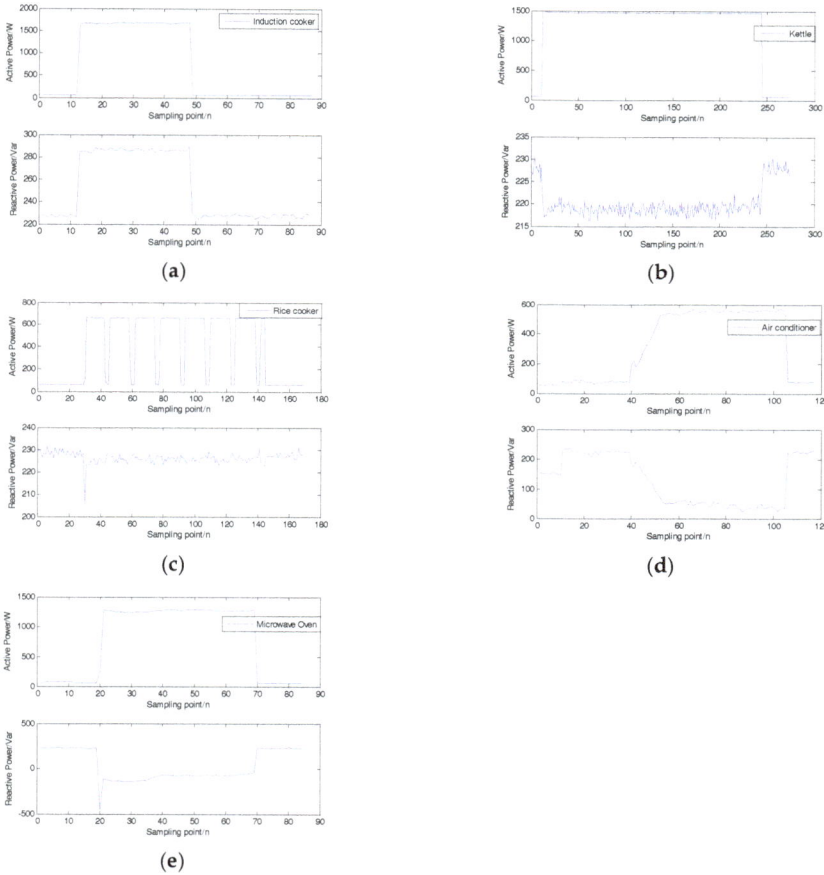

Figure 5. Active and reactive power diagrams of different devices: (**a**) induction cooker, (**b**) kettle, (**c**) rice cooker, (**d**) air conditioner, and (**e**) microwave oven.

Figure 6 shows the above load during its state of turning on and/or turning off in the actual test. The order sequence is air conditioner on, induction cooker on, kettle on, microwave oven on, rice cooker on, then kettle off, induction cooker off, microwave oven off, rice cooker off, and air conditioner off, respectively, as shown in Table 2. Table 3 shows the timing of the change point obtained by the original CUSUM load event detection [23]. It can be seen that there was a certain multi-detection phenomenon due to the increase slowly of active power after turning on the air conditioner. Nevertheless, our method could detect this type of load and could get better result accuracy, as shown in Table 4. So the resulting change points is approximated by the real-time change points.

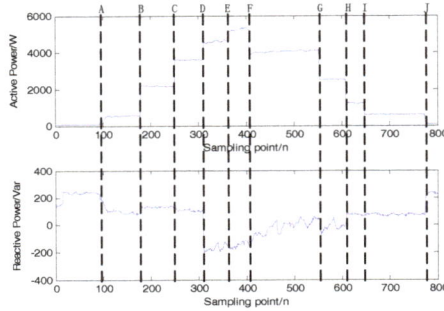

Figure 6. The active and reactive power during the load turning on or off.

Table 2. The description of the load event for the test.

Order Sequence	Event Description
A	Air conditioner on
B	Induction cooker on
C	Kettle on
D	Microwave oven on
E	Rice cooker on
F	Kettle off
G	Induction cooker off
H	Microwave oven off
I	Rice cooker off
J	Air conditioner off

Table 3. Load event change point obtained by the original CUSUM method.

Devices On	Change Points	Devices Off	Change Points
Air conditioner	93, 98, 105	Air conditioner	773, 778
Induction cooker	178	Induction cooker	552
Kettle	248, 253	Kettle	402, 409
Microwave oven	308, 313	Microwave oven	608
Rice cooker	333, 343, 353, 362	Rice cooker	647

Further, Table 4 shows the difference of P and Q in the case of the time interval $T = 1$, using the change point detected by our method. It was found that the obtained load identification features had a certain distance from the corresponding actual load characteristic information shown in Table 1. In particular, the active power with slow increases, such as the air conditioner, greatly differed from the actual characteristic values, which may fail to identify the load. In addition, these loads have different load characteristics between on and off, such as electric kettles and induction cookers. In this paper, the particle resampling algorithm was used to extract features. By retaining the particles with larger weights, some off-center particles were discarded and the best load characteristics could then be found.

Table 4. Differences on the load switching point.

Devices	State	Change Point	Difference of P/W	Difference of Q/var
Air conditioner	on	103	26.1	−8.4
	off	774	−500.07	146.66
Induction cooker	on	178	1562	51.2
	off	552	−1635	−86.51
Kettle	on	250	1566	−14.1
	off	406	−1211	18.5
Microwave oven	on	310	755	−291.86
	off	608	−1279	66.9
Rice cooker	on	362	523	−2.7
	off	647	−614.6	16.43

Taking the air conditioning load switching on and off as an example, Figure 7 shows the whole process of particle resampling. It can be seen that, in Figure 7a,b, there are a few particles far away from the center point obtained by the whole particles and that the distribution is not uniform. So, these particles can be replaced by the resampling method, as seen in Figure 7c,d. These particles almost converge to a state after resampling. In order to describe the state clearly, Table 5 illustrates the fluctuation range of active power and reactive power. After resampling, the fluctuation range is lower than before resampling. So, the load feature can be obtained properly. Here, we used the difference of the center of resampling particles after the device was switched on and the center of the resampling particles before the device was switched on as the load feature. Finally, the obtained active power was 462.17 W and the reactive power was −150.55 Var. Referring to Table 1, it can be found that the value of the obtained active power and reactive power was in the range of air conditioning, rather than the range of other devices. So we can easily identify this device.

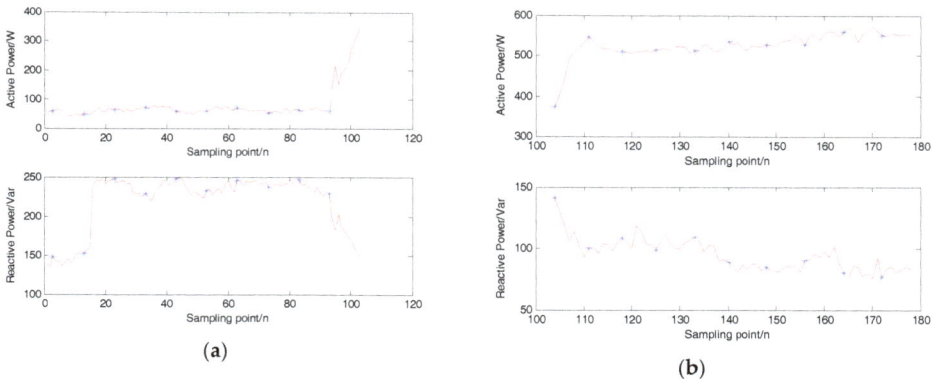

Figure 7. *Cont.*

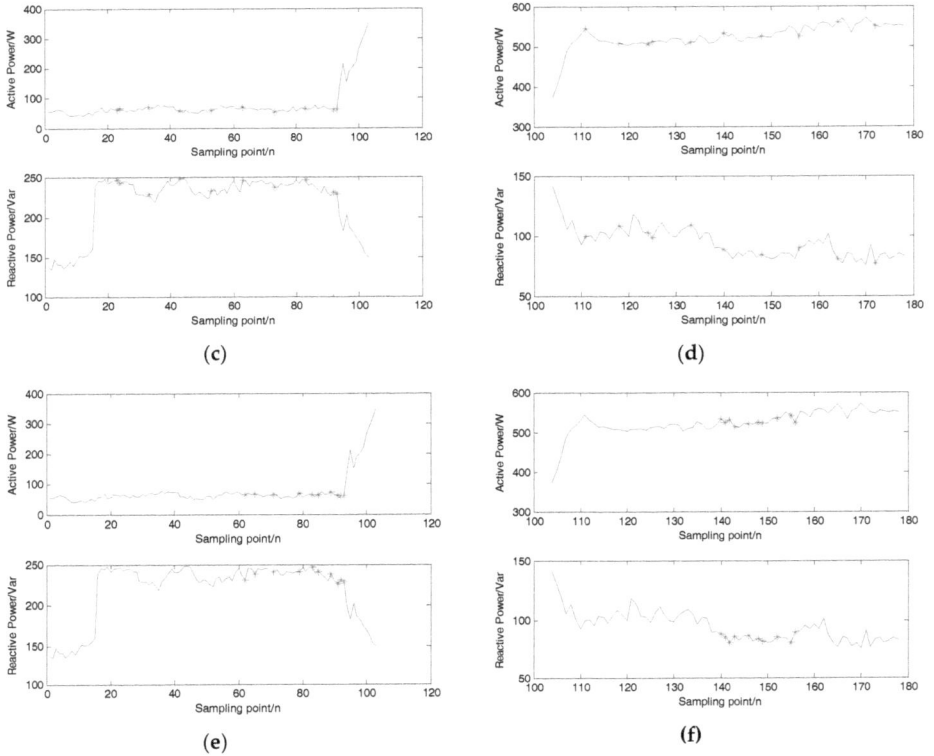

Figure 7. Air-conditioning resampling process: (**a**) particles before resampling (before change point); (**b**) particles before resampling (after change point); (**c**) particles after one resampling (before change point); (**d**) particles after one resampling (after change point); (**e**) particles after resampling (before change point); and (**f**) particles after resampling (after change point).

Table 5. Air conditioning load switching and resampling results.

Scenes	Before Device Turning on P/W	After Device Turning on P/W	Before Device Turning on Q/Var	After Device Turning on Q/Var
Before resampling	48.1–68.9	374.3–558.3	147.1–247.6	76.5–141.2
After one resampling	52.8–68.9	505.1–558.3	227.8–247.6	76.5–109.2
After resampling	60.5–71.6	513.8–540.5	226.5–246.0	80.7–89.8

In addition, Table 6 shows the obtained P and Q features using our method. It can be seen that the value of P and Q had a better correlation with the load features in Table 1. Notably, the obtained value did not change the sign if the status of the load was turning off.

Table 6. Load characteristic information after resampling.

Turn on	P/W	Q/Var	Turn off	P/W	Q/Var
Event A	461.2	−150.6	Event F	−1325.5	−29.8
Event B	1622.2	46.7	Event G	−1568.6	−54.92
Event C	1410.0	−25.3	Event H	−1242.2	119.9
Event D	975.4	−283.6	Event I	−609.6	1.3
Event E	687.7	33.5	Event J	−556.0	156.77

In order to verify the effectiveness of the above-obtained features, here, an induction cooker, a kettle, an air conditioner, a rice cooker, and a microwave oven are represented by A_1, A_2, A_3, A_4, and A_5. In addition, nine levels of fuzzy language are used for active and reactive features:

$$\Psi = \{EB, VB, B, MB, F, MG, G, VG, EG\}, \tag{15}$$

respectively, denoted as extremely bad, very bad, bad, more or less bad, fair, more or less good, good, very good, and extremely good, and they are linked to the finite chain $L = \{1, 2, \dots, 9\}$. According to the previous two-dimensional fuzzy membership evaluation method, the results of the final evaluation are shown in Table 7. It is not difficult to see that the load characteristics obtained after the load event detection could basically match the actual switched electrical equipment.

Table 7. The value of 2-D fuzzy membership about each event.

Event	Induction Cooker	Kettle	Rice Cooker	Air Conditioner	Microwave Oven
A	3	4	6	8.69	4.43
B	8.17	6.06	3.78	3.01	4.30
C	5.48	8.24	5.50	4	5.5
D	5	5.5	5.22	5.5	7.57
E	5.70	4.64	6.92	4.85	3.5
F	5.57	7.46	5.42	4.13	5.63
G	8.72	6.02	3.59	3	4.10
H	4.89	6.5	4.5	5.34	7
I	4.18	6.16	8.95	5.82	3.68
J	3.5	4	6.54	8.25	5

3.2. Validated on REDD

The reference energy disaggregation dataset (REDD) is a freely available data presented by J. Z. Kolter and M. Johnson [24]. The REDD contains detailed power usage information from several homes, which provides circuit-level data, rather than plug-level data. Here, we used it for demonstrating the performance of the event detection and feature extraction.

3.2.1. Event Detection

This part presents the performance of the proposed bilateral CUSUM event detection method and makes a comparison with the original CUSUM method over the REDD. In this work, the confusion matrix-based metrics are taken as the evaluation metric [25]. So, the detection events were divided into four categories: true positive (TP), true negative (TN), false positive (PF), and false negative (FN). Only TP, FP, and FN are considered in event detection performance evaluation because TN is usually infinite. On the basis of the confusion matrix, the evaluation can be carried out by using the following measurements, as seen in Table 8.

Table 8. The measurements to evaluate event detection.

Symbols	Math Equation	Description
TPR	$TPR = \frac{TP}{TP+FN} \in [0,1]$	True positive rate: This is the ratio that is correctly judged to be positive in all samples that are actually positive.
FPR	$FPR = \frac{FP}{FP+TN} \in [0,1]$	False positive rate: The ratio that is erroneously judged to be positive in all samples that are actually negative.
Acc	$Acc = \frac{TP+TN}{TP+TN+FP+FN}$	Accuracy: The ratio of the number of correct decisions to the total output of the system is used to measure the frequency of correct decisions made by the system.
Score	$Score = \frac{TP}{TP+FP+FN}$	Score: This definition is meaningless in event detection as there is no available TN count. To solve this problem, Dixon [26] proposed the definition of score.

Taking data from House3 in the REDD as an example, Table 9 illustrates the main metrics using the bilateral CUSUM, original CUSUM, and BIC (Bayesian information criterion) detector [27]. This is a day-of-event detection statistics. From the results, it can be seen that the bilateral CUSUM had the highest Score. The original CUSUM event detection method had the highest true positive rate (TPR), but it often had multiple FPs. The BIC detector had the lowest TPR and Score. By comparison, the bilateral CUSUM method performed better than the other two methods.

Table 9. The performance of different event detection algorithms on the reference energy disaggregation dataset (REDD).

Algorithm	TP	FP	FN	TPR	FPR	Acc	Score
Bilateral CUSUM	115	1	5	95.83%	--	--	95.04%
Original CUSUM	118	35	2	98.33%	--	--	76.13%
BIC detector	108	39	12	90%	--	--	67.92%

3.2.2. Feature Extraction on REDD

The data in the REDD in House 3 contained 10 appliances: lighting2/4/5, refrigerator, furnace, washer1/2, microwave, bathroom_gfi, and kitchen outlets2. The characteristics of a single electrical equipment were obtained from the statistics of the information from a single channel. This paper mainly focuses on active and reactive power information. In order to demonstrate the effectiveness of the feature extraction method, the resampling and average value method [27] were compared. Figures 8 and 9 illustrate the two-dimensional discrete fuzzy number. It can be seen that the particles obtained by the resampling method had higher overall accuracy in hypothesis and identification. Although the particles selected by the average value method and resampling method had a good performance on furnace, washer1, lighting4, and lighting5, the particles selected by the average value method made mistakes in identifying bathroom_gfi and kitchen_outlets2 as these devices having similar features. This demonstrates that our method has advantages in terms of feature extraction.

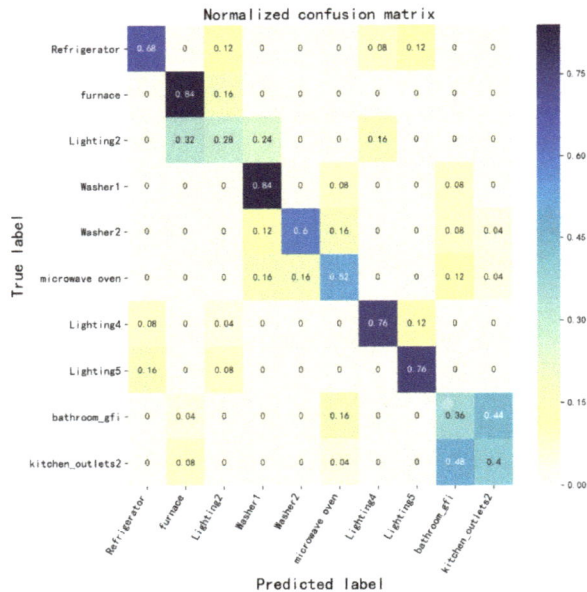

Figure 8. Features performance on REDD based on the average value method.

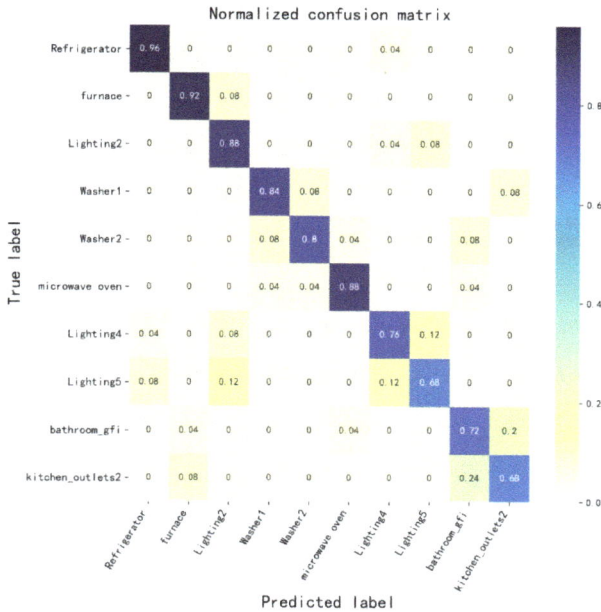

Figure 9. Features performance on REDD based on the resampling method.

4. Conclusions and Future Works

In this paper, a load identification features selection method was proposed and validated by using the laboratory dataset and REDD. The experiment showed that our method has the ability to extract the desired features, which may accurately match the features of the device recorded in the database. In the experiment, the performance of load event detection was also carried out. A bilateral CUSUM combined with the minimum inner-class variance approach that we proposed also promoted the detection effectiveness, especially for the climbing character of a load event. Besides, the resampling method incorporated in our method was used to find the stable characteristics of the load event, thus obtaining the desired features. Through the 2-D fuzzy membership measurement, it was found that the feature extracted by the resampling method is closer to that of the actual device, and can be applied to load identification. However, this resampling method requires a relatively stable switching period of the device. In practice, the device being turned on or off is random, thus the obtained features may fail to match the features of devices in the database.

In the future, the limitations mentioned above should be carefully considered. For example, some devices have a short duration after state switching, and the corresponding steady-state time is also very short. In this case, the method combining the transient features may work better. Moreover, different algorithms, based on different features, may be valid for certain types of devices. Therefore, it is necessary and feasible to integrate the proposed method with other complementary NILM models.

Author Contributions: Methodology, H.H.; Validation, Y.X.; Formal analysis, B.Q.; Data curation, X.L. and H.Z.

Funding: This research received no external funding.

Conflicts of Interest: The authors declare no conflict of interest.

References

1. Hosseini, S.S.; Agbossou, K.; Kelouwani, S.; Cardenas, A. Non-intrusive load monitoring through home energy management systems: A comprehensive review. *Renew. Sustain. Energy Rev.* **2017**, *79*, 1266–1274. [CrossRef]
2. Hart, G.W. Nonintrusive appliance load monitoring. *Proc. IEEE* **1992**, *80*, 1870–1891. [CrossRef]
3. Giri, S.; Bergs, M. An error correction framework for sequences resulting from known state-transition models in Non-Intrusive Load Monitoring. *Adv. Eng. Inform.* **2017**, *32*, 152–162. [CrossRef]
4. Zeifman, M.; Roth, K. Nonintrusive appliance load monitoring: Review and outlook. *IEEE Trans. Consum. Electron.* **2011**, *57*, 76–84. [CrossRef]
5. Yang, C.C.; Soh, C.S.; Yap, V.V. A systematic approach in appliance disaggregation using k-nearest neighbours and naive Bayes classifiers for energy efficiency. *Energy Effic.* **2018**, *11*, 239–259. [CrossRef]
6. Bonfigli, R.; Felicetti, A.; Principi, E.; Fagiani, M.; Squartini, S.; Piazza, F. Denoising autoencoders for Non-Intrusive Load Monitoring: Improvements and comparative evaluation. *Energy Build.* **2018**, *158*, 1461–1474. [CrossRef]
7. Chang, H.H.; Lian, K.L.; Su, Y.C.; Lee, W.J. Power-Spectrum-Based Wavelet Transform for Nonintrusive Demand Monitoring and Load Identification. *IEEE Trans. Ind. Appl.* **2014**, *50*, 2081–2089. [CrossRef]
8. Lin, Y.H.; Tsai, M.S. Development of an Improved Time–Frequency Analysis-Based Nonintrusive Load Monitor for Load Demand Identification. *IEEE Trans. Instrum. Meas.* **2014**, *63*, 1470–1483. [CrossRef]
9. Zoha, A.; Gluhak, A.; Imran, M.; Rajasegarar, S. Non-Intrusive Load Monitoring Approaches for Disaggregated Energy Sensing: A Survey. *Sensors* **2012**, *12*, 16838–16866. [CrossRef]
10. Dinesh, C.; Perera, P.; Godaliyadda, R.I.; Ekanayake, M.P.B.; Ekanayake, J. Non-intrusive load monitoring based on low frequency active power measurements. *AIMS Energy* **2016**, *4*, 414–443. [CrossRef]
11. De Baets, L.; Ruyssinck, J.; Develder, C.; Dhaene, T.; Deschrijver, D. On the Bayesian optimization and robustness of event detection methods in NILM. *Energy Build.* **2017**, *145*, 57–66. [CrossRef]
12. Ruyi, L.; Mingshan, H.; Dongguo, Z.; Hong, Z.; Wenshan, H. Optimized nonintrusive load disaggregation method using particle swarm optimization algorithm. *Power Syst. Prot. Control* **2016**, *44*, 30–36.
13. Zheng, Z.; Chen, H.N.; Luo, X.W. A Supervised Event-Based Non-Intrusive Load Monitoring for Non-Linear Appliances. *Sustainability* **2018**, *10*, 1001. [CrossRef]
14. Yang, C.C.; Soh, C.S.; Yap, V.V. A systematic approach to ON-OFF event detection and clustering analysis of non-intrusive appliance load monitoring. *Front. Energy* **2015**, *9*, 231–237. [CrossRef]
15. Berges, M.; Goldman, E.; Matthews, H.S.; Soibelman, L.; Anderson, K. User-Centered Non-Intrusive Electricity Load Monitoring for Residential Buildings. *J. Comput. Civ. Eng.* **2011**, *25*, 471–480. [CrossRef]
16. Norford, L.K.; Leeb, S.B. Non-intrusive electrical load monitoring in commercial buildings based on steady-state and transient load-detection algorithms. *Energy Build.* **1996**, *24*, 51–64. [CrossRef]
17. Leeb, S.B.; Shaw, S.R.; Kirtley, J.L. Transient event detection in spectral envelope estimates for nonintrusive load monitoring. *IEEE Trans. Power Deliv.* **1995**, *10*, 1200–1210. [CrossRef]
18. Gordon, N.J.; Salmond, D.J.; Smith, A.F. Novel Approach to Nonlinear/Non-Gaussian Bayesian State Estimation. *Radar Signal Process. IEE Proc. F* **1993**, *140*, 107–113. [CrossRef]
19. Martino, L.; Elvira, V.; Camps-Valls, G. Group Importance Sampling for Particle Filtering and MCMC. *Digit. Signal. Process.* **2018**, *82*, 133–151. [CrossRef]
20. Yang, C.; Shi, Z.; Han, K.; Zhang, J.J.; Gu, Y.; Qin, Z. Optimization of Particle CBMeMBer Filters for Hardware Implementation. *IEEE Trans. Veh. Technol.* **2018**, *67*, 9027–9031. [CrossRef]
21. Zadeh, L.A. Fuzzy sets. *Inf. Control.* **1965**, *8*, 338–353. [CrossRef]
22. Wang, G.; Shi, P.; Xie, Y.; Shi, Y. Two-dimensional discrete fuzzy numbers and applications. *Inf. Sci. Int. J.* **2016**, *326*, 258–269. [CrossRef]
23. Niu, L.; Jia, H. Transient Event Detection Algorithm for Non-intrusive Load Monitoring. *Autom. Electr. Power Syst.* **2011**, *35*, 30–35.
24. Kolter, J.Z.; Johnson, M. REDD: A public data set for energy disaggregation research. In Proceedings of the SustKDD Workshop Data Min. Appl. Sustain., San Diego, CA, USA, 2011; pp. 59–62.
25. Benitez, D.; Anderson, K.D.; Bergest, M.E.; Ocneanut, A.; Benitez, D.; Moura, J.M.P. Event detection for Non Intrusive load monitoring Event Detection for Non Intrusive Load Monitoring. In Proceedings of the IECON 2012—38th Annual Conference on IEEE Industrial Electronics Society, Montreal, QC, Canada, 25–28 October 2012; pp. 3312–3317.

26. Dixon, S. On the computer recognition of solo piano music. In Proceedings of the Australasian Computer Music Conference, Brisbane, Australia, 23–27 July 2000; pp. 31–37.
27. Xiao, J.; François, A.; Jing, Z.; Sarra, H. Non-intrusive load event detection algorithm based on Bayesian information criterion. *Power Syst. Prot. Control.* **2018**, *46*, 8–14.

Article

Analysis of Heat Transfer and Thermal Environment in a Rural Residential Building for Addressing Energy Poverty

Yiyun Zhu [1], Xiaona Fan [1,*], Changjiang Wang [2] and Guochen Sang [1]

[1]   Faculty of Civil Engineering, Xi'an University of Technology, Xi'an 710048, China; zyyun@xaut.edu.cn (Y.Z.); sangguochen@xaut.edu.cn (G.S.)

[2]   Department of Engineering and Design, University of Sussex, Brighton BN1 9RH, UK; C.J.Wang@sussex.ac.uk

*   Correspondence: 15109270561@163.com; Tel.: +86-151-0927-0561

Received: 7 October 2018; Accepted: 25 October 2018; Published: 28 October 2018

Abstract: Reducing energy consumption and creating a comfortable thermal indoor environment in rural residential buildings can play a key role in fighting global warming in China. As a result of economic development, rural residents are building new houses and modernizing existing buildings. This paper investigated and analyzed a typical rural residential building in the Ningxia Hui Autonomous Region in Northwest China through field measurements and numerical simulation. The results showed that making full use of solar energy resources is an important way to improve the indoor temperature. Reasonable building layout and good thermal performance of the building envelope can reduce wind velocities and convective heat loss. Insulation materials and double-glazed windows should be used to reduce energy loss in new buildings, although it is an evolution process in creating thermally efficient buildings in rural China. This research provides a reference for the design and construction of rural residential buildings in Northwest China and similar areas for addressing energy poverty.

Keywords: rural residential building; solar energy; heat transfer; wind velocities; field test and numerical simulation

1. Introduction

Climate change has become a worldwide issue and buildings account for over 40% of global energy consumption, a figure which is still rising [1,2]. Building sectors can potentially make significant reductions in greenhouse gas emissions compared with other sectors. Energy efficiency in the built environment can make great contributions to a sustainable economy [3]. In addition to minimizing energy requirements, sustainable buildings should also be designed and constructed to reduce water consumption, use low environmental impact materials, reduce wastage, protect the natural environment, and safeguard human health and wellbeing [4,5]. In China, there are about 600 million people living in rural areas. With the economic development over the last several decades, people in these areas have been building new houses, and a great amount of energy consumption is expected as a consequence of the growing living standard [6]. A report of building energy efficiency in rural China by Evans et al. [7] found that most of these buildings are very energy inefficient. Shan et al. [8] and Liu et al. [9] also reported energy and environmental situations, challenges, and intervention strategies in Chinese rural buildings. They found that the walls are typically built of solid bricks and single-layer glass windows with large window/wall ratios being commonly used in northern China. The energy consumption per household in northern China can be 10 times as much as that in southern China. Although a large amount of energy is consumed for space heating in northern China, the indoor thermal environment is still poor and does not meet the thermal comfort requirement of the occupants [10,11]. Rural energy inefficient buildings, however, are not just a concern in China, which is a developing country; as reported by Roberts et al. [12] and Bouzarovski et al. [13], the level of fuel poverty in the United Kingdom increased rapidly from 2003 to 2010 due to the dramatic increase in electricity and gas prices. Fuel poverty is when people are unable to adequately heat their homes due to a lack of resources and because of the inefficiency of house insulation and heating. Fuel poverty in China should be addressed because poor thermal comfort can lead to respiratory problems, circulatory problems, pneumonia, etc. [14]. To increase building energy efficiency, common measures such as the cavity wall, roof insulation, double glazing, low-emissivity glass, and draught proofing can be used [15]. Boeck et al. [16] reviewed many methods which can be used to improve the energy performance of residential buildings. These measures can potentially solve energy efficiency problems in buildings; however, they need to be adapted to the local environment, building types, and occupants' habits. To achieve an optimal building design, the overall concept of the construction needs to respond to the local environment and the intended use of the building. As pointed out by Mitterer et al. [17] and Wang et al. [18], a profound understanding of the reaction of a building to the specific climate and user's behavior is important in holistic building climate designs.

Building form can affect energy consumption. Hemsath and Bandhosseini [19] highlighted that the vertical and horizontal geometric proportions are sensitive factors related to building energy use. Larger surface-to-volume ratios increase heat transfer through the building envelope by conduction and convection. Montazeri et al. [20] conducted research on the effect of the ratio of building width to height on the convective heat transfer coefficient at the windward facades. They found that the convective heat transfer coefficient reduces when the building's width/height ratio increases. It was explained that the wind blocking effect is more pronounced for wider buildings, and the time that air is in contact with the upstream building facades increases, which therefore decreases the temperature difference between the air and the windward facades.

Solar radiation affects the surface temperature of walls [21], and it can be explored in building space heating. Pisello et al. [22] continuously monitored indoor and outdoor thermal conditions in two types of buildings which had different envelopes and a window/wall ratio of 0.17 for the south facade. Because of the different construction of the envelopes, they found that the difference of radiant temperature was more than $1\ ^\circ$C.

According to the climate regions of architecture in China, most areas in Northwest China are in cold or severe cold zones with a fragile ecological environment and lagging economic development. A large number of rural residents have built many widely distributed rural buildings, yet the design and construction of rural residential buildings still lack the guidance of scientific theory. It is an indisputable fact that the indoor thermal environment is poor in winter and there is high heating energy consumption. Therefore, it is very important to understand the climate characteristics, building types, and thermal performance of the enclosure structure in this area, which is particularly important to improve the indoor thermal environment quality and reduce building energy consumption in rural residential areas in the Northwest and similar areas of China.

Ningxia Hui Autonomous Region, in the hinterland of Northwest China, is located at the intersection zone of Ningxia, Gansu, and Mongolia provinces and has climate characteristics and residential forms typical of Northwest China [23]. Therefore, in this paper, a typical rural residential building in Zhongwei, Ningxia was studied to show the effect of the building's construction, layout, occupant habits, and solar radiation on the building's energy efficiency. The aim of the paper is to foster an evolution process for enhancing building thermal efficiency in rural buildings. Heat flow rate and heat flux through the building envelopes were analyzed with ANSYS finite element simulations and field measurements. Because the convective heat transfer coefficient plays a major role in the heat transfer of buildings, the effects of the building enclosures on the heat transfer coefficient in the rural residential buildings were studied in this paper as well. ANSYS CFX was employed to simulate air velocities around the building.

2. Methods

2.1. Object Selection

During the period from 16 to 20 January 2015, we carried out a field survey targeting rural residential buildings in the Ningxia Zhongwei areas. It was found that in the local area, there exist two kinds of residential buildings: earth houses and brick houses, which account for 13% and 87% of the residential structures, respectively. The survey also found that most of the new brick houses face south, with a long east–west and short north–south layout. The outer wall is often made of 370-mm solid clay brick, the roof is usually flat or in double slope, and most of the windows are in single frame and made of aluminum alloy or plastics. Based on the above initial analysis, in order to investigate the indoor thermal environment and thermal efficiency in local dwellings, a representative brick concrete building, as shown in Figure 1, was selected in this paper for subsequent measurements and analysis. The inside of the main bedroom was also being used as living room, as shown in Figure 2.

Figure 1. A typical residential building in Ningxia.

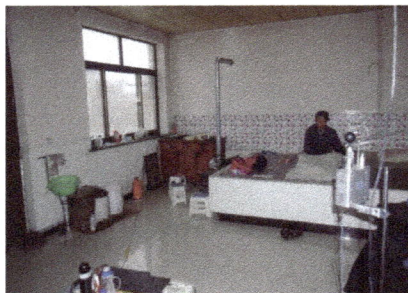

Figure 2. The inside situation of the main bedroom.

2.2. Data Acquisition

The outdoor measurement parameters included air temperature and solar radiation intensity, and indoor measurement parameters included air temperature and air velocity.

The intensity of solar radiation was measured by a solar radiometer, and the measuring points were arranged outdoors with no shelter, as denoted by the symbol "■" in Figure 3. The air temperature was measured with a thermometer and hydrometer. The measuring points of outdoor air temperature were under outdoor shades, and the measuring points of the indoor temperature were at 1.5 m above the floor in the center of the room, as denoted by the symbol "●" in Figure 3. The interior wall surface temperature was measured by an enclosure heat transfer coefficient field detector, and the measuring point was located at the middle of the measured wall surface, as denoted by the symbol "▲" in Figure 3. However, due to the concern that it would be inconvenient to place a hot wire anemometer in the room for a long time, the measurements of average wind speed were done by intermittent tests, and the measuring point was placed at the height of 1.0 m above the floor in the center of the room, as denoted by the symbol "⊕" in Figure 3. The monitoring period for the above measuring points was 24 h, and the acquisition time interval was 10 min. The models and parameters of the instruments are listed in Table 1, and the layout of the measuring points are shown in Figure 3.

Table 1. Models and parameters of the measuring instruments.

Test Instrument	Type	Test Parameters	Range	Accuracy
Solar radiometer	JTDL-4	Solar radiation intensity	0–2000 W/m^2	±0.2 °C
Thermometer and hydrometer	TESTO175-H	Air temperature	−20 to 70 °C	±0.1 °C
Envelope structure heat transfer coefficient detector	JTNT-C	Interior wall surface temperature	−20 to 85 °C	±0.2 °C
Hot wire anemometer	Testo425	Air velocity	0–20 m/s	±0.03%

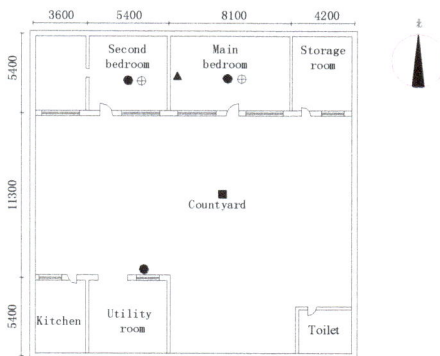

Figure 3. Layout of the residential building.

To conduct numerical heat transfer simulations, the following material properties of the building envelope listed in Table 2 were used. In the numerical simulations, heat transfer by convection and conduction were considered. Heat transfers by radiation were studied by analyzing the heat transfer rates from the measured temperature data and the simulated results.

Table 2. Thermal conductivities of materials.

Material	Thermal Conductivity (W/m·K)
Solid clay brick	0.81
Mortar plasters	0.93
Glass	1.3
Door (PVC)	0.19
Air	0.024

In the convective heat transfer analysis, the convective heat transfer coefficients were affected by the temperature difference between wall and fluid. As reported by Obyn and Van [24], the selection of surface convective heat transfer coefficient affects the evaluation of energy consumption of buildings. In the work carried out by Awbi and Hatton [25], convective heat transfer coefficients of buildings in the range of 1–6 W/m^2·K were obtained.

According to a number of intermittent field tests, the average wind velocity of the indoor movement area is 0.06 m/s at maximum, far below the human body's sensory threshold to air (0.2 m/s). Therefore, the convective heat transfer coefficient was calculated using the free convection method. This was also done because the measured indoor temperature was mainly in the single figures and the mean outdoor temperature was −2.98 °C according to the field tests. Air properties at 0 °C (273 K) were used to calculate the convective heat transfer coefficient:

Prandtl number $Pr = 0.715$ $\mu \cdot c_p/k$, thermal conductivity $k = 0.0243$ W/m·K, dynamic viscosity $\mu = 1.720 \times 10^{-5}$ N·s/m^2, specific heat $c_p=1005$ J/kg·K, and density $\rho = 1.293$ kg/m^3.

For the vertical walls of this house, the characteristic length was 3.0 m, which was the height of the room. Thus, the Grashof number Gr was

$$Gr = \frac{\rho^2 L^3}{\mu^2} g\beta(T - T_\infty) = 1.07 \times 10^{10} \tag{1}$$

Rayleigh number Ra is $Ra = Gr \times Pr = 7.63 \times 10^9 > 10^9 \tag{2}$

From Equation (2), the Ra is greater than 10^9; therefore, the flow of the air inside the room was turbulent. The average Nusselt number Nu_{avg} was

$$Nu_{avg} = 0.677(0.952 + Pr)^{-\frac{1}{4}} Pr^{\frac{1}{2}} Gr_L^{\frac{1}{4}} = 148.9 \tag{3}$$

The average heat transfer coefficient h_{avg} was

$$h_{avg} = \frac{Nu_{avg}k}{L} = 1.21 \text{ W/m}^2 \cdot \text{K} \tag{4}$$

The calculated convective heat transfer coefficient was 1.21 W/m^2·K, which was in the range of data reported by Awbi and Hatton [25].

Based on the building layout and dimensions in Figure 3, a 2D model was created in ANSYS software to analyze the heat flux through the building. The model is shown in Figure 4a. Because of the thin glass panel in the windows, a mesh size of 0.006 m was used, and the ANSYS 2D plane thermal element type 55 was used for meshing the walls, windows, and doors. The calculated convective heat transfer coefficient of 1.21 W/m^2·K was applied to the wall, window, and door surfaces. The mesh of the internal partition wall with different material layers are shown in Figure 4b, for which there was one layer of mortar plasters on each side of the clay brick internal wall. The external surface of the front wall had a layer of ceramic tiles.

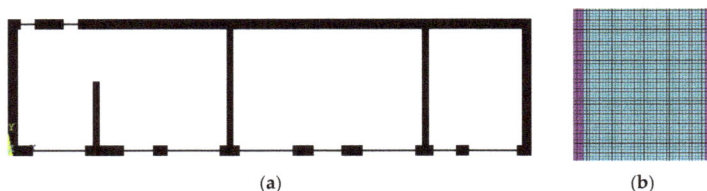

(a)   (b)

Figure 4. (a) Finite element model of the building (plan view) and (b) finite element mesh of the internal partition wall.

To analyze heat transfer by convection and minimize heat loss via the convection mode, the relationships between the external convective heat transfer coefficient and wind velocity were studied. Hagishima and Tanimoto [26] measured the convective heat transfer coefficient for building envelopes, which is influenced by wind speed. Mirsadeghi et al. [27] reviewed prediction models of external convective heat transfer coefficients in building energy simulation programs and pointed out that different models can result in 30% deviations in the yearly energy demand. It is recommended by Liu and Harris [28] that the model should be used for one-story buildings; hence, it was adopted in this paper.

External convective heat transfer coefficient $h_{c,ext}$ for total wind velocity at 0.5 m away from the wall is shown in Equations (5) and (6) [28].

$$h_{c,ext} = 2.08V + 2.97 \text{ (Windward)} \tag{5}$$

$$h_{c,ext} = 1.57V + 2.64 \text{ (Leeward)} \tag{6}$$

In order to analyze the effect of wind on the heat transfer coefficient and understand the effect of the building layout on the wind velocities, a numerical model was created using ANSYS CFX software to obtain wind velocities near the buildings. Fine meshes were used in the modelling; an element size of 0.2 m was used for meshing the surfaces of the building. To simulate the mean flow characteristics for turbulent flow conditions, the most common k-ε model was used as it is suitable for a wide range of simulations. A no-slip wall boundary condition was applied to all surfaces, and the standard wall function in ANSYS CFX was used with zero roughness height, as used by Ramponi and Blocken [29]. The model and a part of the numerical mesh are shown in Figure 5a,b, respectively. When the north and west winds were simulated separately, a typical wind velocity of 10 m/s was applied to the model to show the variations of wind velocities near the building envelopes.

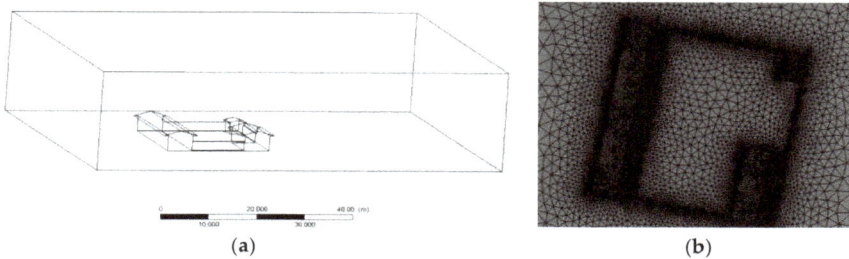

(a)

(b)

Figure 5. (**a**) Fluid domain and (**b**) meshes around the building in the simulation.

3. Results

The tests were all done on fine days and the outdoor weather conditions were similar. Therefore, the indoor thermal environment parameters on 17 January were analyzed.

The changes of indoor and outdoor temperatures of the residential building during the tests are shown in Figure 6.

It can be concluded from Figure 6 that the outdoor temperature varied from −7.6 to 2.5 °C and the mean temperature was −2.7 °C. The indoor average temperature was 9.48 °C in the main function room, and the average temperature in the secondary function room was 4.02 °C. The average surface temperature of the west wall of the main bedroom was 9.45 °C. To conclude, in this region it is cold and the indoor temperature is low in winter.

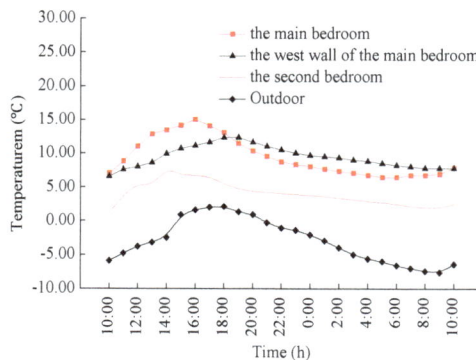

Figure 6. Indoor and outdoor temperatures of the residential building.

The changes of solar radiation intensity during the tests are as shown in Figure 7.

Figure 7 shows that the local sunshine duration was about 11 h, the solar radiation intensity was 286 W/m² on average and came to the peak value of 544.8 W/m² at around 4:00 pm, the scattered radiation intensity was 124.9 W/m², and the direct solar radiation intensity accounted for about 80% of the total radiation intensity. In summary, the sunshine in this area lasts longer and the solar radiation intensity is higher. The solar radiation certainly needs to be considered and explored further in the building designs.

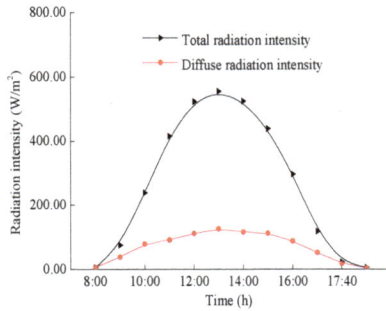

Figure 7. Solar radiation intensity outside the residential building.

Figure 8 shows that when the air temperature of the main bedroom was 9.48 °C, the surface temperatures of the west and east side walls of the main bedroom were about 7.23 °C and the inside surface temperature of the north wall was 4.87 °C, which is about half of the room temperature.

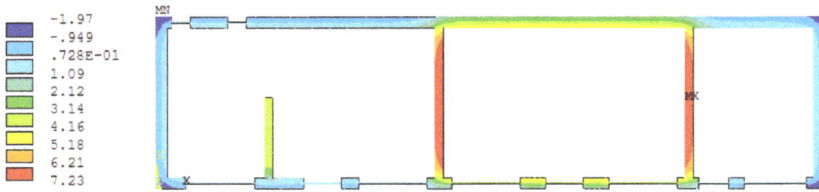

Figure 8. Temperature distributions in the walls with natural convection on the surface.

It can be seen from Figure 9 that heat flux at the middle section of the partition wall between the main bedroom and the second room was 3.04 W/m². The relatively large heat flux occurred at the window corners at south and north walls of the main bedroom. From the simulation results, the flow rate from the main bedroom to the second bedroom through the partition wall was 55.3 W, and the heat flow through the north, south, and west side enclosures of the second bedroom was 260.5 W. The difference between the two heat flow rates is 205.2 W.

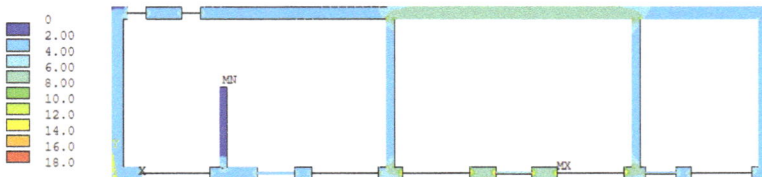

Figure 9. Heat flux through the walls under the free convection simulation.

The above heat flow rates were calculated based on the existing windows, which have a single layer of glass. If two layers of glass with a 12-mm air gap between them were installed, the heat flow rate through the north, south, and west enclosures of the second bedroom would be reduced to 242.2 W, with a reduction of 18.2 W or about 7% of the current heat flow rate out of the second bedroom.

North and west wind velocities around the buildings are shown in Figure 10a,b. Also, a sectional view of the velocity distribution in the north wind simulation is shown in Figure 10c.

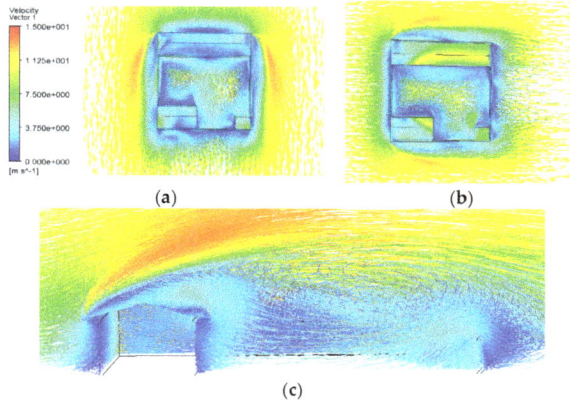

(a)   (b)

(c)

Figure 10. Wind velocity around the building: (**a**) north wind, (**b**) west wind, and (**c**) sectional view of velocity distribution in the north wind.

It can be seen from Figure 10 that the wind velocities around the building in the north wind is smaller than that in the west wind. Wind velocities on the top of the roof are much higher in the west wind than that in the north wind.

The wind velocities at 0.5 m from the south and north wall surfaces are presented in Figure 11a,b in the north wind simulations, respectively. The velocities at 0.5 m from the south wall vary from 1.1 to 4.0 m/s. The velocities at 0.5 m from the north wall vary from 1.2 to 6 m/s. The wind velocities at 0.5 m from the south wall surface in the west wind are also shown in Figure 11c, which vary from 2.5 to 3.5 m/s. However, the wind velocities at 0.5 m from the north wall surface in the west wind, as shown in Figure 11d, vary from 2.2 to 5.8 m/s.

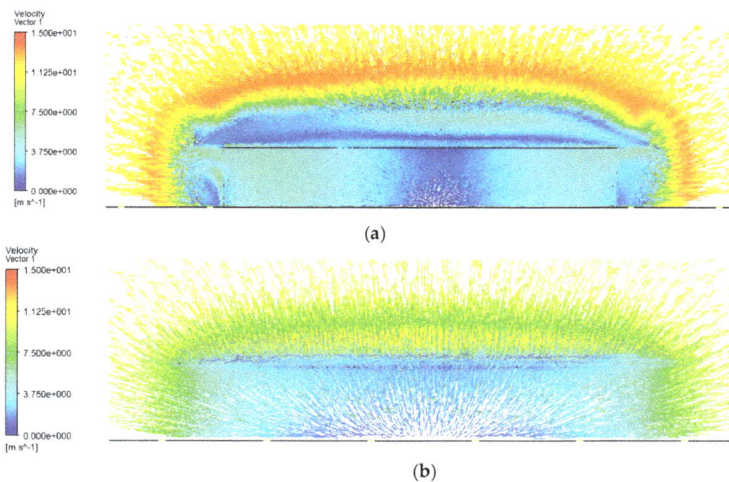

(a)

(b)

Figure 11. *Cont.*

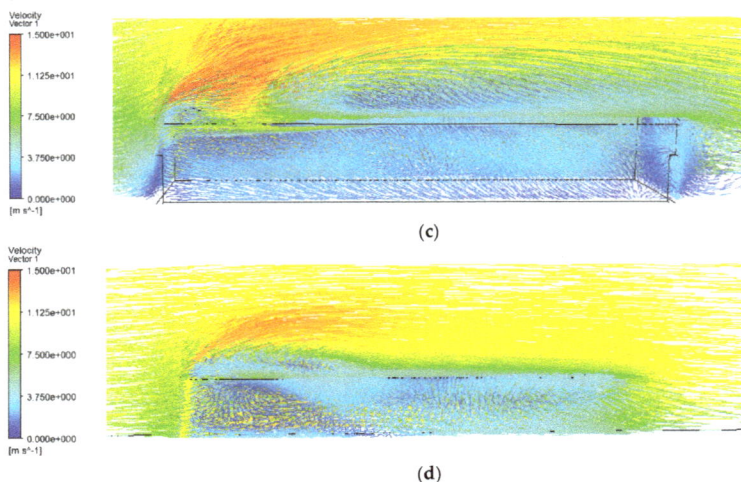

Figure 11. Wind velocities at 0.5 m to the walls of the building: (**a**) south wall in the north wind, (**b**) north wall in the north wind, (**c**) south wall in the west wind, and (**d**) north wall in the west wind.

4. Discussion

As Shan et al. [8] reported, a comfortable indoor temperature is around 15 °C in rural areas in northern China in contrast to 20 °C in urban areas because rural occupants wear thick clothing and move in and out rooms more frequently. From the measured data shown in Figure 6, the indoor temperature is much lower than 15 °C, except at 4:00 p.m. The indoor average temperature of the main bedroom is 9.48 °C, which is only 63% of 15 °C, in secondary rooms, and the indoor average temperature is 4.02 °C, which is only 26.8% of 15 °C. As reported by Santamouris et al. [30], there are several national and international standards which define comfortable indoor temperatures, and they are in the range of 18–21 °C. Therefore, the indoor temperature was quite low in the current study, and low indoor temperatures have a great impact and effect on various illness. In 2000, Clinch and Healy [31] studied housing standards and excess winter mortality in Ireland and Norway. They reported that relative excess winter mortality from cardiovascular and respiratory diseases in Ireland was higher than that in Norway. A possible explanation for this may be due to poorer Irish housing standards than those in Norway and that the indoor temperature was greatly impacted by falls of outdoor temperature. Zhao et al. [32] recently studied the effect of cold temperatures on clinical visits for cardiorespiratory diseases in the Ningxia Hui Autonomous Region, the same region as this study. They collected cardiovascular and respiratory illness data from 203 villages between 1 January 2012 and 31 December 2015. The average temperature in the 203 villages was 8.5 °C. They concluded that the overall suboptimal temperatures were responsible for 13.1% of total clinic visits for cardiovascular illness, and 25.9% of total clinic visits for respiratory disorders. From Figures 8 and 9, it can be seen that most of the heat loss from the main bedroom is through the north and south walls, windows, and doors. Paolini et al. [33] studied the hydrothermal performances of residential buildings at urban and rural sites and found that the most significant differences between urban and rural indoor conditions were related to the moisture levels, as computed by the indoor Humidex index. With the lower average temperature of 9.48 °C in the main bedroom and a variation of about 10 °C within 24 h, the relative humidity in the indoor environment fluctuates, and the effect of this will be studied in future work.

It can be concluded from Figure 6 that from 8:30 am to 5:00 pm, the outdoor temperature is on the rise, while from 5:00 pm to 8:30 am the next day, the outdoor temperature shows a decreasing trend. From 10:00 a.m. to 4:00 p.m., the temperature in the main bedroom increased from 7 to

15.0 °C in 6 h, and the temperature change rate was 1.0 °C/h. From 12:00 to 6:00 a.m., the indoor temperature of the main bedroom reduced from 8.0 to 6.4 °C in 6 h, and the temperature change rate was −0.27 °C/h. However, during the same period, the corresponding outdoor temperature change rates were 1.25 and −0.75 °C/h, respectively. The higher indoor temperature rise rate between 10:00 a.m. and 4:00 p.m. can be explained by the heating contribution from solar radiation, which is shown in Figure 7. The slower indoor temperature decreasing rate is due to the energy stored in the wall. The surface temperature of the west side wall of the main bedrooms was lower than the room temperature between 10:00 a.m. and 6:00 p.m., however, it was reversed between 7:00 p.m. and 10:00 a.m. the following day. It is evident that the walls transfer their stored energy to the room. As reviewed by Navarro et al. [34], high thermal mass materials in buildings can provide thermal stability and smooth thermal fluctuations. Yang et al. [35] pointed out that using a thermal storage medium to utilize solar energy is a relatively simple, economical, and reliable way to improve the building thermal environment.

It is interesting to note here that there was no coal stove in the second bedroom; the heat sources were solar radiation energy and the heat from the internal walls adjacent to the main bedroom. The solar radiation intensity is shown in Figure 7. It can be seen that the maximum of solar radiation occurred at 1:00 p.m. and the higher values of radiation were between 12:00 and 2:00 p.m. The air temperature in the second bedroom increased from 1.2 to 7.8 °C, a notable increase of 6.8 °C, between 10:00 a.m. and 2:00 p.m. This building is in one of the richest regions in terms of solar resources in China [23], so solar radiation certainly needs to be considered and explored further in building designs.

To further analyze the contribution of solar radiation in room space heating, the analysis of the heat flow rate of the second bedroom shows that there is more heat leaving the second room than is gained from the main bedroom through the internal wall. From the calculations of heat flow rate, the net heat flow rate out of the second bedroom is 205.2 W. Because the average indoor and outdoor temperatures were used in the simulation, the average solar energy gained by the second bedroom should be at least 205.2 W, which is about 3.7 times the energy gained from the internal wall which is adjacent to the main bedroom. This means that solar radiation plays a big role in maintaining the higher temperature in the second bedroom than the outdoor temperature.

The average surface temperature of the west side wall of the main bedroom is 9.45 °C, which is very close to the average air temperature of 9.48 °C in the main bedroom. This can be explained by the location of the coal stove which was close the west side of the main bedroom, as seen in Figure 2. The simulated wall surface temperature as shown in Figure 8 was about 2 °C less than that of the main bedroom temperature; this is due to the omission of the radiation effect of the coal stove on the wall surface temperatures.

Rural buildings in northern China have a unique style. Courtyards are often open and main buildings face south, which can maximize the exposure of the walls and windows of the building to the sun and reduce the velocity of cold northern winds. It is evident from the simulations and analysis that a great amount of solar energy is absorbed by the building. From the simulation results shown in Figures 10 and 11, the building and the enclosures reduce the wind velocities in the courtyard. When the building is south facing, not only is the solar radiation energy absorbed by the room, it can be seen that the wind velocity on the roof of the building in the north wind is much less than that in the west wind, and the wind velocity is reduced greatly in the front of the main bedroom. The north wind is the dominate wind in the region; therefore, south-facing buildings should be constructed.

When a north wind velocity of 10 m/s was simulated, it was found that the wind velocities at 0.5 m from the building in the south of the building varied from 1.1 to 4 m/s. The convective heat transfer coefficients obtained from Equation (6) are 4.37 to 8.92 W/m²·K, which will affect the heat transfer rate in individual rooms in the building. Therefore, to accurately predict the heat flow rate in the building, numerical simulations can help determine the convective heat transfer coefficients at various points in the building. The simulation results echo the findings by Montazeri et al. [20] that a wider building has more impact on wind blocking.

It can be seen from Figure 6 that the temperature of the main bedroom dropped below 10 °C between 9:00 p.m. and 10:00 a.m. on the following day. To maintain a comfortable indoor temperature in a rural residential building, more energy needs to be consumed. With growing concerns about energy consumption, builders and owners want to design and build energy efficient buildings in rural China. Because most of the heat is lost through the south and north walls of the main bedroom, it is important to show the builders and owners the benefits of using insulation materials and increasing wall thickness, even if this may incur additional costs to the initial budget. Studies suggest that insulation board or double-glazed windows could be installed to reduce heat loss, as shown in the paper. Since energy is so important in maintaining a comfortable thermal environment for occupants, to minimize the building energy used while choosing building envelopes and insulation materials, their environmental impact also needs to be considered. Huedo et al. [36] provided a sustainability evaluation model based on a lifecycle assessment for different envelope assemblies, building orientations, and climate zones. Making buildings more energy efficient will be an evolution process in rural China since people are not used to changing windows or installing insulation materials on existing buildings. Santamouris et al. [30] showed that the indoor temperature in dwellings of very-high-deprivation residents in Athens, Greece was very low, with an average temperature of 12.2 °C, and that the thermal quality of the building envelope was low. They suggested improving the thermal performance of low-income houses to improve indoor environmental quality. To address energy poverty and improve building energy efficiency, as suggested by He et al. [37], some easy methods could be used in rural cold regions, Including improving the tightness of doors and windows and reducing window and door cold bridges such as by coating wood doors with heat preservation materials. Heat transfer through the roof and floors are usually overlooked in rural regions, and a sloping roof tends to have better insulation effect than a flat roof. Further, moisture-proofing and insulation design under the floors can reduce heat loss. Other measures for reducing energy consumption could be optimizing the length/with ratio and shape coefficient of rural buildings.

5. Conclusions

The typical rural residential building studied in this paper has its merits. Its south-facing layout is not only able to absorb solar radiation during the winter, it also can reduce north wind velocities around the building. The wind blocking effect can then reduce the convective heat transfer coefficients and minimize heat loss. The computer simulations can predict the velocity distributions of the wind around the building, from which the function of the rooms or the layout of the building could be optimized. The good practice of constructing south-facing buildings and enclosures in rural areas should be promoted with the demonstrated benefits in this study. The solar energy absorbed by the room raised the room temperature around 7 °C above the outdoor average temperature, so solar energy should be maximized in the area for space heating. Heat loss from the room also can be minimized by using double-glazed windows and insulation boards, which should be considered and used in new residential buildings. To conduct heat flow rate calculations in the building energy efficiency analysis, computational fluid dynamics should be used to obtain the convective heat transfer coefficient at different locations.

Author Contributions: Funding acquisition, Y.Z.; Methodology, Y.Z. and G.S.; Investigation, X.F. and G.S.; Modelling, C.W.; Preparing the manuscript, X.F. and C.W.

Acknowledgments: This research was funded by the National Natural Science Foundation of China, grant number [51678483], and the APC was funded by [51678483].

Conflicts of Interest: The authors declared no conflicts of interest.

References

1. UNEP SBCI. *Buildings and Climate Change, Summary for Decision-Makers*; United Nations Environment Programme: Paris, France, 2009.
2. Aksoezen, M.; Daniel, M.; Hassler, U.; Kohler, N. Building age as an indicator for energy consumption. *Eng. Build.* **2015**, *87*, 74–86. [CrossRef]

3. Clarke, J.A.; Johnstone, C.M.; Kelly, N.J.; Strachan, P.A.; Tuohy, P. The role of built environment energy efficiency in a sustainable UK energy economy. *Energy Policy* **2008**, *36*, 4605–4609. [CrossRef]

4. Akadiri, P.O.; Chinyio, E.A.; Olomolaiye, P.O. Design of a sustainable building: A conceptual framework for implementing sustainability in the building sector. *Buildings* **2012**, *2*, 126–152. [CrossRef]

5. Ortiz, O.; Castells, F.; Sonnemann, G. Sustainability in the construction industry: A review of recent of recent developments based on LCA. *Build. Mater.* **2009**, *23*, 28–39. [CrossRef]

6. Li, K.Q. Inclusive Development: A Better World for All. *British Think Tanks*, 18 June 2014.

7. Evans, M.; Yu, S.; Song, B.; Deng, Q.Q.; Liu, J.; Delgado, A. Building energy efficiency in rural China. *Energy Policy* **2014**, *64*, 243–251. [CrossRef]

8. Shan, M.; Wang, P.S.; Li, J.; Yue, J.; Yang, X. Energy and environment in Chinese rural buildings: Situations, challenges, and intervention strategies. *Build. Environ.* **2015**, *91*, 271–282. [CrossRef]

9. Liu, J.P.; Wang, L.Y.; Yoshino, Y.; Liu, Y.F. The thermal mechanism of warm in winter and cool in summer in China traditional vernacular dwellings. *Build. Environ.* **2011**, *46*, 1709–1715. [CrossRef]

10. Liu, Y.Y.; Jiang, J.; Wang, D.J.; Liu, J.P. The indoor thermal environment of rural school classrooms in Northwestern China. *Indoor Built Environ.* **2017**, *25*, 631–641. [CrossRef]

11. Liu, Y.Y.; Song, C.; Zhou, X.J.; Liu, J.P. Thermal requirements of the sleeping human body in bed warming conditions. *Eng. Build.* **2016**, *130*, 709–720. [CrossRef]

12. Roberts, D.; Vera-Toscano, E.; Phimister, E. Fuel poverty in the UK: Is there a difference between rural and urban areas? *Energy Policy* **2015**, *87*, 216–223. [CrossRef]

13. Bouzarovski, S.; Petrova, S.; Sarlamanov, R. Energy poverty policies in the EU: A critical perspective. *Energy Policy* **2012**, *49*, 76–82. [CrossRef]

14. Healy, J.D.; Clinch, J.P. Fuel poverty thermal comfort and occupancy: Results of a national household-survey in Ireland. *Appl. Energy* **2002**, *73*, 339–343. [CrossRef]

15. Fan, X.N.; Zhu, Y.Y.; Sang, G.C. Optimization design of enclosure structure of solar energy building in Northwest China based on indoor zoning. In Proceedings of the 2017 6th International Conference on Energy and Environmental Protection, Zhuhai, China, 30 June–2 July 2018; Atlantis Press: Hong Kong, China, 2017; Volume 143, pp. 30–40.

16. Boeck, L.D.; Verbeke, S.; Audenaert, A.; Mesmaeker, L.D. Improving the energy performance of residential buildings: A literature review. *Renew. Sustain. Energy Rev.* **2016**, *52*, 960–975. [CrossRef]

17. Mitterer, C.; Kunzel, HM.; Herkel, S.; Holm, A. Optimizing energy efficiency and occupant comfort with climate specific design of the building. *Front. Archit. Res.* **2012**, *1*, 229–235. [CrossRef]

18. Wang, D.J.; Jiang, J.; Liu, Y.F.; Wang, Y.Y.; Xu, Y.C. Student responses to classroom thermal environment in rural primary and secondary schools in winter. *Build. Environ.* **2017**, *115*, 104–117. [CrossRef]

19. Hemsath, T.L.; Bandhosseini, K.A. Sensitivity analysis evaluating basic building geometry's effect on energy use. *Renew. Energy* **2015**, *76*, 526–538. [CrossRef]

20. Montazeri, H.; Blocken, B.; Derome, D.; Carmeliet, J.; Hensen, J.L.M. CFD analysis of forced convective heat transfer coefficients at windward building facades: Influence of building geometry. *J. Wind Eng. Ind. Aerodyn.* **2015**, *146*, 102–116. [CrossRef]

21. Hassanain, A.A.; Hokam, E.M.; Mallick, T.K. Effect of solar storage wall on the passive solar heating construction. *Eng. Build.* **2011**, *43*, 737–747. [CrossRef]

22. Pisello, A.L.; Pignatta, G.; Piselli, C.; Castaldo, V.L.; Cotana, F. Effect of dynamic characteristics of building envelope on thermal-energy performance in winter conditions: In field experiment. *Energy Build.* **2014**, *80*, 218–230. [CrossRef]

23. Zhang, Q.Y.; Joe, H. *Chinese Standard Weather Databank for Buildings*; Machine Industry Press: Beijing, China, 2014. (In Chinese)

24. Obyn, S.; Van, M.G. Variability and impact of internal surfaces convective heat transfer coefficients in the thermal evaluation of office buildings. *Appl. Therm. Eng.* **2015**, *97*, 258–272. [CrossRef]

25. Awbi, H.B.; Hatton, A. Natural convection from heated room surfaces. *Energy Build.* **1999**, *30*, 233–244. [CrossRef]

26. Hagishima, A.; Tanimoto, J. Field measurements for estimating the convective heat transfer coefficient at building surface. *Build. Environ.* **2003**, *38*, 873–881. [CrossRef]

27. Mirsadeghi, M.; Costola, D.; Blocken, B.; Hensen, J.L.M. Review of external convective heat transfer coefficient models in building energy simulation programs: Implementation and uncertainty. *Appl. Therm. Eng.* **2013**, *56*, 134–151. [CrossRef]

28. Liu, Y.; Harris, D.J. Full-scale measurements of convective coefficient on external surface of a low-rise building in sheltered conditions. *Build. Environ.* **2007**, *42*, 2718–2736. [CrossRef]

29. Ramponi, R.; Blocken, B. CFD simulation of cross-ventilation for a generic isolated building: Impact of computational parameters. *Build. Environ.* **2012**, *53*, 34–48. [CrossRef]

30. Santamouris, M.; Alevizos, S.M.; Aslanoglou, L.; Mantzios, D.; Milonas, P.; Sarelli, I.; Karatasou, S.; Cartalis, K.; Paravantis, J.A. Freezing the poor-indoor environmental quality in low and very low income households during the winter period in Athens. *Energy. Build.* **2014**, *70*, 61–70. [CrossRef]

31. Clinch, J.P.; Healy, J.D. Housing standards and excess winter mortality. *J. Epidemiol. Community Health* **2000**, *54*, 719–720. [CrossRef] [PubMed]

32. Zhao, Q.; Zhao, Y.; Li, S.; Zhang, Y.; Wang, Q.; Zhang, H.; Qiao, H.; Li, W.; Huxley, R.; Williams, G.; et al. Impact of ambient temperature on clinical visits for cardio-respiratory diseases in rural villages in northwest China. *Sci. Total Environ.* **2018**, *612*, 379–385. [CrossRef] [PubMed]

33. Paolini, R.; Zani, A.; MeshkinKiya, M.; Castaldo, V.L.; Pisello, A.L.; Antretter, F.; Poli, T.; Cotana, F. The hydrothermal performance of residential buildings at urban and rural sites: Sensible and latent energy loads and indoor environmental conditions. *Energy Build.* **2017**, *152*, 792–803. [CrossRef]

34. Navarro, L.; Gracia, A.D.; Colclough, S.; Browne, M.; Mccormack, S.J. Thermal energy storage in building integrated thermal systems: A review. Part 2. Integration as passive system. *Renew. Energy* **2016**, *85*, 1334–1356. [CrossRef]

35. Yang, L.; He, B.J.; Ye, M. The application of solar technologies in building energy efficiency: BISE design in solar-powered residential buildings. *Technol. Soc.* **2014**, *38*, 111–118. [CrossRef]

36. Huedo, P.; Mulet, E.; Lopez-Mesa, B. A model for the sustainable selection of building envelope assemblies. *Environ. Impact Assess. Rev.* **2016**, *57*, 63–77. [CrossRef]

37. He, B.J.; Yang, L.; Ye, M. Building energy efficiency in China rural areas: Situation, drawbacks, challenges, corresponding measures and policies. *Sustain. Cities Soc.* **2014**, *11*, 7–15. [CrossRef]

![applied sciences logo] *applied sciences*

MDPI

Article

Effects of Configurations of Internal Walls on the Threshold Value of Operation Hours for Intermittent Heating Systems

Shuhan Wang * and Ke Zhong

School of Environmental Science and Engineering, Donghua University, Shanghai 201620, China; zhongkeyx@dhu.edu.cn
* Correspondence: shuhanwang@mail.dhu.edu.cn; Tel.: +86-216-779-2554; Fax: +86-2167-792-522

Received: 28 January 2019; Accepted: 18 February 2019; Published: 21 February 2019

Abstract: The heating load of intermittent heating is not always lower than that of continuous heating for heat storage and release of internal walls. Therefore, the threshold value of daily operation hours exists, and is affected by the configuration of internal walls. A comparative study is performed between continuous and intermittent heating modes to investigate the threshold value of daily operation hours for different internal wall configurations by employing computational fluid dynamic (CFD) models. Meanwhile, field tests on the temperature distribution within a thermal mass was carried out to validate the simulation. The results show that the heating load index of intermittent heating is larger than that of continuous heating with increased amplitude ranging from 31.58% to 152.63%. The threshold value of daily operation hours is, respectively, 18.04 h, 15.80 h, 14.59 h, and 13.46 h for four internal wall configurations. Moreover, with the increase in the insulation level of internal walls, the threshold value of daily operation hours decreases. In addition, the results indicate that it is more economical to use continuous heating when the daily operation hours are more than the threshold values.

Keywords: threshold value of daily operation hours; intermittent heating; configurations of internal wall; heat storage and release; hot summer and cold winter climate zone

---

1. Introduction

Residential heating issues in the hot summer and cold winter (HSCW) climate zone of China have attracted increasing attention for the low outdoor temperature and absence of district heating [1–3]. The HSCW climate zone of China, located near the lower reaches of the Yangtze River, is one of the most economically developed regions, and has the highest population density in China. The mean temperature of January is between 2 °C and 7 °C, which is about 8 °C lower than other places of the same latitude in the world [4]. According to Chinese heating policy [5], central heating systems are not provided in the HSCW zone. To improve the poor indoor thermal environment, more than 90% of households have been equipped with individual air conditioners for heating. The survey performed by Yoshino [6] and Hu [7] showed that the operation of heating devices in the HSCW zone is intermittent, and occupants run heating devices only in the room they are using.

For intermittent heating mode, envelopes release heat during the heating cessation period. Therefore, more heat is absorbed a stored by building walls during the heating period, while it is not required for continuous heating mode. For this reason, the heating load of the unit floor area per unit time for intermittent heating mode is larger than that of continuous heating mode. The situation where daily heating load of intermittent heating is larger than that of continuous heating may occur, which leads to a threshold value of daily operation hours. Dreau and Heiselberg [8] investigated the effect of heating duration on energy consumption under two residential buildings with different levels of insulation. The results showed that the amount of energy reached a peak value for heating durations longer than 18 h. Slightly larger threshold values of heating duration of about 20 h were obtained by Badran [9] for a local residential building in Jordan with several levels of insulation thickness.

However, the main objective of those studies [8,9] was only to investigate the dynamic thermal performance of external walls. In the HSCW zone, usually just one room in the house is heated at a given time, which leads to the temperature difference between indoor air and the adjoining room air being slightly lower than that between indoor air and outdoor air. Moreover, the inner surface area of the internal walls is approximately 5 times larger than that of the external wall. Therefore, the effect of dynamic thermal performance of internal walls on the threshold value of daily operation hours need to be analyzed.

Much research has been performed on the dynamic thermal behavior of walls for intermittent heating mode. Tsilingiris [10] investigated the effect of thermal resistance and heat capacity as well as thermal constant on the heat exchange through the building envelope by implicit finite-difference method. Meng [11] and Zhang [12] analyzed the temperature response rate and the heat flow of different wall insulation forms under intermittent and continuous operation in summer by simulative and experimental method, respectively. Yuan et al. [13] analyzed the effects of insulation and thermal resistance on dynamic heat transfer of building walls by combining mathematical models and numerical solutions. Their results showed an evident process of heat storage and release of walls was found for intermittent heating mode, while the key points of those studies are still the heat transfer of external walls. The thermal performance of internal walls plays a vital role in affecting the threshold value of daily operation hours, but little research has been carried out.

According to the above problems, the aim of this study is to explore the effect of internal wall configurations on the threshold value of operation hours for intermittent heating mode. Therefore, four typical internal wall configurations commonly used in the HSCW region are considered, two of which are uninsulated walls with different thermal mass, and others are integrated with insulation layer of different materials. Moreover, five intermittent heating duration cases according to occupants' behavior in residential buildings are selected in the present work. Computational fluid dynamic (CFD) techniques are employed to perform a series of unsteady cases. The analysis of the threshold value of operation hours for different internal wall configurations is carried out by comparing the heating load between the intermittent and continuous heating mode.

2. Model Setup and Validation

2.1. Computational Domain

In the present study, a room located at the middle of an intermediate floor is selected as the model prototype, as shown in Figure 1. The room's internal dimensions are 4 m (length) × 4.3 m (width) × 2.7 m (height) and the floor area is 17.2 m². The insulated external wall consists of 0.2 m reinforced concrete layer and 0.035 m extruded polystyrene (XPS) layer inside of the reinforced concrete layer. A single-glazing window in the external wall has dimensions of 2.5 m × 1.2 m. The ceiling and floor are made of a 0.1 m reinforced concrete layer. Moreover, all the walls are plastered on both sides with 0.015 m cement layer. Two pieces of furniture, with the sizes of 2.2 m × 0.8 m × 0.3 m + 0.6 m × 0.6 m × 0.3 m (L × W × H, furniture 1) and 1.6 m × 0.55 m × 0.3 m (L × W × H, furniture 2), respectively, are placed inside the room. The warm air inlet, with dimensions of 0.7 m × 0.1 m, is at a height of 2 m from the floor level, while the air outlet, with the same dimensions, is 0.3 m above the air inlet.

Four different configurations of internal walls commonly used in the HSCW zone are considered in this paper. A detailed schematic of internal wall configurations is shown in Figure 2. Figure 2a shows the Wall 1 containing thermal mass layer of 0.2 m reinforced concrete and 0.015 m plaster layers. Figure 2b shows the Wall 2, which consists of thermal mass layer of 0.2 m reinforced concrete, 0.03 m insulation layers of thermal mortar and 0.015 m plaster layers. Figure 2c shows a lightweight wall (Wall 3), which contains base wall layer of 0.2 m aerated concrete block and 0.015 m plaster layers.

Figure 2d shows the Wall 4 containing thermal mass layer of 0.2 m reinforced concrete, insulation layers of 0.02 m XPS board and 0.015 m plaster layers. Thermal properties of the construction materials are given in Table 1.

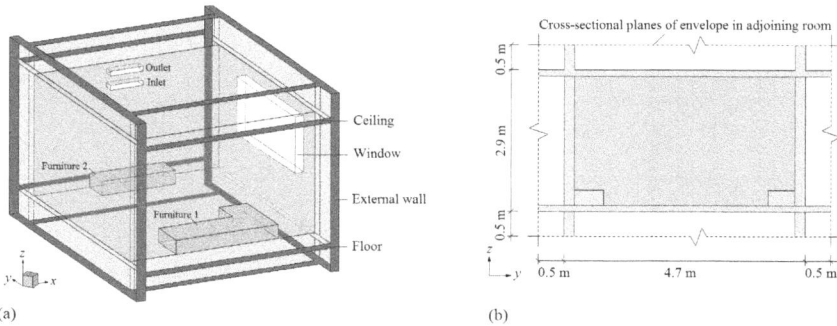

Figure 1. Schematic of the room: (**a**) three-dimensional schematic of the room model; (**b**) side view (plane *y-z*) of the model.

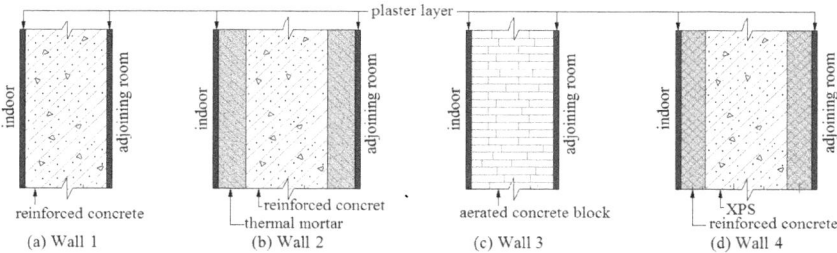

Figure 2. Sections of four kinds of internal wall structures.

Table 1. Thermo-physical properties of building construction materials [5].

Material	Density ρ (kg/m^3)	Specific Heat Capacity c [J/(kg·°C)]	Thermal Conductivity λ [W/(m·°C)]	Thermal Storage S [W/(m^2·°C)]	Thermal Diffusivity α [m^2/s]
Plaster layer	1700	1050	0.87	10.75	4.87×10^{-7}
Reinforced concrete	2500	920	1.74	17.2	7.57×10^{-7}
Thermal mortar	600	1050	0.18	2.87	2.86×10^{-7}
Aerated concrete block	700	1050	0.18	3.10	2.45×10^{-7}
XPS	35	1380	0.036	0.32	7.45×10^{-7}

The thermal resistance and thermal diffusivity of the four walls are also given in Table 2 to better understand the heat conduction and storage process of walls.

Table 2. Thermal resistance and thermal diffusivity of walls.

Internal Wall Configuration	Thermal Resistance R [m^2·°C/W]	Thermal Diffusivity α [m^2/s]
Wall 1	0.15	6.89×10^{-7}
Wall 2	0.48	3.16×10^{-7}
Wall 3	1.15	2.30×10^{-7}
Wall 4	1.26	1.12×10^{-7}

Due to the effect of the thermal bridge, the heat transfer of connections between external and internal walls as well as the floor is a three-dimensional unsteady process. The heat transfer through the external wall is considered to be one-dimensional along the *x*-axis away from the connectors.

For this reason, the envelopes of 0.5 m height in the four adjoining rooms on the directions of $\pm y$, $\pm z$ based on the investigated room are also contained in the computational domain (seen in Figure 1). Moreover, the cross-sectional planes of envelope in the adjoining rooms are adiabatic surfaces.

The indoor thermal comfort environment is achieved by operating individual heating facilities with 600 m^3/h supply air volume. The velocity of the air inlet is 2.38 m/s. The heating device is operated with high running power (3,500 W added with additional 1000 W electric auxiliary heat) when the average room temperature is lower than 18 °C. When the indoor air temperature rises to 20 °C, the running power of the heating device drops to 1800 W, and then the air temperature will decrease. Based on this system, the indoor air temperature varies between 18 and 20 °C.

2.2. Governing Equations

The airflow and temperature distribution in the room are set to maintain conservation laws of mass, momentum and energy. The airflow is simulated as unsteady state, incompressible, and turbulent. Based on the above assumptions, the governing time-averaged equations are given by [14]:

$$\frac{\partial u_i}{\partial x_i} = 0 \tag{1}$$

$$\frac{\partial u_i}{\partial t} + \frac{\partial (u_i u_j)}{\partial x_j} = -\frac{1}{\rho}\frac{\partial p}{\partial x_i} + \frac{\partial}{\partial x_j}\left(\nu\frac{\partial u_i}{\partial x_j} - \overline{u'_i u'_j}\right) \tag{2}$$

$$\frac{\partial (C_p T)}{\partial t} + \frac{\partial (C_p T u_i)}{\partial x_i} = \frac{\partial}{\partial x_j}\left[\left(\frac{\lambda}{\rho} + \frac{C_p \nu_t}{Pr_t}\right)\frac{\partial T}{\partial x_j}\right] \tag{3}$$

$$-\overline{u'_i u'_j} = \nu_t\left(\frac{\partial u_i}{\partial x_j} + \frac{\partial u_j}{\partial x_i}\right) - \frac{2}{3}k\delta_{ij} \tag{4}$$

$$\nu_t = C_\mu \frac{k^2}{\varepsilon} \tag{5}$$

where x_i represents the Cartesian coordinates while the subscripts i and j range between 1 and 3, separately, refer to the (x, y, z) directions in space. The term u_i is the velocity of component i. p and T are the pressure and temperature, respectively. ρ and λ are the air density and thermal conductivity, respectively. ν is the kinematic viscosity, ν_t is the turbulent eddy viscosity, k is turbulent kinetic energy, and δ_{ij} is the Kronecker delta (i = j, δ_{ij} = 1; i $\neq$ j, δ_{ij} = 0).

The renormalization group (RNG) k-ε turbulence model is used to simulate the three-dimensional turbulent airflow, as it has been shown to accurately describe the flow field of the near wall region and relatively better predict of the indoor environment than the standard k-ε model by introducing an additional term in the ε-equation [15–17].

The transfer equations of turbulent kinetic energy k and turbulence dissipation rate ε are given by:

$$\frac{\partial k}{\partial t} + \frac{\partial}{\partial x_i}(k u_i) = \frac{\partial}{\partial x_j}\left[\left(\nu + \frac{\nu_t}{\sigma_k}\right)\frac{\partial k}{\partial x_j}\right] - \overline{u'_i u'_j}\frac{\partial u_j}{\partial x_i} - \varepsilon \tag{6}$$

$$\frac{\partial \varepsilon}{\partial t} + \frac{\partial}{\partial x_i}(\varepsilon u_i) = \frac{\partial}{\partial x_j}\left[\left(\nu + \frac{\nu_t}{\sigma_\varepsilon}\right)\frac{\partial \varepsilon}{\partial x_j}\right] - C_{1\varepsilon}\frac{\varepsilon}{k}\overline{u'_i u'_j}\frac{\partial u_j}{\partial x_i} - C^*_{2\varepsilon}\frac{\varepsilon^2}{k} \tag{7}$$

$$C^*_{2\varepsilon} = C_{2\varepsilon} + \frac{C_\mu \rho \eta^3 (1 - \eta/\eta_0)}{1 + \zeta \eta^3} \tag{8}$$

where $\eta = S\ (k/\varepsilon)$ is the ratio between the time scales of the turbulence and the mean flow, S is the coefficient of surface tension and is defined by $S = \sqrt{2S_{ij}S_{ij}}$, $S_{ij} = (\partial u_i/\partial x_j + \partial u_j/\partial x_i)/2$.

The constants appearing in the RNG k-ε model are: $C_\mu = 0.0845$, $C_{1\varepsilon} = 1.42$, $C_{2\varepsilon} = 1.68$, $\sigma_k = \sigma_\varepsilon = 0.7178$, $\zeta = 0.012$, $\eta_0 = 4.38$.

To observe the dynamic heat transfer process in envelope, a three-dimensional system of wall with thickness direction x, width direction y and height direction z is established (see Figure 3a). δ, W and H are the thickness, width, and height of the wall, respectively.

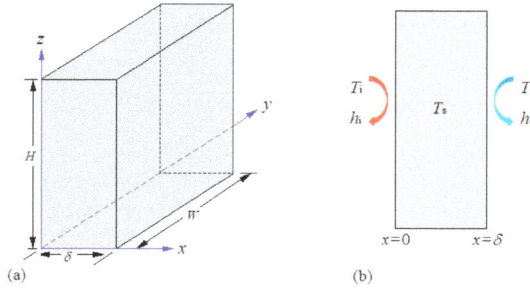

Figure 3. Simulation model of internal wall: (**a**) three-dimensional schematic of internal wall; (**b**) heat transfer boundary conditions on the wall surfaces.

The heat transfer across the wall is described in the following equation [18,19]:

$$\frac{\partial T_s}{\partial t} = \frac{\lambda_s}{\rho_s C_s}\left[\frac{\partial}{\partial x}\left(\frac{\partial T_s}{\partial x}\right) + \frac{\partial}{\partial y}\left(\frac{\partial T_s}{\partial y}\right) + \frac{\partial}{\partial z}\left(\frac{\partial T_s}{\partial z}\right)\right] \qquad (9)$$

Convective heat transfer boundary conditions are adopted on the wall's inner and outer surfaces. For the inner surface, it is expressed as:

$$-\lambda_s\frac{\partial T_s}{\partial x}\bigg|_{x=0} = h_i(T_{x=0} - T_i) \qquad (10)$$

For the outer surface, it can be expressed as:

$$-\lambda_s\frac{\partial T_s}{\partial x}\bigg|_{x=\delta} = h_o(T_a - T_{x=\delta}) \qquad (11)$$

Moreover, adiabatic boundary conditions are used on the cross-sectional planes ($y = 0$ and $y = W$, $z = 0$ and $z = H$) of the envelope [20]:

$$\frac{\partial T_s}{\partial y}\bigg|_{y=0,W} = \frac{\partial T_s}{\partial z}\bigg|_{z=0,H} = 0 \qquad (12)$$

where ρ_s and λ_s are the density and thermal conductivity of the wall, respectively. T_a refers to the temperature of the surrounding air (i.e., the outdoor air or the adjoining air). T_i is the indoor air temperature. h_i and h_o are the heat transfer coefficients of inner and outer surfaces of walls, respectively.

2.3. Numerical Aspects and Boundary Conditions

The commercial program of ANSYS 6.3.26 with the finite volume method is employed to solve the three-dimensional numerical simulations. To improve the accuracy of the numerical simulations, second-order discretization schemes are applied to the momentum, turbulent kinetic energy, and turbulent dissipation rate equations. Semi-Implicit Method for Pressure Linked Equations (SIMPLE) algorithm is used to evaluate pressure-velocity coupling in the continuity equations. Meanwhile, the standard wall-function method is applied to model the near wall regions with low Reynolds number [21].

According to the "Design standard for energy efficiency of residential buildings in hot summer and cold winter zone" (JGJ 134-2010) [22], the cold air infiltration is 1 h^{-1}. The heat transfer coefficient for the window should be less than 4.7 W/(m$^2\cdot$°C), and the window: external wall area ratio should be less than 0.4 (north-facing) and 0.45 (south-facing), respectively. Thus, the area of window in the computational model is 3 m^2, and the heat transfer coefficient of window is 3.0 W/(m$^2\cdot$°C).

At the inlet of the computational domain, uniform velocity and temperature are imposed, and the inlet values of turbulent kinetic energy (k_0) and dissipation rate (ε_0) profiles [23] are set as follows:

$$T_u = 0.16(Re_L)^{-1/8} \tag{13}$$

$$\varepsilon_0 = C_\mu^{3/4}\frac{k^{3/2}}{l} \tag{14}$$

where V, T_u and Re_L, respectively, denote the mean velocity, the turbulence intensity, and the Reynolds number at the inlet. l is a length scale defined as $l = 0.07 L$, where L is the hydraulic diameter at the inlet. Furthermore, outflow boundary condition is applied at the outlet of the computational domain and non-slip conditions are imposed on the inner surfaces of solid walls.

2.4. Grid Independency

The computational domain is discretized with tetrahedral elements, while non-uniform computational grids are used for the present simulation. To describe the heat transfer and flow status more accurately, grids are further refined near walls, inlet and outlet, while relatively coarse grids are used for zones away from solid surfaces. The expansion rate between two consecutive cells is no more than 1.12.

To ensure the first numerical point is located inside the logarithmic layer, the variable y^+, termed as the dimensionless wall distance, is controlled in the range of 90~220, satisfy the requirement range of $30 \leq y^+ \leq 300$ for RNG k-ε model using standard wall functions [24].

$$y^+ = \frac{y}{\nu}\sqrt{\frac{\tau_w}{\rho}} \tag{5}$$

where, y is the vertical distance from wall to the first cell center. ν is the kinematic viscosity. τ_w is the wall shear stress.

Grid independency is carried out to achieve converged results of the simulations. A set of pre-simulations are carried out to check the same physical parameters by using different grid densities (coarse, normal, and fine), and the grid number is increased until the numerical results will not be affected by the grid size. In this study, the root-mean-square error in the temperature, shown in Equation (16), $\varepsilon_{r,m,s}$ of less than 2% is used as the criterion for grid independence [25].

$$\varepsilon_{r,m,s} = \sqrt{\frac{1}{N}\sum_1^N\left(\frac{T_{i,f} - T_{i,c}}{T_r}\right)^2} \tag{16}$$

where, $T_{i,f}$ is the temperature in the former grid number, $T_{i,c}$ is the temperature in the current grid number, the subscript i represents the ith sampling point, $T_r = 19$ °C is the reference temperature, and N is the number of the examined sample points. Based on the different configurations of building envelope, the resulting unstructured grid numbers of 4.03~5.02 million are used for the simulations.

2.5. Model Validation

Validation exercises were performed to ensure the reliability of the results from the CFD simulations. The temperature fields in a test chamber were investigated by experimental and numerical simulation. Figure 4 gives the detailed arrangement of the test system.

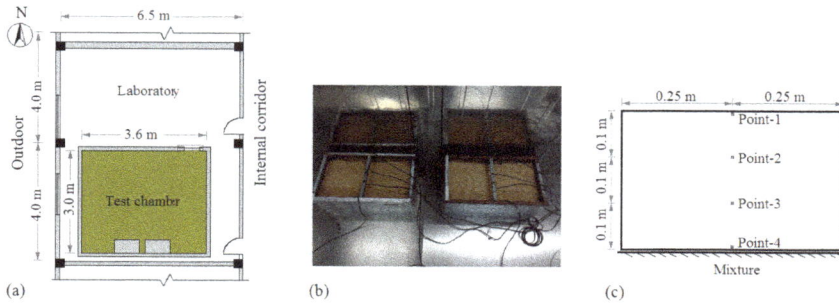

Figure 4. Schematic of the experimental system: (**a**) layout of the laboratory, (**b**) dry sand and steel reinforcing rods in the boxes, (**c**) measuring points in the mixture.

The experiment was carried out in an environment chamber with dimensions of 3.6 × 3.0 × 2.6 m (L × W × H), which is placed in a large laboratory, as shown in Figure 4a. The walls of the chamber were made of 0.01 m thickness colored steel plates which were hollow and filled with 0.07 m thickness insulated rock wool. The heat transfer coefficient of the envelopes was about 0.9 W/(m^2·°C). The dimension of door was 0.7 m × 2 m. The door was open, and the area of which was covered with plastic film. The fresh air ratio was regulated to 10%, and the condition of slightly positive pressure was maintained in the chamber during the experiment. Two boxes with dimensions of 0.5 m × 0.3 m × 0.28 m (see Figure 4b), containing mixture of dry sand and steel reinforcing rods were placed at the floor near one internal wall. Moreover, the mass ratio of the dry sand to the steel reinforcing rods was about 0.91, and the thermal conductivity of the mixture was 0.7 (W/(m·°C)).

The air conditioning system was used to warm the chamber. The supply velocity and temperature of the warm air inlet is 1.62 m/s and 36.9 °C, respectively. The air inlet was placed at the height of 0.15 m with the dimension of 0.3 m × 0.1 m. The exhaust device was mounted at the center of the ceiling with the dimension of 0.3 m × 0.2 m.

The temperature was measured by Type-K thermocouples with an accuracy of ±0.1 °C. The temperature of each wall was measured at the central point by thermocouples. Because the walls were made of colored steel plates with uniform temperature distribution, the value of the points can represent the corresponding surface temperature. In addition, four thermocouples were placed uniformly in each mixture, the turn of which from external to internal side is measuring point 1, 2, 3 and 4. The distribution of four measuring points in the mixture is shown in Figure 4c. All the thermocouples were covered with aluminum foil tape to minimize radiation effects.

Before the tests, the air temperature of the chamber and the outside laboratory had been kept at 10 °C for two hours by adequately ventilating. Then the air conditioning system was operated, and the air temperature of the outside laboratory was kept at 10 ± 1 °C during the experiment.

Moreover, the RNG *k*-*ε* model was verified by experimental data to simulate the thermal performance of heating room and the dynamic thermal behavior of thermal mass. Figure 5 gives the comparisons of temperature distribution of the mixture 1 between CFD simulations and the experimental data.

Figure 5a–d indicates that the numerical predictions of the temperature distribution in Mixture 1 are in good agreement with the measured results. The temperature difference of Point 2 is the maximum of the four points, while which is less than 0.3 °C. Moreover, the temperature difference of the other three points is less than 0.1 °C. Hence, it can be concluded that the numerical simulation used in the present study is appropriate for simulating the temperature fields in the heating room.

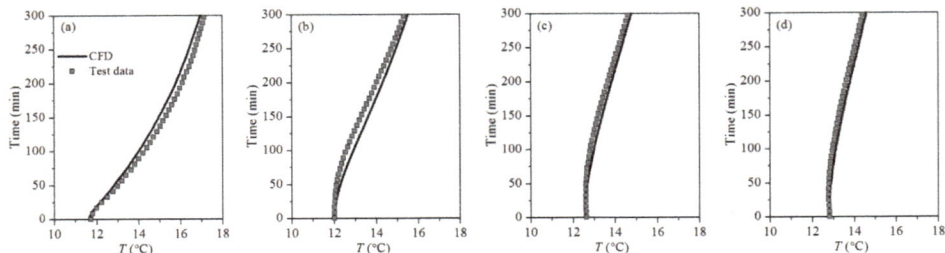

Figure 5. Validation of temperature distribution in a thermal mass: (**a**) Point-1; (**b**) Point-2; (**c**) Point-3; (**d**) Point-4.

2.6. Study Cases

In the present study, five intermittent heating durations have been considered, ranging from 0.5 h to 8 h, according to the occupants' behavior in residential buildings [26]. A heating cycle contains heating duration and heating cessation duration, in which the heating cessation duration is 2 h. The results of four heating cycles is given in this paper. The detailed heating operation cases are summarized in Table 3.

Table 3. Information of operation cases.

Operation Case	Heating Duration per Operation τ_0 (h)	Hours Contained in a Heating Cycle τ (h)
$C_{T=0.5}$	0.5 h	2.5 h
$C_{T=1}$	1 h	3 h
$C_{T=2}$	2 h	4 h
$C_{T=4}$	4 h	6 h
$C_{T=8}$	8 h	10 h

3. Results and Discussion

3.1. Comparison of Surface Temperature of Four Internal Wall Configurations

To observe the dynamic thermal behavior of four internal walls for different heating durations, Figure 6 shows the variations of inner surface temperature $T_{in,i}$ of the internal walls, taking heating duration of 1 and 8 h as examples.

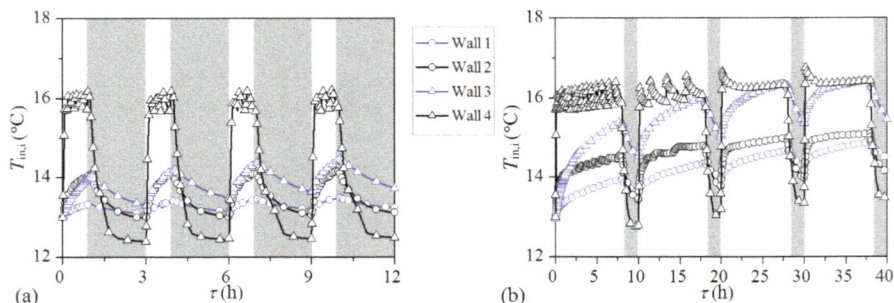

Figure 6. Variations of inner surface temperature $T_{in,i}$ of the internal walls with time: (**a**) $C_{T=1}$, (**b**) $C_{T=8}$. Note: the light gray zone in Figure 6 represents the heating cessation period.

It is found in Figure 6 that at the starting time of heating cycle, the $T_{in,i}$ of Wall 4 (insulted with XPS) is increasing rapidly at the beginning of the heating period, and then fluctuates at 16 °C, approaching the indoor air temperature T_i (18 °C to 20 °C). However, $T_{in,i}$ of the other three internal walls increases slowly and is obviously lower than the $T_{in,i}$ of Wall 4 during the whole heating period. Once the heating device is shut off, $T_{in,i}$ drops sharply for Wall 4, but drops slowly for the other three internal walls. The reason is the heat capacity of the inner layer of Wall 4 is the lowest, which leads to nearly no time lag between changes of $T_{in,i}$ and T_i (indoor air temperature).

Comparing Figure 6a,b, it can be seen that the $T_{in,i}$ of Wall 2 is equal to that of Wall 3 during heating period in Figure 6a. However, when it comes to the longer heating duration, as shown in Figure 6b, significant temperature difference is found between the inner surface of Wall 2 and Wall 3. This is because the thermal resistance of Wall 2 is smaller than Wall 3, which leads to more heat transferring to the inside of Wall 2 with the heating duration increases.

The internal walls absorb and store heat from indoor air during the heating period. Then, some heat transfers through the internal walls, which leads to the fluctuation of surface temperature of internal walls close to the adjoining room. The outer surface temperature ($T_{in,o}$) of internal walls close to the adjoining room is shown in Figure 7 for short and long heating duration.

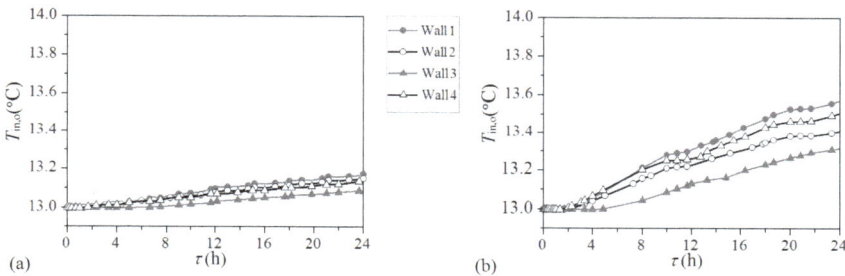

Figure 7. Variations of outer surface temperature $T_{in,o}$ of the internal walls with time: (a) $C_{\tau=1}$, (b) $C_{\tau=8}$.

As shown in Figure 7a, the $T_{in,o}$ of four internal walls is similar to each other and close to the adjoining air temperature. It indicates that almost all the heat absorbed by the inner layer of internal walls is stored and little energy transfer to the outer surface of internal walls for short heating duration.

An apparent increase of $T_{in,o}$ of four internal walls is found in Figure 7b for long heating duration, and the difference of $T_{in,o}$ between four walls increases with the increase of heating duration. The raise of $T_{in,o}$ ranks from high to low with the configuration of Wall 1, Wall 4, Wall 2, and Wall 3. This is dependent on the heat capacity of inner layer and thermal resistance of four internal walls. $T_{in,o}$ of the Wall 1 is the highest due to its higher thermal capacity and lower thermal resistance.

3.2. Comparison of Surface Heat Flow of Four Internal Walls

Figure 8 shows the variation of inner surface heat flow (q_i) of four internal walls with time under short and long heating durations.

As shown in Figure 8a, the inner surface heat flow q_i of Wall 4 is the lowest of the four walls, and the heat flow q_i of the other three walls are similar to each other during heating period. This is because Wall 4 gives the smallest temperature difference between the inner surface and indoor air, and hence the smallest heat flow q_i. It can also be seen from Figure 8a that the internal walls release heat to the indoor air during heating cessation period, and the heat flow achieves a similar effect by the configuration of internal walls.

The results in Figure 8b indicate that Wall 4 still demonstrates the smallest heat flow among the four internal wall types, which is consistent with the situation presented in Figure 8a. However, the heat flow of Wall 3 drops gradually with the heating duration and is almost equal to the q_i of Wall 4. This is because of the similar temperature different between indoor air and inner surface temperature of Wall 3 and Wall 4.

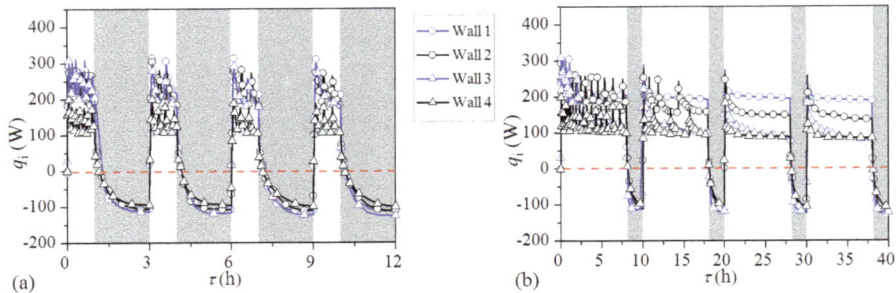

Figure 8. Variation of inner surface heat flow q_i of four internal walls with time: (**a**) $C_{\tau=1}$, (**b**) $C_{\tau=8}$.

To evaluate the heat transferring to the adjoining room, the variation of the heat flow q_o between outer surface (close to the adjoining room) of internal wall and adjoining air are shown in Figure 9.

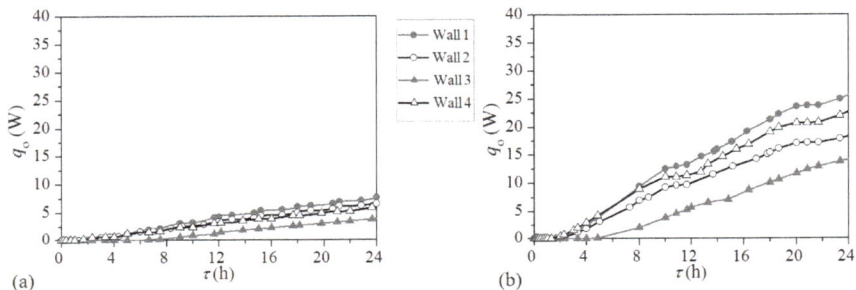

Figure 9. Variation of outer surface heat flow q_o of four internal walls with time: (**a**) $C_{\tau=1}$, (**b**) $C_{\tau=8}$.

It can be observed from Figure 9a that the heat flow q_o of four internal walls is always small for heating case $C_{\tau=1}$ (less than 5 W), and the difference of q_o between four walls is slight. With the increase of heating duration, the heat flow q_o increases obviously (as shown is Figure 9b). Moreover, a significant difference of q_o between four walls is observed. This is because of the temperature difference of outer surface between the four internal walls.

The results in Figure 9b also shows that the heat flow is nearly zero when the heating duration is less than 5 h. It means the process of heat absorption and storage of internal walls lasts 5 h before heat transferring to the adjoining room, no matter what the configuration of internal wall is.

3.3. Threshold Value of Daily Operation Hours Under Four Internal Wall Configurations

It is obvious that the extent of building walls heated by indoor air is different under various heating durations, which will lead to different heating loads of the unit floor area per unit time. Therefore, the heating load index Q_i, as expressed in Equation (17), is introduced to analyze the characteristics of the heating load during the heating period. Also, the increasing rate η, as expressed in Equation (18), is employed to compare the heating load index between intermittent heating mode and continuous heating mode.

$$Q_i = \int_0^{\tau_o} pd\tau_o / (F \times \tau_o) \tag{17}$$

$$\eta = \frac{Q_{i,m} - Q_{i,c}}{Q_{i,c}} \times 100\% \tag{18}$$

where p is the running power of heating device. τ_o is the heating duration. F is the floor area of the investigated room. $Q_{i,m}$ is the heating load index for intermittent heating. $Q_{i,c}$ is the heating load index for 24 h continuous heating (C_{con}).

Figure 10 demonstrates the effect of intermittent heating duration on the heating load index Q_i for four internal wall configurations, and the case of 24 h (C_{con}) heating duration is also given.

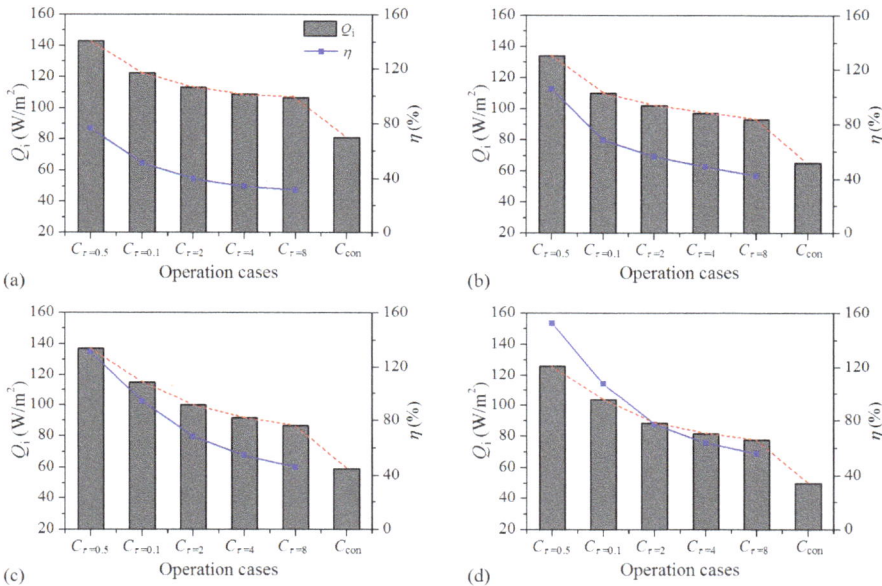

Figure 10. Variations of Q_i and η with operation cases: (**a**) Wall 1, (**b**) Wall 2, (**c**) Wall 3, (**d**) Wall 4.

Figure 10 shows that Q_i of intermittent heating is significantly larger than that of C_{con}. This is because the inner layer of walls release heat during heating cessation time, which leads to more energy being required to heat the inside layer of the walls at the next heating cycle for intermittent heating. Compared with Figure 10a–d, it can be found that the heating load index with the same heating duration drops in turn under the Wall 1, Wall 2, Wall 3, and Wall 4. The heating load index depends on the heat convection between the inner surface of walls and indoor air. Wall 1 gives the largest temperature difference between the inner surface of walls and indoor air. Therefore, the largest Q_i is obtained by Wall 1.

It can be seen in Figure 10a that the increasing rate η ranges from 31.58% to 76.35% for five heating durations. Moreover, the heating cessation time ratio of $C_{\tau=8}$ is 20%, while the increasing rate η is 31.58%, which is larger than 20%. This indicates that the daily heating load of heating time ratio 19.2/24 is larger than that of 24 h continuous heating and a threshold value of daily operation hour exists for intermittent heating mode. By comparing Figure 10a–d, it also can be seen that with various internal wall configurations, the increasing rate η is different. This will lead to various threshold values of daily operation hours.

The daily heating load is calculated by Equation (19):

$$Q = Q_i \times \frac{24\tau_o}{\tau} \tag{19}$$

where τ is the hours contained in a heating cycle (see Table 3).

The variation of daily heating load (*Q*) with operation hour for internal wall configurations is shown in Figure 11.

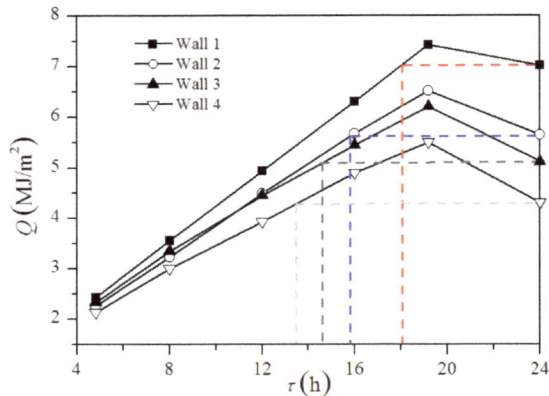

Figure 11. Variations of daily heating load (*Q*) with operation hour.

As illustrated in Figure 11, the daily heating load (*Q*) is increasing continuously at first and then decreasing linearly with the operation hour per day. Comparing the heating load of intermittent (Q_m) and 24 h continuous heating (Q_c), Q_m is larger than Q_c when the heating time ratio is 19.2/24 and Q_m is smaller than Q_c when the heating time ratio is 12/24.

The threshold value of operation hours 18.04 h, 15.80 h, 14.59 h and 13.46 h, respectively, for Wall 1, Wall 2, Wall 3, and Wall 4. The threshold value of Wall 1 is 18.04 h, which is close to 18 h obtained in the research [8]. This is because the thermal performance of envelope investigated in two papers is similar. With the operation hours lower than the threshold value, the daily heating load is clearly less than that of continuous heating. Nevertheless, it seems to be more economical to use continuous heating mode when the operation hour is more than the threshold value.

It also can be found that the variation of heating load index with internal wall configurations is consistent with the variation of threshold value with internal wall configurations, which drops in turn under Wall 1, Wall 2, Wall 3, and Wall 4. This means that with the increase of insulation level of the internal wall, the optimum heating duration decreases.

4. Conclusions

The relatively low outdoor temperature and absence of district heating leads to the intermittent heating operation mode in the HSCW climate zone of China. More heat is absorbed by building walls during the heating period due to the heat release and storage of the internal wall, while it is not required for continuous heating mode. Therefore, the heating load of the unit floor area per unit time for intermittent heating mode is larger than that of continuous heating mode. This means that the daily heating load of intermittent heating will be larger than that of continuous heating, and a threshold value of daily operation hours occur for intermittent heating.

In the present work, a three-dimensional simulation method is established to study the heating load of buildings with different intermittent heating duration and configuration of internal walls. The threshold value of daily operation hours is analyzed under four configurations of internal walls by comparing the heating load of continuous and intermittent heating. Thorough experiments on temperature distributions of thermal mass are carried out to validate the simulation model.

In the intermittent heating room, the temperature of building envelope increases by storing energy during heating period, and the inside layer cools to a lower temperature during heating cessation period. As a result, more energy is used to heat the inside layer of the walls at the starting time of the heating cycle. This leads to a lager heating load index for intermittent heating system than continuous heating system with increasing rate ranging from 31.58% to 152.63%.

By comparing the daily heating load with operation hours, it is found that the daily heating load increases to a peak value as the operation hours increase, and then starts to decrease. This indicates that the threshold value of operation hours is 18.04 h, 15.80 h, 14.59 h and 13.46 h for four internal wall configurations, respectively. Moreover, with the increase of insulation level of internal walls, the optimum heating duration decreases. In addition, the results indicate that it is more economical to use continuous heating when the daily operation hours are more than the threshold values.

It should be noted that the threshold value of daily operation hours given in this paper is obtained according to the present situation of heating in the HSCW zone. However, the heating duration per day and the amount of heated rooms are increasing with the rapid growth of the economy and continuous increase of disposable personal income. This leads to a higher average air temperature of adjoining rooms; thus, a smaller threshold value of daily operation hours is obtained, which will be discussed in future works. Moreover, just four configurations of internal wall commonly used in the HSCW zone are investigated in this paper, and the threshold value of daily operating hours varies for other configurations of the internal wall.

Author Contributions: Conceptualization, S.W. and K.Z.; Formal analysis, S.W.; Funding acquisition, K.Z.; Methodology, S.W. and K.Z.; Writing—original draft, S.W. and K.Z.; Writing—review & editing, S.W.

Funding: This research was funded by National Natural Science Foundation of China (Grant No. 51478098) and the Fundamental Research Funds for the Central Universities (Grant No. CUSF-DH-D-2017098).

Conflicts of Interest: The authors declare no conflict of interest.

Nomenclature

C_p	Specific heat capacity of air (J/(kg·°C))
Pr_t	Turbulent Prandtl number
Tu	Turbulence intensity at the inlet (%)
Re_L	Reynolds number at the inlet
k_0	Turbulent kinetic energy (m^2/s^2)
V	Local air velocity (m/s)
T	Local air temperature (°C)
T_r	Reference temperature (°C)
T_o	Outdoor air temperature (°C)
T_i	Indoor air temperature (°C)
P	Power of heating device (kW)
l	Length scale (m)
y+	Non-dimensional distance
$T_{in,i}$	Inner surface temperature of the internal wall (°C)
$T_{in,o}$	Outer surface temperature of the internal wall (°C)
h_i	Heat transfer coefficient of inner surface of the internal wall (W/(m^2·°C))
h_o	Heat transfer coefficient of outer surface of the internal wall (W/(m^2·°C))
V_0	Volume of the investigated room (m^3)
n	Air change rate of infiltration (h^{-1})
F	Floor area of the room (m^2)
Q_i	Heating load index of room (W/m^2)
Q	Daily heating load (MJ/m^2)
q_i	Inner surface heat flow of internal wall (W)
q_o	Outer surface heat flow of internal wall (W)
Greek symbols	
p	Air density (kg/m^3)
λ	Thermal conductivity (W/(m·°C))
τ_o	Heating duration per operation (h)

τ	Hours contained in a heating cycle (h)
ν	Kinematic viscosity (m^2/s)
ν_t	Turbulent eddy viscosity (m^2/s)
δ_{ij}	Kronecker delta
$\varepsilon_{r,m,s}$	Root-mean-square error in temperature
η	Increasing rate (%)

References

1. Yoshino, H.; Yoshino, Y.; Zhang, Q.; Mochida, A.; Li, N.; Li, Z.; Miyasaka, H. Indoor thermal environment and energy saving for urban residential buildings in China. *Energy Build.* **2006**, *38*, 1308–1319. [CrossRef]
2. Hu, T.; Yoshino, H.; Jiang, Z. Analysis on urban residential energy consumption of Hot Summer & Cold Winter Zone in China. *Sustain. Cities Soc.* **2013**, *6*, 85–91.
3. Guo, S.; Yan, D.; Peng, C.; Cui, Y.; Zhou, X.; Hu, S. Investigation and analyses of residential heating in the HSCW climate zone of China: Status quo and key features. *Build. Environ.* **2015**, *94*, 532–542. [CrossRef]
4. Yu, J.; Yang, C.; Tian, L. Low-energy envelope design of residential building in hot summer and cold winter zone in China. *Energy Build.* **2008**, *40*, 1536–1546. [CrossRef]
5. *Code for Thermal Design of Civil Building*; The People's Republic of China National Standard GB 50176-2016; China Architecture and Building Press: Beijing, China, 2016. (In Chinese)
6. Yoshino, H.; Guan, S.; Lun, Y.F.; Mochida, A.; Shigeno, T.; Yoshino, Y.; Zhang, Q.Y. Indoor thermal environment of urban residential buildings in China: Winter investigation in five major cities. *Energy Build.* **2004**, *36*, 1227–1233. [CrossRef]
7. Hu, S.; Yan, D.; Cui, Y.; Guo, S. Urban residential heating in hot summer and cold winter zones of China-Status, modeling, and scenarios to 2030. *Energy Policy* **2016**, *92*, 158–170. [CrossRef]
8. Le, J.; Heiselberg, P. Energy flexibility of residential buildings using short term heat storage in the thermal mass. *Energy* **2016**, *111*, 991–1002.
9. Badran, A.A.; Jaradat, A.W.; Bahbouh, M.N. Comparative study of continuous versus intermittent heating for local residential building: Case studies in Jordan. *Energy Convers. Manag.* **2013**, *65*, 709–714. [CrossRef]
10. Tsilingiris, P.T. Wall heat loss from intermittently conditioned space-The dynamic influence of structural and operational parameters. *Energy Build.* **2006**, *38*, 1022–1031. [CrossRef]
11. Meng, X.; Luo, T.; Gao, Y.; Zhang, L.; Huang, X.; Hou, C.; Shen, Q.; Long, E. Comparative analysis on thermal performance of different wall insulation forms under the air-conditioning intermittent operation in summer. *Appl. Therm. Eng.* **2018**, *130*, 429–438. [CrossRef]
12. Zhang, L.; Luo, T.; Meng, X.; Wang, Y.; Hou, C.; Long, E. Effect of the thermal insulation layer location on wall dynamic thermal response rate under the air-conditioning intermittent operation. *Case Stud. Therm. Eng.* **2017**, *10*, 79–85. [CrossRef]
13. Yuan, L.; Kang, Y.; Wang, S.; Zhong, K. Effects of thermal insulation characteristics on energy consumption of buildings with intermittently operated air-conditioning systems under real time varying climate conditions. *Energy Build.* **2017**, *155*, 559–570. [CrossRef]
14. *ANSYS FLUENT, 14.0 Theory Guide*; ANSYS: Canonsburg, PA, USA, 2011.
15. Gebremedhin, K.G.; Wu, B.X. Characterization of flow field in a ventilation space and simulation of heat exchange between cows and their environment. *J. Therm. Biol.* **2003**, *28*, 301–319. [CrossRef]
16. Coussirat, M.; Guardo, A.; Jou, E.; Egusquiza, E.; Cuerva, E.; Alavedra, P. Performance and influence of numerical sub-models on the CFD simulation of free and forced convection in double-glazed ventilated façades. *Energy Build.* **2008**, *40*, 1781–1789. [CrossRef]
17. Rohdin, P.; Moshfegh, B. Numerical predictions of indoor climate in large industrial premises. A comparison between different k-ε models supported by filed measurements. *Build. Environ.* **2007**, *42*, 3872–3882. [CrossRef]
18. Meng, X.; Yan, B.; Gao, Y.; Wang, J.; Zhang, W.; Long, E. Factors affecting the in situ measurement accuracy of the wall heat transfer coefficient using the heat flow meter method. *Energy Build.* **2015**, *86*, 754–765. [CrossRef]
19. Ozel, M. Thermal performance and optimum insulation thickness of building walls with different structure materials. *Appl. Therm. Eng.* **2011**, *31*, 3854–3863. [CrossRef]

20. Casalegno, A.; Antonellis, S.D.; Colombo, L.; Rinaldi, F. Design of an innovative enthalpy wheel based humidification system for polymer electrolyte fuel cell. *Int. J. Hydrogen Energy* **2011**, *36*, 5000–5009. [CrossRef]
21. Ye, X.; Kang, Y.; Zuo, B.; Zhong, K. Study of factors affecting warm air spreading distance in impinging jet ventilation rooms using multiple regression analysis. *Build. Environ.* **2017**, *120*, 1–12. [CrossRef]
22. Ministry of Housing and Urban-Rural Development of the People's Republic of China. *Design Standard for Energy Efficiency of Residential Buildings in Hot Summer and Cold Winter Zone (JGJ134-2010)*; China Architecture and Building Press: Beijing, China, 2010. (In Chinese)
23. *ANSYS FLUENT 14.0 User's Guide*; ANSYS: Canonsburg, PA, USA, 2011.
24. Hussain, S.; Oosthuizen, P.H.; Kalendar, A. Evaluation of various turbulence models for the prediction of the airflow and temperature distributions in atria. *Energy Build.* **2012**, *48*, 18–28. [CrossRef]
25. Ye, X.; Zhu, H.; Kang, Y.; Zhong, K. Heating energy consumption of impinging jet ventilation and mixing ventilation in large-height space: A comparison study. *Energy Build.* **2016**, *130*, 697–708. [CrossRef]
26. Lin, B.; Wang, Z.; Liu, Y.; Zhu, Y.; Ouyang, Q. Investigation of winter indoor thermal environment and heating demand of urban residential buildings in China's hot summer-cold winter climate region. *Build. Environ.* **2016**, *101*, 9–18. [CrossRef]

![applied sciences logo] *applied sciences*

MDPI

Article

Risk-Constrained Optimal Chiller Loading Strategy Using Information Gap Decision Theory

Er Shi [1], Farkhondeh Jabari [2], Amjad Anvari-Moghaddam [3,*], Mousa Mohammadpourfard [4] and Behnam Mohammadi-ivatloo [2]

[1]   School of Energy and Power Engineering, Changsha University of Science and Technology, Changsha 410114, China; shier@csust.edu.cn
[2]   Faculty of Electrical and Computer Engineering, University of Tabriz, Tabriz 5166616471, Iran; f.jabari@tabrizu.ac.ir (F.J.); bmohammadi@tabrizu.ac.ir (B.M.-i.)
[3]   Department of Energy Technology, Power Electronic Systems, 9220 Aalborg, Denmark
[4]   Faculty of Chemical and Petroleum Engineering, University of Tabriz, Tabriz 5166616471, Iran; mohammadpour@tabrizu.ac.ir
*   Correspondence: aam@et.aau.dk

Received: 14 February 2019; Accepted: 8 May 2019; Published: 10 May 2019

Abstract: This paper presents a novel framework for economic cooling load dispatch in conventional water-cooled chillers. Moreover, information gap decision theory (IGDT) is applied to the optimal chiller loading (OCL) problem to find the optimum operating point of the test system in three decision-making modes: (a) risk-neutral approach, (b) risk-aversion or robustness approach, and (c) risk-taker or opportunistic approach. In the robustness mode of the IGDT-based OCL problem, the system operator enters a desired energy cost value in order to find the most appropriate loading points for the chillers so that the total electricity procurement cost over the study horizon is smaller than or equal to this critical value. Meanwhile, the cooling load increase is maximized to the highest possible level to find the most robust performance of the benchmark grid with respect to the overestimated load. Similarly, the risk-taker optimization method finds the on/off status and the partial load ratio (PLR) of the chillers in order to keep the total energy cost as low as the given cost function. In addition, the minimum value of cooling load decrease can be found while satisfying the refrigeration capacity of the chiller and the load-generation balance constraint. Thus, a mixed-integer non-linear programming problem is solved using the branch and reduce optimization (BARON) tool of the generalized algebraic mathematical modeling system (GAMS) for a five-chiller plant, to demonstrate that IGDT is able to find a good solution in robustness/risk-taker OCL problem.

Keywords: optimal chiller loading (OCL); uncertain cooling demand; information gap decision theory (IGDT); mixed-integer non-linear programming problem (MINLP)

---

1. Introduction

In summer, different end users, such as residential and commercial sectors, consume more electricity for building space cooling. This may lead to an energy crisis and cascading power outages [1,2]. Therefore, the economic operation of electrical air conditioners is important to reduce the energy demand of interconnected power systems. The optimal short-term scheduling of multiple-chiller systems is a cost-effective tool to minimize the total energy cost and power consumption of multiple-chiller plants [3,4]. The main objective of the economic chiller dispatch problem is to minimize the power consumption of the chillers while satisfying the cooling load-generation balance constraint and the refrigeration capacity of the chiller units [5,6]. The partial load ratio, refrigeration production, and electrical power consumption of the chillers have been selected as the decision variables of the optimization problem. Moreover, the binary variables that show the on or off status of the chillers are used to determine which ones are turned on at each operating time interval. The variable climatic conditions affect the building cooling demand, the optimum value of the partial load ratios (PLRs), and the cooling capability of the chillers, as well as their power consumption [7]. Therefore, the uncertainties associated with the cooling load should be modeled by short-term scheduling of electrical air conditioners [8,9].

Recently, researchers have presented fast optimization algorithms for solving the optimal chiller loading (OCL) problem. In [10], the branch and bound method was proposed to find the best values of the PLR of the chillers and minimize the power consumption of the water coolers. Chang et al. [11] proved that the gradient method achieves the optimum scenario with less calculation time and better objective function than the Lagrangian approach. The authors of [12] demonstrated that if the particle swarm optimization (PSO) is integrated with the neural networks, the power consumption of the chillers will be 18% less than that achieved by linear regression and the equal loading distribution method. The simulated annealing method, which is used for heating a specific metal to its melting temperature, reducing shape defects, and cooling the modified metal, provides more accurate solutions than the Lagrangian approach [13]. Coelho and Mariani [14] solved the OCL problem by using the Gaussian distribution function coupled with the firefly search algorithm. The firefly search algorithm is a well-known search strategy that was inspired by the behavior of fireflies, which attract mating partners based on light intensities. In this model, it is assumed that all fireflies except one are of the same sex and only the firefly of the different sex can be attracted by the others. Other search algorithms, such as evolution strategy [15], teaching learning procedure [16], cuckoo search algorithm [17], differential evolution method [18], exchange market strategy [19,20], and basic open-source non-linear mixed-integer programming [21], provide global optimal solutions with a lower computational burden in less time than the genetic algorithm [22,23]. Lo et al. [24] introduced a novel OCL strategy based on non-linear ripple weight indices and self-adaption repulsion factors, known as the ripple bee swarm optimization technique, but invasive weed optimization [25] is able to find better operating points in the three test systems than those obtained by this technique. Saeedi et al. [26] applied an interval robust optimization algorithm to the OCL problem in order to model the uncertainty of the cooling demand. Minimum and maximum forecasted values of cooling load over a 24-h study horizon were considered in order to minimize the total electricity requirement of a three-chiller standard grid. The partial load ratio, refrigeration production, and power consumption of the chillers in three scenarios—(a) minimum cooling load profile, (b) forecasted demand, and (c) maximum load level—were compared. Recent studies on the possible use of thermoacoustic refrigerators have provided quite significant results, which may lead to a considerable reduction in environmental pollutants and, in the future, a reduction in costs [27] due to the use of new smart-window technologies that reduce the cooling load [28].

Different search methods have been proposed for solving the OCL problem and saving energy in air conditioning systems, but the uncertainty of the cooling demand has only been discussed in [26]. Two consecutive intervals have been considered for modeling the underestimated and overestimated cooling loads in the robust optimization approach. Meanwhile, information gap decision theory (IGDT) could be applied to economic dispatch problems in order to model both risk-aversion or robustness and risk-taker or opportunistic aspects of risk-constrained OCL strategy. This paper implements the IGDT algorithm on a benchmark five-chiller plant in order to solve a mixed-integer non-linear programming problem with a generalized algebraic mathematical modeling system (GAMS). In the risk-aversion OCL problem, the system operator enters a critical cost value. The cooling load increase is then maximized in such a way that the sum of the refrigeration production of the chillers is higher than or equal to the overestimated cooling demand, and the daily energy cost of the test system is smaller than the given cost function. In other words, the total electricity cost over the study horizon is not considered to be the objective function. It is restricted so that it is smaller than the critical cost function. The maximum value of the cooling demand increase that can be satisfied by chillers is then calculated as the objective function. In the risk-taker or opportunistic decision-making process, the minimum value of the cooling load decrease is calculated in order to reduce the total energy cost so that it is as low as the given cost function. The refrigeration capacity constraints of the chillers and load-generation balance criterion are also modeled in robustness and risk-taker modes. The on/off status, partial load ratio, cooling generation, and electricity consumption of each chiller are selected as the decision variables that are found in the base case study and the robustness and opportunistic OCL problem. In all reviewed works, the total power consumption of the chiller plants was minimized as the objective function, but real-time electricity prices should be considered for minimization of the total energy cost of multi-chiller systems. The novel contributions of this paper can be summarized as follows:

- The IGDT method was applied to the OCL problem to model the uncertainty of the cooling demand.
- The robustness or risk-aversion decision-making approach was used for maximizing the value of the cooling load increase while minimizing the daily energy cost so that it was as low as the critical energy cost.
- The opportunistic or risk-taker OCL strategy was used to find the best operating point of the chillers so that the minimum value of the cooling load decrease that results in target cost saving was found.

2. Proposed IGDT-Based OCL Strategy

Figure 1 shows the single-line diagram of a conventional multi-chiller plant. In this system, the total electricity cost is minimized over the T-hour study horizon, as given by Equation (1), where λ_t and P_i^t are the electricity price and the power consumption of chiller i at operating time interval t, which can be calculated from Equations (2)–(4), where u_i^t is a binary decision variable that is equal to 1 if the ith chiller is on at hour t; otherwise, it will be 0. Based on Equation (3), the partial load ratio of chiller i is selected as another decision variable and represents the cooling capacity constraint. As is evident from Equation (4), when chiller i is on at hour t, its electrical power consumption depends on the constant coefficients a_i, b_i, c_i, and d_i and its cooling production; otherwise, it will be 0. Equation (5) demonstrates that the refrigeration capability of the chillers should be equal to or larger than the cooling demand at time interval t. Note that CL_t and RT_i denote the cooling load at hour t and the refrigeration capacity of chiller i, respectively [26].

$$\text{Energy cost} = \sum_{t=1}^{T} \sum_{i=1}^{N} \left(\lambda_t \times P_i^t \right) \tag{1}$$

$$u_i^t = \left\{ \begin{array}{ll} 0 & \text{if chiller } i \text{ is off} \\ 1 & \text{if chiller } i \text{ is on} \end{array} \right\} \tag{2}$$

$$PLR_i^t = \left\{ \begin{array}{ll} 0 & \text{if chiller } i \text{ is off} \\ \frac{\text{Cooling load of chiller } i \text{ at time } t}{\text{Refrigeration capacity of chiller } i} = \text{rand} & \text{if chiller } i \text{ is on} \end{array} \right\} \tag{3}$$

$$P_i^t = \left\{ \begin{array}{ll} 0 & \text{if } u_i^t = 0 \\ a_i + b_i PLR_i^t + c_i \left(PLR_i^t \right)^2 + d_i \left(PLR_i^t \right)^3 & \text{if } u_i^t = 1 \end{array} \right\} \tag{4}$$

$$CL_t \leq \sum_{i=1}^{N} PLR_i^t \times RT_i \tag{5}$$

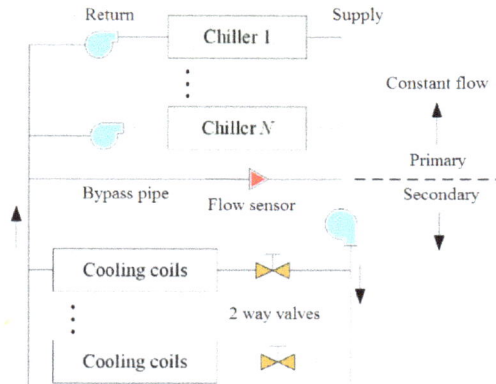

Figure 1. Single-line diagram of a conventional multi-chiller plant.

2.1. IGDT-Based OCL Problem

The main objective of the IGDT approach is to maximize the horizon of the cooling demand uncertainty, while minimizing the daily energy cost of the multiple-chiller plant to be as low as the given value. The IGDT method enables the system operator to make the appropriate and most cost-effective decisions regarding the probable fluctuations of the cooling demand. In the economic dispatching of chiller units, the cooling load may behave adversely and lead to a higher energy cost or it may behave desirably and lead to a lower electricity cost. In other words, the IGDT strategy assesses the robustness and opportunistic aspects of the air conditioning process by modeling the unexpected variations of the cooling load using three components: (a) system model, (b) performance requirement, and (c) uncertainty model.

2.1.1. System Model

It is presumed that the cooling demand, CL_t, is an uncertain parameter, and it may be increased or decreased at each time interval t. Moreover, the input–output model of the multi-chiller system is shown as the energy cost function $F\left(x_i^t, CL_t\right)$, where x_i^t denotes the decision variables of the optimization problem, which include the on/off status, partial load ratio, cooling production, and electrical power consumption of ith chiller at hour t. The daily energy cost function, $F\left(x_i^t, CL_t\right)$, should be minimized to be as low as possible.

2.1.2. Performance Requirement

The expectations of the system operator regarding the energy cost function are evaluated using the robustness and opportunistic strategies, Equations (6) and (7), respectively. According to Equation (6), the robust optimization problem is formulated with an aim to maximize the uncertainty variable, α, while the energy cost is less than the given cost, F_k. The uncertainty variable, α, is maximized as per Equation (8). The system operator makes the robustness decision with less sensitivity to the variations of the uncertain parameter, CL_t. Moreover, the energy cost over the study time interval T is smaller than the predefined critical cost F_k. The robustness function, $\hat{\alpha}$, investigates how robust the multi-chiller system is against the possible increase in cooling demand. A risk-taker decisionmaker requests a lower energy cost by implementing the opportunity mode of the IGDT algorithm. As formulated by Equation (9), the variable β is the minimum value of α with the aim of achieving lower costs for the decision variables, x_i^t. Note that F_w represents the maximum cost of the opportunity strategy, which is determined by the system operator, for paying less under the favorable changes of the uncertain cooling demand, CL_t.

$$\hat{\alpha} = \underset{\alpha}{\text{Max}}\{\text{Maximum energy cost is lower than a predefined critical cost}\} \tag{6}$$

$$\hat{\alpha} = \underset{\alpha}{\text{Max}}\{\text{Maximum energy cost is lower than a predefined critical cost}\} \tag{7}$$

$$\hat{\alpha}\left(x_i^t, F_k\right) = \underset{\alpha}{\text{Max}}\left\{\alpha : \underset{x_i^t}{\text{Max}}F\left(x_i^t, CL_t\right) \le F_k\right\} \tag{8}$$

$$\hat{\beta}\left(x_i^t, F_w\right) = \underset{\beta}{\text{Min}}\left\{\alpha : \underset{x_i^t}{\text{Min}}F\left(x_i^t, CL_t\right) \le F_w\right\} \tag{9}$$

2.1.3. Uncertainty Model in Risk-Aversion or Robustness Mode

The robustness variable, $\hat{\alpha}\left(x_i^t, F_k\right)$, is defined for the risk-aversion decision-making approach and evaluates the greatest value of the uncertainty variable, α, when the maximum cost is smaller than the predefined cost, F_k. The larger $\hat{\alpha}\left(x_i^t, F_k\right)$ indicates higher robustness against uncertainty. Therefore, $\hat{\alpha}\left(x_i^t, F_k\right)$ will increase as F_k increases, and vice versa. The uncertain cooling demand can be calculated by Equation (10). The increasing rate of the cooling load causes an increase in the energy cost function, which is evident from Equation (10). According to Equation (11), the objective is to maximize α for the given critical cost, F_k. The parameter CL_t^0 represents the forecasted cooling demand at operating time interval t.

$$CL_t = (1 + \alpha) \times CL_t^0 \tag{10}$$

$$\hat{\alpha}\left(Q_i^t, P_i^t, F_k\right) = \underset{\alpha}{\text{Max}}\left\{\alpha : \underset{Q_i^t, P_i^t}{\text{Max}} \sum_{t=1}^{T} \sum_{i=1}^{N} \left(\lambda_t \times P_i^t\right) \le F_k\right\} \tag{11}$$

2.1.4. Uncertainty Model in Opportunistic Decision-Making Strategy

The opportunity function, $\hat{\beta}\left(x_i^t, F_w\right)$, assesses the feasibility of the lower costs. Therefore, a small value of $\hat{\beta}\left(x_i^t, F_w\right)$ is desirable. According to Equation (12), the opportunity variable is the lowest value of α for minimization of energy cost as low as F_w. Therefore, it is expected that $\hat{\beta}\left(x_i^t, F_w\right)$ increases with the reduction of F_w for the energy cost minimization approach as in Equations (13) and (14). As expected from Equation (13), the decreasing rate of the cooling demand causes a decrease in electricity cost.

$$\hat{\beta}\left(x_i^t, F_w\right) = \underset{x_i^t}{\text{Min}}\hat{\alpha}\left(x_i^t, F_w\right) \tag{12}$$

$$CL_t = (1 - \alpha) \times CL_t^0 \tag{13}$$

$$\hat{\beta}\left(Q_i^t, P_i^t, F_k\right) = \underset{\alpha}{\text{Min}}\left\{\alpha : \underset{Q_i^t, P_i^t}{\text{Min}} \sum_{t=1}^{T} \sum_{i=1}^{N} \left(\lambda_t \times P_i^t\right) \le F_w\right\} \tag{14}$$

3. Numerical Result and Discussions

A mixed-integer non-linear program (MINLP) was developed by GAMS software [29] and was solved using a branch and reduce optimization navigator (BARON) tool [30]. A benchmark multi-chiller system with two 550 and three 1000 RT chillers [31] was used for simulations. The power consumption coefficients and the cooling capacity of the chillers are presented in Table 1. The forecasted value of the cooling load over a 24-h study horizon [31] is illustrated in Figure 2. The hourly variations of the electricity prices [32] are shown in Figure 3.

Table 1. Power consumption coefficients and cooling capacity of chillers.

Chiller	a_i	b_i	c_i	d_i	Q_i^{max} (RT)
1	57.2	329.73	0.05	7.85	550
2	50.09	419.28	−123.8	76.36	550
3	−76.29	1226.94	−709.37	296.93	1000
4	−72.56	1100.42	−145.77	−137.1	1000
5	−186.18	1817.08	−1755.59	847.43	1000

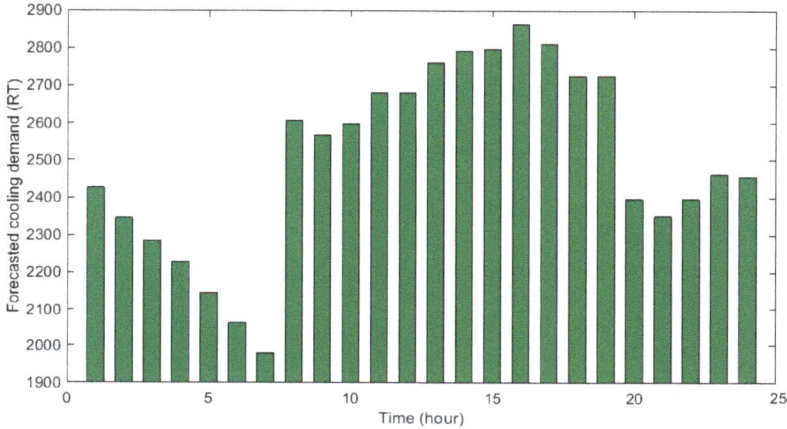

Figure 2. Forecasted value of cooling demand over a sample 24-h time interval.

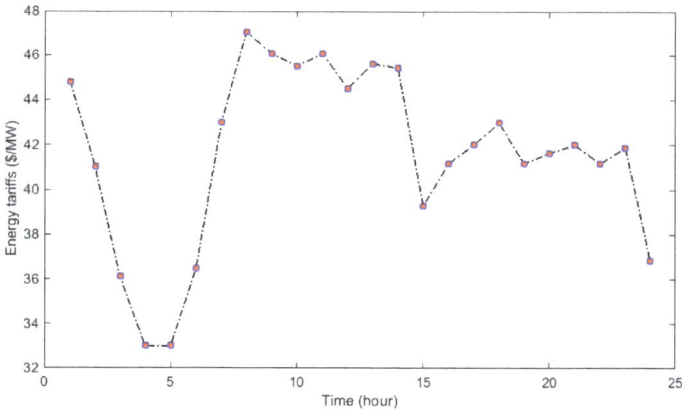

Figure 3. Hourly variations of energy rates.

First, the robustness OCL problem was solved using the GAMS optimization software in conjunction with the different values of the critical energy costs. The risk-aversion optimization problem was performed on the test system, taking Equations (1)–(5), (10), and (11) into consideration. The system operator then determined the value of the known parameter F_k for 12 iterations ($k = 12$). As mentioned in Section 2.1.3, the main objective of the robustness mode at each scenario k was to maximize the percentage of the cooling load increase by increasing α in such a way that the maximum value of the daily energy cost is smaller than or equal to the predefined critical cost F_k. Meanwhile, the cooling capacity of the chillers was satisfactory at each solution. In addition, the total refrigeration production of the chiller units was more than or equal to the overestimated cooling demand, or

$(1 + \alpha) \times CL_t^0$. Figure 4 shows the optimum value of the cooling demand increase at each iteration of the robustness economic dispatch problem. The vertical axis represents the predefined critical energy cost, F_k, at each iteration k. The horizontal axis refers to the maximum value of the cooling load increase, which is obtained by solving the robust short-term scheduling problem. As expected from Equations (3)–(5), (10), and (11), when the building cooling load increased, the value of the refrigeration production of the chillers increased, as did the electrical power consumption of the chillers. Figure 4 shows the optimum value of the cooling load increase at each scenario of the robust economic dispatch problem. It is clear that the daily electricity cost increased, a result which is to be expected as per Equation (1). In the opportunistic optimal cooling energy procurement strategy, the target energy cost, F_w, is lower than that obtained from solving the base case study (Equations (1)–(5)) without considering the uncertainty of the cooling load. In other words, the system operator wants the daily electricity purchasing cost to be smaller than the energy cost of the multi-chiller system under $\alpha = \beta = 0$ or the base optimization problem. The system owner wants to reduce the daily energy cost of the five-chiller plant so that it is as low as the target energy cost, F_w. Hence, the minimum value of the cooling load decrease, which reduces the energy cost to predefined values F_w, must be found by solving the optimization problem (Equations (1)–(5), (13) and (14)). As shown in Figure 5, eight scenarios were considered for the risk-taker decision-making approach ($w = 8$). As expected from Equations (5) and (13) and displayed in Figure 5, when the cooling demand decreased, the value of the cooling production of the chillers also decreased. Therefore, they consumed less electrical power while supplying this underestimated cooling demand. The day-ahead energy cost was also reduced, as predicted by Equations (1)–(4).

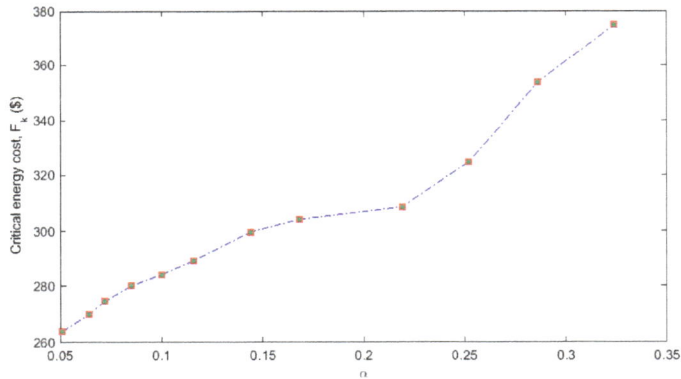

Figure 4. The optimum value of the cooling load increase at each scenario of the robust economic dispatch problem.

Figures 6 and 7 show the daily electricity consumption of the chillers in three cases: (a) risk-neutral, (b) risk-aversion or robustness OCL problem, and (c) risk-taker or opportunistic economic dispatch process. As mentioned before, when the values of the robustness and opportunity factors are equal to zero, $\alpha = \beta = 0$ and the base OCL problem is solved in accordance with the forecasted cooling load profile (Figure 2). Hence, the daily electricity cost will be equal to $254.5 as a result of 6097.69 kW power consumption by the chillers, as seen in Figures 6 and 7. If the system owner wants to pay the higher energy costs and procure greater electrical power for the chillers, the maximum value of the cooling demand increase can be found, as shown in Figure 6. This figure enables the system operator to know how much the cooling demand can be increased to meet the specific value of the energy cost. It is also possible to find the minimum value of the cooling load decrease in order to save the electrical power a certain value, as demonstrated in Figure 7.

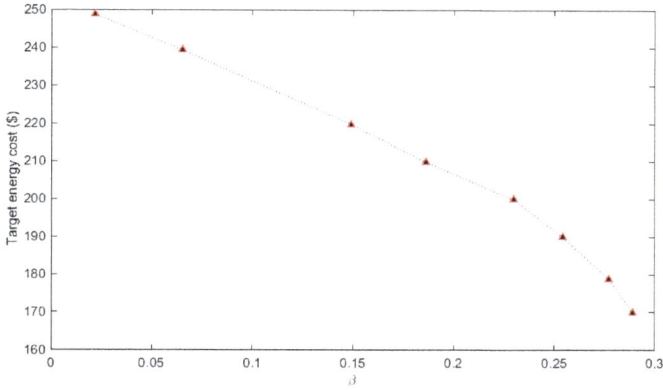

Figure 5. The minimum value of the cooling load decrease at each iteration of the opportunistic optimal chiller loading (OCL) strategy.

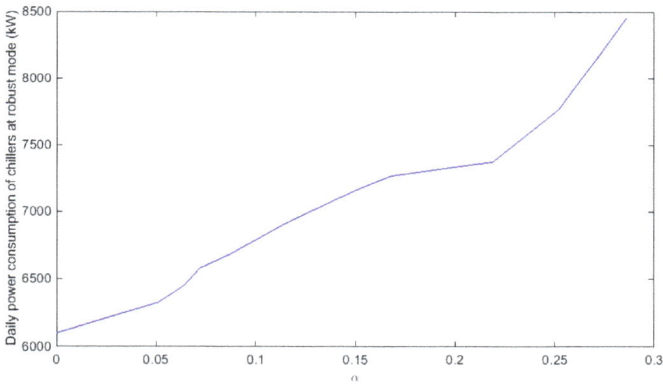

Figure 6. Daily power consumption of chillers versus α.

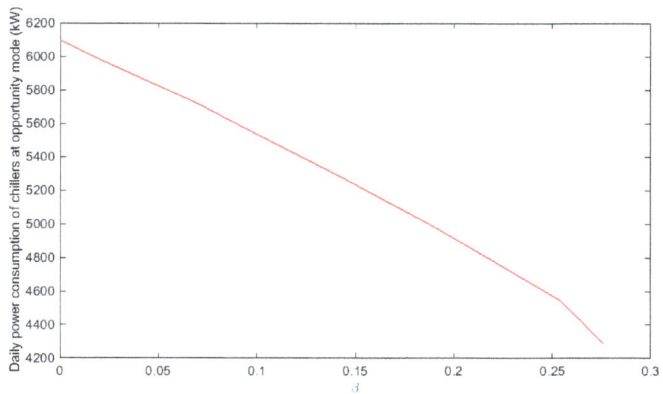

Figure 7. Daily energy requirement of chiller units versus β.

The PLR cooling generation and the electrical power consumption of the chillers in the three cases are shown in Figure 8a–c, respectively. Figure 8a depicts the PLRs of the chillers versus the critical (F_k) and target (F_w) costs. The orange line marks the boundary between the opportunistic and robustness modes. At hour 7, the PLR and the power consumption of chiller 2 will be equal to zero in all robustness and risk-taker scenarios. The optimization problem detects that if chiller 2 is off and does not operate at hour 7, the total energy cost of the system will be less than when it turns on at this hour. It may be economic to turn on chiller 2 at hours with different cooling demand values. Figure 8 demonstrates how the cooling demand changes at hour 7, as well as how much electrical power is consumed by the chillers to supply this value. This information can be utilized when it comes to making robustness/opportunistic decisions with respect to the cooling load increase/decrease. Figure 8 also allows the system operator to see the value of the power consumption of the chillers and daily energy cost when the cooling load at hour 7 changes from 1400 to 2600 RT. As seen in Figure 2, the forecasted cooling demand at hour 7 in the deterministic or risk-neutral optimization problem, without considering the uncertainty of the cooling demand, is equal to 1979 RT, which is shown with a vertical line. If the cooling demand is less than 1979 RT (opportunistic zone), the system owner gains more energy cost savings from the underestimated cooling load. In the same manner, when the cooling demand increases from 1979 RT (robust mode), the daily energy cost increases. Hence, the value of the power consumption of the chillers will change according to the left and right sides of the vertical boundary line, designated as "Opportunity" and "Robust". For example, if the opportunity factor is equal to 0.289 and the cooling demand is $(1 - 0.289) \times 1979 = 1407$ RT in opportunistic operating mode, the value of the power consumption of chillers 1, 3, 4, and 5 will be equal to 65, 68, 107, and 0 kW, respectively. However, if the cooling demand at this hour increases to 1431 RT, their power consumption will change to 0, 68, 0, and 101 kW, respectively. Even though the cooling demand increases, it will still be more economic to turn off chillers 1 and 4 and turn on chiller 5. Similarly, if the cooling demand increases from 1979 RT in the deterministic or risk-neutral optimization problem to 2478 RT ($\alpha = 0.252$, $CL_{t=7} = (1 - 0.252) \times 1979 = 2478$) in the robust economic dispatch strategy, the electricity consumption of chiller 1 changes from 0 to 65 kW. The IGDT-based uncertainty modeling approach enables the operator to make appropriate decisions in order to optimize the system's operation with respect to possible changes in cooling load. Similar analysis can be considered for other hours with different values of cooling load.

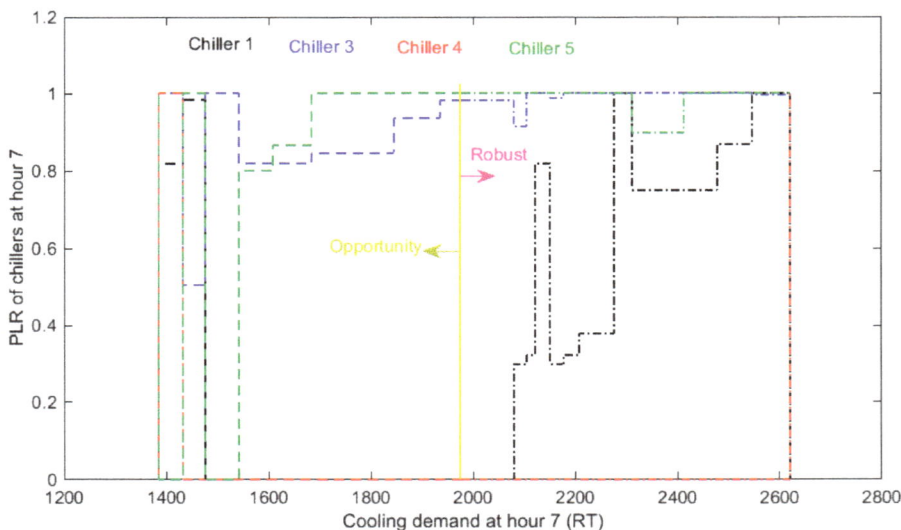

(**a**) Partial load ratio (PLR) of chillers at hour 7.

Figure 8. *Cont.*

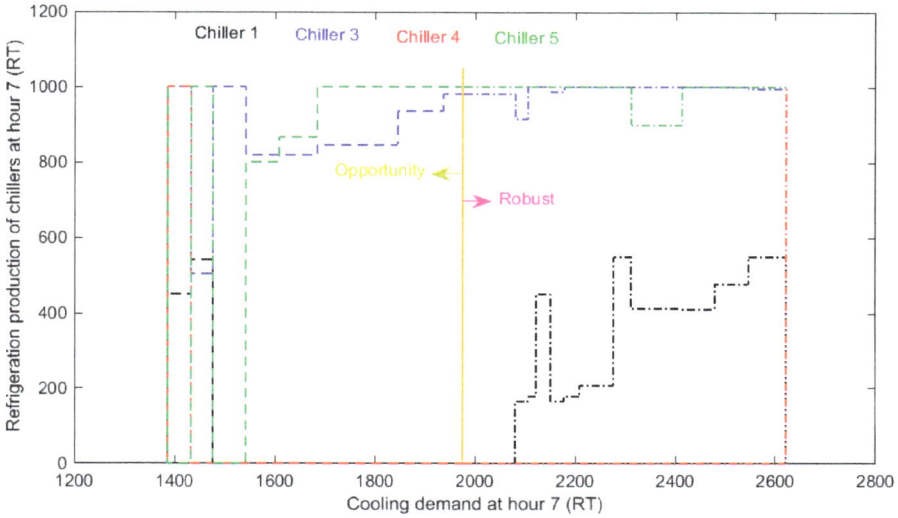

(b) Refrigeration production of chillers at hour 7.

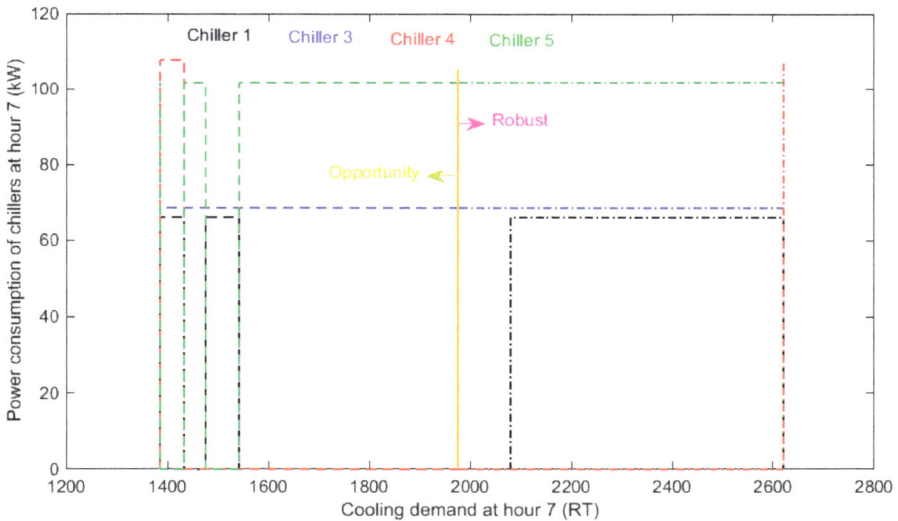

(c) Electrical power consumption of chillers at hour 7.

Figure 8. The optimal operating points of the chillers at hour 7 (a) PLR, (b) cooling generation and (c) power consumption.

4. Conclusions

This paper proposed a novel framework for modeling the uncertain nature of cooling demand in the day-ahead optimal scheduling of multiple-chiller systems. The information gap decision theory approach was used for finding the robustness or risk-aversion solution against the maximum cooling load increase. Moreover, the opportunity for cost and energy saving under the minimum value of the load decrease was maximized. The presented model was shown to enable a system operator to input the desired energy cost parameter and minimize the daily energy cost to this critical value and also to determine the maximum cooling demand increase in the robustness model. In addition, a risk-taker

Appl. Sci. **2019**, *9*, 1925

decisionmaker can schedule the daily refrigeration generation patterns of the chillers so that the daily energy cost of the plant is smaller than the predefined target cost. The minimum value of the cooling demand can also be calculated and reduced for favorable cost saving.

Author Contributions: Conceptualization, F.J.; methodology, B.M.-i; software, F.J.; validation, M.M.; formal analysis, F.J.; investigation, M.M.; resources, A.A.-M. and B.M.-i; writing—original draft preparation, F.J.; writing—review and editing, A.A.-M. and E.S.; visualization, E.S.; and supervision, B.M.-i and M.M.

Conflicts of Interest: The authors declare no conflict of interest.

References

1. Zheng, Z.-X.; Li, J.-Q.; Duan, P.-Y. Optimal chiller loading by improved artificial fish swarm algorithm for energy saving. *Math. Comput. Simul.* **2019**, *155*, 227–243. [CrossRef]
2. Sulaiman, M.H.; Ibrahim, H.; Daniyal, H.; Mohamed, M.R. A new swarm intelligence approach for optimal chiller loading for energy conservation. *Procedia Soc. Behav. Sci.* **2014**, *129*, 483–488. [CrossRef]
3. Wang, Y.; Jin, X.; Du, Z.; Zhu, X. Evaluation of operation performance of a multi-chiller system using a data-based chiller model. *Energy Build.* **2018**, *172*, 1–9. [CrossRef]
4. Powell, K.M.; Cole, W.J.; Ekarika, U.F.; Edgar, T.F. Optimal chiller loading in a district cooling system with thermal energy storage. *Energy* **2013**, *50*, 445–453. [CrossRef]
5. Wang, Y.; Jin, X.; Shi, W.; Wang, J. Online chiller loading strategy based on the near-optimal performance map for energy conservation. *Appl. Energy* **2019**, *238*, 1444–1451. [CrossRef]
6. Chang, Y.-C. Optimal chiller loading by evolution strategy for saving energy. *Energy Build.* **2007**, *39*, 437–444. [CrossRef]
7. Hallegatte, S. Strategies to adapt to an uncertain climate change. *Glob. Environ. Chang.* **2009**, *19*, 240–247. [CrossRef]
8. Huang, P.; Huang, G.; Augenbroe, G.; Li, S. Optimal configuration of multiple-chiller plants under cooling load uncertainty for different climate effects and building types. *Energy Build.* **2018**, *158*, 684–697. [CrossRef]
9. Ho, H.C.; Knudby, A.; Sirovyak, P.; Xu, Y.; Hodul, M.; Henderson, S.B. Mapping maximum urban air temperature on hot summer days. *Remote Sens. Environ.* **2014**, *154*, 38–45. [CrossRef]
10. Chang, Y.-C.; Lin, F.-A.; Lin, C.H. Optimal chiller sequencing by branch and bound method for saving energy. *Energy Convers. Manag.* **2005**, *46*, 2158–2172. [CrossRef]
11. Chang, Y.-C.; Chan, T.-S.; Lee, W.-S. Economic dispatch of chiller plant by gradient method for saving energy. *Appl. Energy* **2010**, *87*, 1096–1101. [CrossRef]
12. Chen, C.-L.; Chang, Y.-C.; Chan, T.-S. Applying smart models for energy saving in optimal chiller loading. *Energy Build.* **2014**, *68*, 364–371. [CrossRef]
13. Chang, Y.-C. An innovative approach for demand side management—Optimal chiller loading by simulated annealing. *Energy* **2006**, *31*, 1883–1896. [CrossRef]
14. Coelho, L.d.S.; Mariani, V.C. Improved firefly algorithm approach applied to chiller loading for energy conservation. *Energy Build.* **2013**, *59*, 273–278. [CrossRef]
15. Chang, Y.-C.; Lee, C.-Y.; Chen, C.-R.; Chou, C.-J.; Chen, W.-H.; Chen, W.-H. Evolution strategy based optimal chiller loading for saving energy. *Energy Convers. Manag.* **2009**, *50*, 132–139. [CrossRef]
16. Rao, R.V. Optimization of Multiple Chiller Systems Using TLBO Algorithm. In *Teaching Learning Based Optimization Algorithm*; Springer: Cham, Switzerland, 2016; pp. 115–128.
17. Coelho, L.d.S.; Klein, C.E.; Sabat, S.L.; Mariani, V.C. Optimal chiller loading for energy conservation using a new differential cuckoo search approach. *Energy* **2014**, *75*, 237–243. [CrossRef]
18. Lee, W.-S.; Chen, Y.-T.; Kao, Y. Optimal chiller loading by differential evolution algorithm for reducing energy consumption. *Energy Build.* **2011**, *43*, 599–604. [CrossRef]
19. Sohrabi, F.; Nazari-Heris, M.; Mohammadi-Ivatloo, B.; Asadi, S. Optimal chiller loading for saving energy by exchange market algorithm. *Energy Build.* **2018**, *169*, 245–253. [CrossRef]
20. Cannistraro, M.; Mainardi, E.; Bottarelli, M. Testing a dual-source heat pump. *Math. Model. Eng. Probl.* **2018**, *5*, 205–210. [CrossRef]

21. Jabari, F.; Mohammadi-Ivatloo, B. Basic Open-Source Nonlinear Mixed Integer Programming Based Dynamic Economic Dispatch of Multi-chiller Plants. In *Operation, Planning, and Analysis of Energy Storage Systems in Smart Energy Hubs*; Springer: Cham, Switzerland, 2018; pp. 121–127.

22. Chang, Y.-C.; Lin, J.-K.; Chuang, M.-H. Optimal chiller loading by genetic algorithm for reducing energy consumption. *Energy Build.* **2005**, *37*, 147–155. [CrossRef]

23. Chang, Y.-C. Genetic algorithm based optimal chiller loading for energy conservation. *Appl. Therm. Eng.* **2005**, *25*, 2800–2815. [CrossRef]

24. Lo, C.-C.; Tsai, S.-H.; Lin, B.-S. Economic dispatch of chiller plant by improved ripple bee swarm optimization algorithm for saving energy. *Appl. Therm. Eng.* **2016**, *100*, 1140–1148. [CrossRef]

25. Zheng, Z.-X.; Li, J.-Q. Optimal chiller loading by improved invasive weed optimization algorithm for reducing energy consumption. *Energy Build.* **2018**, *161*, 80–88. [CrossRef]

26. Saeedi, M.; Moradi, M.; Hosseini, M.; Emamifar, A.; Ghadimi, N. Robust optimization based optimal chiller loading under cooling demand uncertainty. *Appl. Ther. Eng.* **2019**, *148*, 1081–1091. [CrossRef]

27. Piccolo, A.; Siclari, R.; Rando, F.; Cannistraro, M. Comparative performance of thermoacoustic heat exchangers with different pore geometries in oscillatory flow. Implementation of experimental techniques. *Appl. Sci.* **2017**, *7*, 784. [CrossRef]

28. Cannistraro, M.; Castelluccio, M.E.; Germanò, M. New sol-gel deposition technique in the Smart-Windows–Computation of possible applications of Smart-Windows in buildings. *J. Build. Eng.* **2018**, *19*, 295–301. [CrossRef]

29. Generalized Algebraic Mathematical Modeling Systems. Available online: https://www.gams.com/ (accessed on 17 July 2015).

30. Available online: https://www.gams.com/latest/docs/S_BARON.html (accessed on 5 June 2015).

31. Lin, W.-M.; Tu, C.-S.; Tsai, M.-T.; Lo, C.-C. Optimal energy reduction schedules for ice storage air-conditioning systems. *Energies* **2015**, *8*, 10504–10521. [CrossRef]

32. Soroudi, A. *Power System Optimization Modeling in GAMS*; Springer: Cham, Switzerland, 2017.

applied
sciences

MDPI

Article

A Novel Method Based on Neural Networks for Designing Internal Coverings in Buildings: Energy Saving and Thermal Comfort

José A. Orosa [1,*], Diego Vergara [2], Ángel M. Costa [1] and Rebeca Bouzón [1]

[1] Department of N. S. and Marine Engineering, Universidade da Coruña, Paseo de Ronda, 51,
     15011 A Coruña, Spain; angel.costa@udc.es (Á.M.C.); rebeca.bouzon@udc.es (R.B.)
[2] Department of Mechanical Engineering, Catholic University of Ávila, C/Canteros, s/n, 05005 Avila, Spain;
     diego.vergara@ucavila.es or dvergara@usal.es
*   Correspondence: jarosa@udc.es or jose.antonio.orosa@udc.es; Tel.: +34-981-167-000 (ext. 4320)

Received: 5 April 2019; Accepted: 22 May 2019; Published: 25 May 2019

Featured Application: This work shows a new methodology to optimize the exact amount of permeable internal coverings in a building, thereby becoming a passive method control system for this indoor ambience.

Abstract: Although several papers define energy saving and thermal comfort optimization with internal coverings materials, none of them deal with predictive models to improve design in building constructions. Thus, artificial intelligence (AI) procedures were applied in this paper. In particular, neural networks (NNs) were designed for indoor ambiences with internal covering materials in different buildings, were trained and employed to predict indoor ambiences (indoor temperature and relative humidity as a function of weather conditions), and, based on these procedures, local thermal comfort conditions and energy consumption, due to the type of internal covering permeability level, were calculated. Results from this original methodology showed a better acceptability of indoor ambiences when permeable coating materials were used, in agreement with previous research works. At the same time, with permeable coverings, a lower energy consumption of 20% in the heating, ventilation, and air conditioning (HVAC) systems was needed to reach more comfortable conditions during the summer season in the first hours of occupation. Finally, all these results suggest an original methodology to optimize indoor ambiences based on the design of internal coverings by NN.

Keywords: novel method; internal coverings; neural networks; energy; thermal comfort; control system

1. Introduction

Internal coverings are now commonly employed in new and old buildings. In particular, public spaces, such as office buildings, are selected for internal coverings based on the principles of a good economy and easily washable surfaces. Consequently, most of the time, impermeable coverings are selected for buildings due to their high durability and ease of cleaning. On the other hand, in the last decade, permeable internal coverings as a passive method to control indoor ambiences have attracted much attention [1–5]. Initially, this passive effect was neglected for a long time [6]; however, under low ventilation rates, this passivity was clearly appreciated [7]. Finally, initial studies in wooden structures [6,8] were employed to analyze the same passive effect on concrete buildings covered with coatings of different materials, such as paper, wood, and paint or plastic [7,9]. Initially, these studies were evaluated in laboratories to define their coefficients, such as diffusion coefficients or water vapor permeability [10], for applications such as inputting data in future modelling processes [11] so as to define the real effect of internal coverings over indoor ambiences after placing these materials in their final constructive position in buildings [12].

The International Energy Agency (IEA) made several different attempts to predict indoor ambiences based on outdoor weather conditions. However, predicting indoor ambiences is a complex objective with a large number of unknown variables, such as the properties of materials in a building construction positions. Despite the fact that materials are tested in a laboratory in their final application place, experiments with different treatments change their expected behavior, [13] such varnishing, painting, covering with paper, or simply dusting over the covering. Recent papers have reported that many of the current numerical models for building energy systems assume empty rooms and do not account entirely for the thermal inertia of objects and materials, such as furniture. This assumption makes the models invalid for dynamic calculations [14]. The issues arising during the simulation processes are too complex to understand and apply realistically to building behavior.

Despite the interest in heat and mass transfer processes in buildings, toward energy saving and thermal comfort improvement, most of the studies are centered on phase change materials (PCMs) that affect buildings [15,16] due to the addition of new materials, reaching a more intense effect over indoor ambiences. Despite this, in accordance with the sustainability, and considering the impossibility of development in most countries, these constructive materials seem expensive, which is why it is necessary to continue the analysis of this effect on the basis of the usual internal covering materials to reach the optimization level in buildings.

In previous research on the effect of permeable internal coverings in the indoor conditions of 25 office buildings [1–5], it was possible to define, based on statistical studies, the effects of internal coverings over real indoor ambiences during an unoccupied period. During the occupied period, the offices attend to clients and the air changes are so high that the covering materials are ineffective in controlling the ambiences [7]. In this sense, internal covering materials like paper, wood, paint, and plastic can be classified as permeable, semi-permeable, and impermeable, in clear agreement with its expected permeability level.

Furthermore, in previous works [1–5], the statistical hourly study of the partial vapor pressure difference between indoor and outdoor ambiences showed that, although this effect is more intense in wooden constructions, the internal covering over concrete walls act as a barrier that influences the building by controlling the indoor ambience. In particular, permeable materials show a tendency to reduce the indoor partial vapor pressure when it is high in the ambience, and vice versa. Simultaneously, impermeable materials only increase the effect of outdoor humidity, reaching a greater number of dissatisfied persons during the first hours of occupation. As a consequence, during the first hours of occupation, the enthalpy was high and the energy consumption increased to reach an acceptable ambience, which was a peak of energy demand during the morning.

Although our results suggest the applicability of permeable materials, the information on optimal material properties and the amount of internal coverings needed in an indoor ambience to act as an adequate mechanical thermal comfort controller, remains unknown, and so a new design methodology towards nearly-zero energy building (NZEB) is needed [17,18]. This is related to the fact that statistical studies do not allow us to model this process and recognize the real material coefficients once placed in the building.

Despite the fact that there are previous works about control systems of indoor ambiences in buildings and its posterior optimization by artificial intelligence, a few of them are centered on wall construction materials. Furthermore, the permeability level of internal coverings was simulated in heat and mass transfer software resources, which, most of time, do not let researchers develop feedback and redesign the building construction characteristics. Furthermore, in the present paper, the input variables considered were outdoor temperature and relative humidity, which could be related to just one output variable per neural network (NN), such as indoor local perception of indoor air quality (PD), indoor air acceptability (ACC) and indoor enthalpy (energy consumption), with the aim of a future optimization of indoor thermal comfort and a reduction of building energy consumption.

NNs are employed when the statistical results are insufficient in revealing great results in most of the research areas, indicating the solution to different problems, such as natural ventilation and thermal comfort in buildings [18,19]. In particular, NNs are a universal approach that allow us to model everything that statistical curve fitting cannot, letting us model some processes and predict their behavior. Once statistics showed these results, it was of interest to train neural networks based on outdoor conditions to predict indoor conditions and reach a quantitative determination of the permeability level of internal coverings. In the present paper, an initial step was performed on the validation of an original modelling procedure based on real sampled data and its predictions, in accordance with the knowledge developed in the last few decades.

2. Materials and Methods

2.1. Office Buildings

In previous research works [1–5], different statistical studies were developed to identify the behavior of indoor conditions in office buildings in the northwest region of Spain. This region is of special interest due to its high relative humidity, which is nearly 80% throughout the year, and its mild climate with a mild temperature.

Offices were selected owing to their usage of the same construction materials and structure, except the internal covering, which let us relate the effect of internal coverings with different ambiences. This internal covering was classified as permeable, semi-permeable, and impermeable, in accordance with indoor ambience behavior under different weather conditions. The behavior of internal coverings was in accordance with the expected permeability level of other materials, such as paper, paint, and plastic.

Two time periods were identified in these offices, in accordance with the working hours, which could be directly related to the high or low ventilation rate and the presence or absence of humidity sources from its metabolic rate. The occupied period was defined as a period when clients and workers were in the office from 09:00 to 19.00, and the unoccupied period was defined as the time-period when nobody was in the office from 19:00 till 09:00 the next day. During the unoccupied period, the ventilation rate was reduced and only the internal coverings could have decreased humidity these ambiences.

2.2. Sampling Temperature and Relative Humidity

Different weather stations from the Environmental Information System of Galicia (SIAM) [20] in the entire Galician region provide us the main climatic variables, such as temperature, relative humidity, pressure, and air velocity, with a time frequency of 10 min. This sampling frequency can be considered to be adequate for our research work.

On the other hand, to sample indoor conditions, different tiny tag data loggers [21] of temperature and relative humidity were placed in each office during the summer and winter seasons, with a sampling frequency of 5 min after calibration, with a precision range of ±1 °C of temperature and 1% of relative humidity, respectively.

3. Calculation

3.1. Local Thermal Comfort Indexes

Indoor temperature and relative humidity can be related to some local thermal comfort indexes, as shown in a previous study by Toftum et al. [21–23] and Simonson et al. [24]. Different research works revealed that the percentage of dissatisfied persons with warm respiratory thermal comfort (PD_{WRC}) or the percentage of dissatisfied persons with indoor air quality (PD_{IAQ}) were employed to identify the acceptable level of an indoor ambience (ACC_{IAQ}), which ranges from −1 (clearly unacceptable) and 0 (just acceptable) to +1 (clearly acceptable), and the expected percentage of dissatisfied persons. As in previous research, PD_{IAQ} was found to be more sensitive to the changes in the indoor air conditions, which is an interesting index for this study, as compared to the PD_{WRC}. The Equations (1) and (2) represent the variables ACC_{IAQ} and PD_{IAQ}:

$$ACC_{IAQ} = -0.033h + 1.662 \tag{1}$$

$$PD_{IAQ} = \frac{\exp\left(-0.18 - 5.28\ ACC_{IAQ}\right)}{1 + \exp\left(-0.18 - 5.28\ ACC_{IAQ}\right)} \times 100 \tag{2}$$

The two indexes shown in Equations (1) and (2) were obtained by Fang et al. [25] and were based on the laboratory studies, where the subjects were facially exposed to clean air in a climatic chamber, and different indoor air temperatures and relative humidity could be modelled with a perception of indoor air quality. In this sense, the first term defined by these laboratory results showed moist air acceptability due to the indoor air quality. As can be seen from Equation (1), this acceptability is clearly a function of moist air enthalpy (h). On the other hand, the second index obtained from the previous index is the percentage of dissatisfied persons expected to be in disagreement with the proposed indoor air (PD_{IAQ}). This index, shown in Equation (2), is the only function of the moist air acceptability (ACC_{IAQ}).

At the same time, due to the moist air, the enthalpy difference respects the comfortable conditions, both in summer or winter seasons, which are proportional to the energy consumption needed to reach this comfortable ambience; the enthalpy was calculated for each indoor ambience under the effect of each internal covering during extreme summer and winter seasons.

3.2. NN Training and Prediction

Matlab NN [26] is the main software resource employed to train and predict the behavior of internal covering materials as a function of weather conditions. In this sense, it is of interest to highlight that the generalized regression neural network (GRNN), due to its main advantages, is related to the lack of need to define the topology of the network. One of the more complex decisions to make when developing this kind of study is the number of nodes needed that relate to the precision of the results and the time and the number of calculations needed to obtain each of the different predictions. As a consequence, an interesting neural net selection of two hidden layers were selected by default, which is considered to be adequate for most of the real processes.

The neuronal model of generalized regression employed in this research work (GRNN) was proposed and developed by Specht in 1991 [27,28]. It possesses the desirable property of not requiring any iterative training, that is, it can approximate any arbitrary function between input vectors (inputs) and output vectors (outputs), taking the estimation of the function directly from the training data. In this sense, the GRNN model relies on nonlinear regression theory. It is, in essence, a method to estimate a function $f(x, y)$ only through the training set, so that the joint probability function, which is unknown, is estimated using the Parzen estimator [27,28]. To do this, we must first define the following distances between "x" and "y".

The output of the GRNN can be defined by Equations (3) and (4):

$$D_i^2 = (x - x_i)^T ((x - x_i)) \tag{3}$$

$$\hat{y}(x) = \frac{\sum_{i=1}^{n} y^i \exp\left(-\frac{D_i^2}{2\sigma^2}\right)}{\sum_{i=1}^{n} \exp\left(-\frac{D_i^2}{2\sigma^2}\right)} \tag{4}$$

where x isthe training sampled, y is the training sampled output, σ is the smoothing parameter of the GRNN, and T is the number of Parzen windows used in the estimation process. Equation (3) shows the Euclidean distance between the input x_i and the training sampled (D_i), and Equation (4) shows the fundamental expression of the neuronal model.

The topology of the GRNN model has four layers. The first layer represents the vector of inputs or inputs. The second layer, called the pattern layer, is equal to the number of observations of the input vector. The value of the neuron in this non-visible layer is obtained by applying the activation function, $h = exp\left(-\frac{D_i^2}{2\sigma^2}\right)$, which is an extension of the Gaussian multivariate function. The third layer includes two types of summations: (i) $S_s = \sum_i h_i$, which represents the denominator ($D(x)$) and (ii) $S_w = \sum_i w_i h_i$, which is the numerator ($N(x)$), where "w_i" are the values of the output used in the learning phase and which act as weights. Finally, in the fourth layer the output is obtained by the following operation, $y = \frac{S_w}{S_s}$, as we can see in the Figure 1.

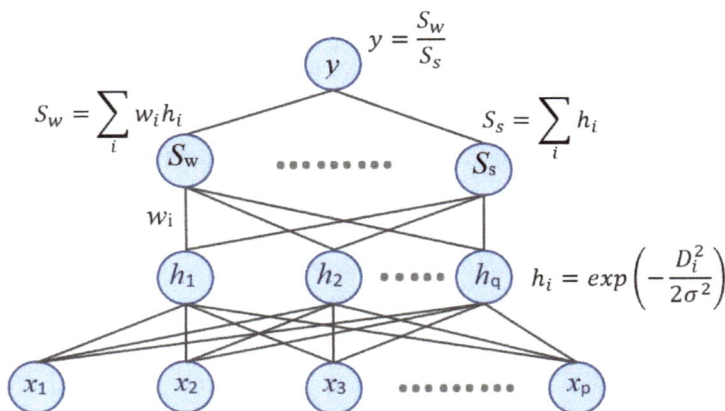

Figure 1. Typology of a generalized regression neural network (GRNN).

Once training data are introduced in the input layer, the training starts and it will stop when a previously marked fitness is reached or, if the values don't improve, when they reach a value close to zero, for a number of consecutive generations. In our case, the average percentage of incorrect predictions for each weather station was selected, as we can see in Table 1.

Table 1. Average percentage of incorrect predictions.

GRNN	Error (%)
Winter permeable	1.59
Winter semi-permeable	1.20
Winter impermeable	1.34
Summer permeable	1.47
Summer semi-permeable	1.21
Summer impermeable	1.68

In the second part of this experimental configuration, the main variables to be employed were selected in accordance with the NN rules. In this sense, NNs can predict only one dependent variable as a function of a wide number of independent variables. The problem appears to adjust this methodology to our case study, where we need to define the indoor conditions as a function of how to minimize the indoor and outdoor temperatures and the relative humidity. As a consequence of this, partial vapor pressure was selected as the main study variable as it represents indoor conditions and due to its difference with the outdoor partial vapor pressure, which used to be employed to analyze heat and mass transfer processes in building construction materials.

However, different results from past studies [1,2] have shown that only outdoor partial vapor pressure, outdoor temperature or relative humidity is needed to train the NNs with an adequate precision. In consequence, these two input variables were selected to train and predict indoor partial vapor pressure, enthalpy, and local thermal comfort indexes during the summer and winter seasons.

Finally, in the second step, to compare and understand the behavior of the internal coverings, each of the trained networks were required to be stimulated under the exact same weather conditions for each season. Figures 2 and 3 depict the outdoor temperature and relative humidity (RH) sampled in a typical Galician night. The data reflected in Figures 2 and 3 are real curves obtained from nearer weather stations that are certificated and calibrated by the Spanish ministry. These are the real weather conditions selected for their usual nearly constant high relative humidity, as is normal in the coastal regions of Galicia. In the winter season relative humidity increases during the night and, in summer, it may reach 100% during long periods of time due to fog, as we can see in Figure 3.

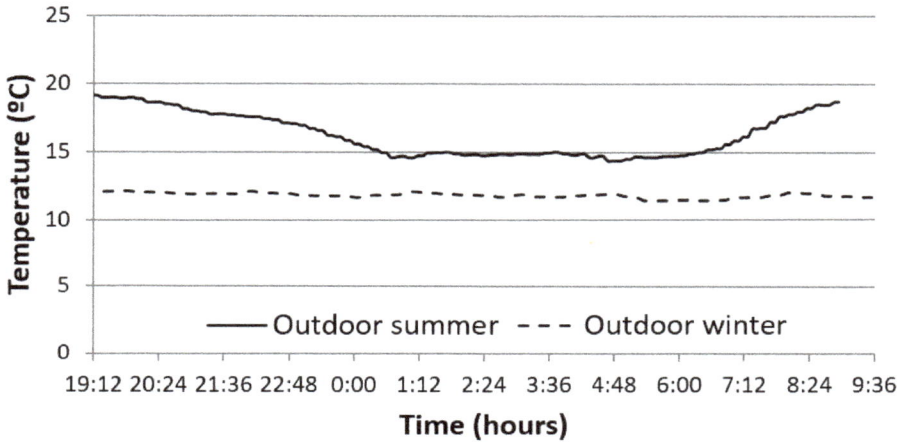

Figure 2. Outdoor temperature during the unoccupied period employed in the prediction process (data collected from different weather stations in Galicia, Spain).

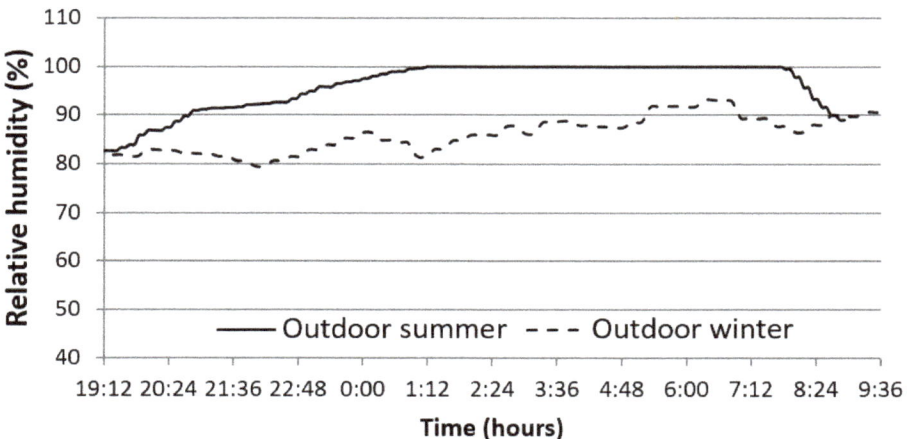

Figure 3. Outdoor relative humidity during the unoccupied period employed in the prediction process (data collected from different weather stations in Galicia, Spain).

An interesting change in temperature from 20–15 °C during the unoccupied period of the summer season can be observed. As we can see in Figure 2, in summer weather the temperate remains constant from 0:00 until 7:00, but during the winter season a nearly constant 12 °C can be observed. This event is typical of the weather conditions that would help understand its effect over indoor ambiences.

4. Results

Using the previous methodology, and with a frequency of 10 min, about 80 samples/day during the unoccupied period over more than 1 week were needed to obtain a minimum training data of 300 samples per office and per season. For NN validation, when the percentage of error is reduced during the training process, it can be modified from the standard 60% of data for training and 40% for validation [29] to a more interesting 75% of data for training the network and 25% to validate it, by comparing the NN results with real sampled data and their derived indexes of thermal comfort and energy saving, obtained from this sampled data, inside each office building.

The stopping criteria was the minimum absolute number of errors obtained in most of the indoor vapor pressure predictions. In particular, the maximum absolute error allowed was fixed to 6 during the training period and 9 during the testing period, with a standard deviation of this error in both cases of 8%, which represented an nearly null percentage of incorrect predictions, providing a clear example of the power of NNs to model this process. Next, as an example of the accuracy obtained to define indoor air variables, such as partial vapor pressure (p_v), the sampled and predicted values were compared during the validation process, as shown in Figure 4.

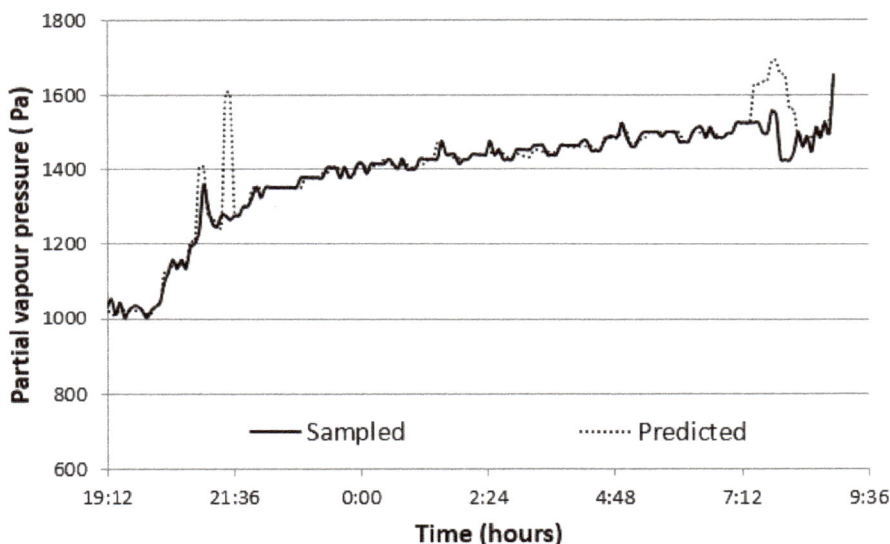

Figure 4. Example of sampled and predicted partial vapor pressure in the winter season.

For the indoor partial vapor pressure prediction, more interesting thermodynamic variables and thermal comfort indexes were predicted. The moist air enthalpy, PD_{IAQ} and ACC_{IAQ}, was calculated for each season and for each internal covering during the unoccupied period from 19:00 to 09:00. Thus, partial vapor pressure, ACC_{IAQ}, PD_{IAQ}, and enthalphy were represented as the three most used materials for internal covering (paper, paint, and plastic) during the winter season in Figures 5–8 and during the summer season in Figures 9–12. Taking into account the permeability of these materials (Table 2) and the results obtained in previous works [1–5], based on a statistical study of indoor ambiences with these wall internal coverings, the following assumption is considered in the present study: Paper represents the permeable materials family, paint represents the semi-permeable materials family, and plastic represents the impermeable materials family.

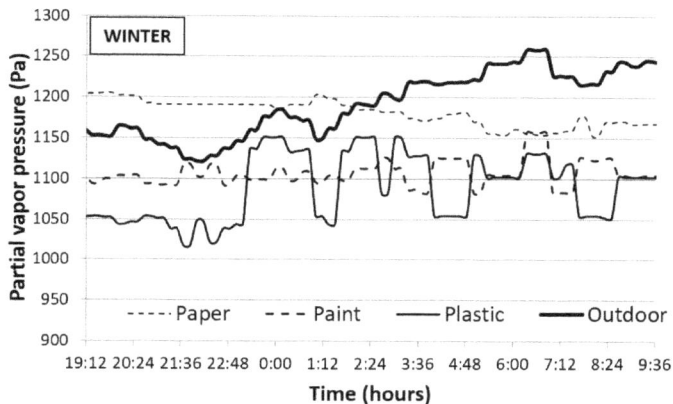

Figure 5. Indoor partial vapor pressure predicted for each internal covering during the winter season.

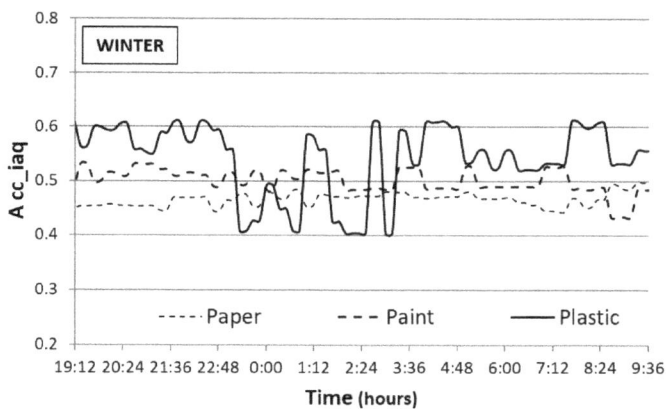

Figure 6. Indoor air acceptability (ACC$_{IAQ}$) predicted for each internal covering during the winter season.

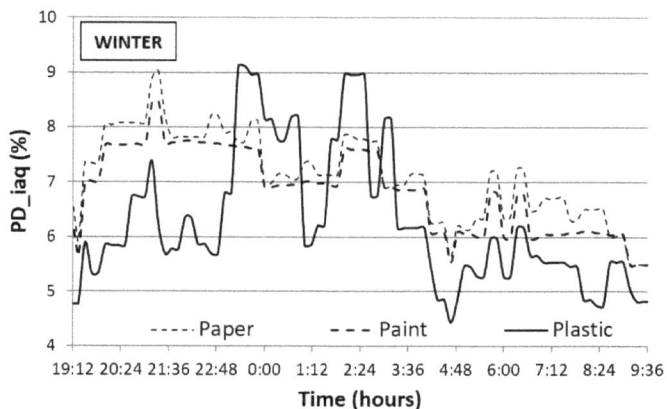

Figure 7. Indoor air percentage of dissatisfied persons (PD$_{IAQ}$) predicted for each internal covering during the winter season.

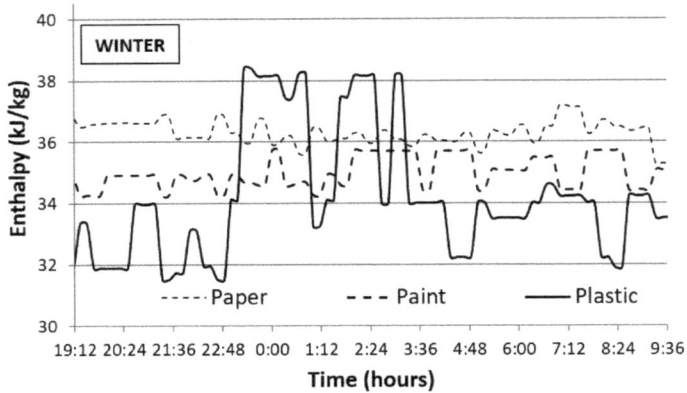

Figure 8. Indoor air enthalpy predicted for each internal covering during the winter season.

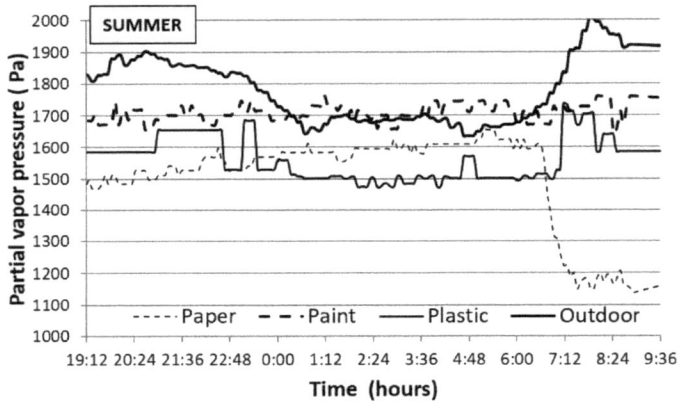

Figure 9. Indoor air partial vapor pressure predicted for each internal covering during the summer season.

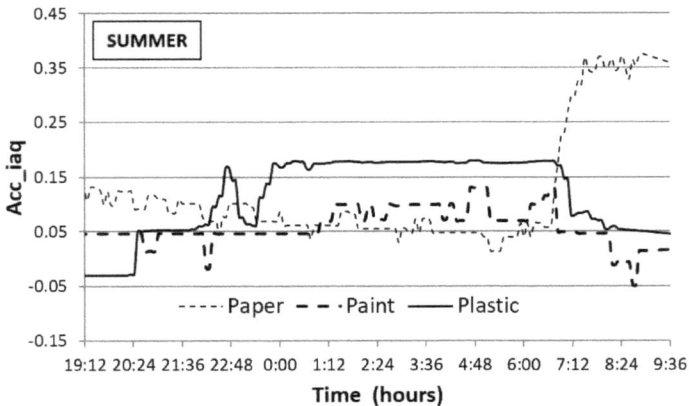

Figure 10. Indoor air acceptability (ACC$_{IAQ}$) predicted for each internal covering material during the summer season.

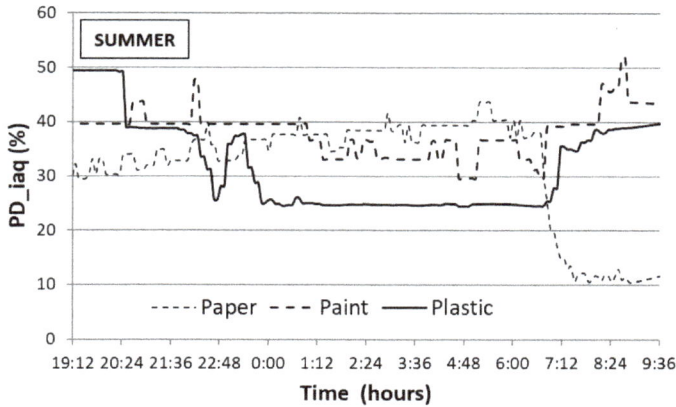

Figure 11. Indoor air percentage of dissatisfied persons (PD$_{IAQ}$) predicted for each internal covering during the summer season.

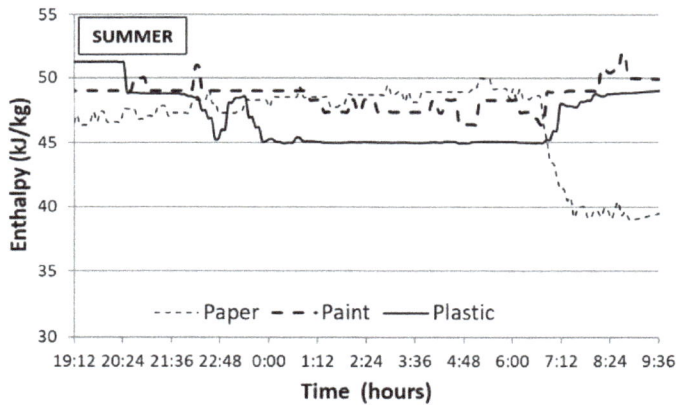

Figure 12. Indoor air enthalpy predicted for each internal covering during the summer season.

Table 2. Permeability values of different types of coverings [1–5].

Covering	Permeability (kg/(m s Pa))
Paper and Plaster	1.44e-10
Paint	1.75e-12
Plastic	0.80e-12

5. Discussion

Each NN was trained based on the respective indoor sampled conditions of each office and its respective outdoor weather conditions, with an adequate margin of error. Each of the different trained networks had predicted indoor conditions as a function of the same outdoor weather (Figures 2 and 3) with the aim to demonstrate the effect of internal coverings over indoor ambiences, as in the laboratory analysis.

Previous to this analysis, it is important to remember that, in previous works [1,2], statistical studies showed that internal covering materials used in several offices such as paper, paint, and plastic, showed a statistical behavior representative of the permeable, semi-permeable, and impermeable materials and, in consequence, the offices buildings with these materials were selected to model the behavior of internal covering materials.

In accordance with the previous works performed in the summer season, for this climatic region, a more interesting period helps to appreciate the more intense effect of internal coverings. In this sense, we can appreciate that, during the winter (Figure 5), permeable coverings tended to exert an opposite effect on the outdoor conditions. In particular, the humidity cumulated and released from permeable materials is expected to be found in the first 4 h (from 19:30 to 11:30) and, after this, only permeable internal coverings will reduce partial vapor pressure more slowly than impermeable internal coverings. Additionally, the opposite effect was appreciated during the last hours of occupation in this season.

This effect was more intense in the summer, as can be seen in Figure 9. As soon as the outdoor partial vapor pressure was reduced from an initial value of 19:12 to 0:00, the indoor partial vapor pressure in offices with permeable internal coverings released humidity and tended to increase its partial vapor pressure. Furthermore, the opposite effect to the increase in the outdoor partial vapor pressure can be appreciated during the last hours of the unoccupied period from 06:00 to 09.00.

We can also see that, during the winter and summer seasons, the indoor ambiences of office buildings with impermeable internal coverings showed a tendency very similar to that of the outdoor ambiences. Finally, the office with semi-permeable materials tended to maintain a nearly constant partial vapor pressure during the night in both the seasons (cf. Figures 5 and 9).

As a consequence of this effect of the internal covering over indoor ambiences, we can appreciate clear consequences to thermal comfort and energy saving. For instance, as seen in Figure 6, the acceptability index (ACC_{IAQ}) during the winter showed a nearly constant value of 0.5, while the impermeable coverings showed a great variability of this index over time. This variability is a clear example of the effect of temperature and relative humidity over indoor partial vapor pressure in office buildings with impermeable coverings.

As seen in Figure 10, this same index revealed higher acceptability during the summer season as a consequence of lower indoor partial vapor pressure during the first and last hours of the unoccupied period. Consequently, a better acceptability of indoor ambiences was expected when these materials (permeable coatings) were used.

The same effect was identified via the expected number of people that were unsatisfied with the perception of indoor air quality, by means of the PD_{IAQ} index (Figures 7 and 11). As expected, the percentage of dissatisfied persons during the winter season tended to be reduced to an extremely low value, which was similar for all the offices. Despite this, the effect of internal coverings was more intense in summer and, as a direct effect of indoor partial vapor pressure, the percentage of dissatisfied persons tended to be nearly 10% during the last hours of inoccupation and the first hours of occupation. Moreover, from these results, it is possible to confirm that, in general, a better indoor ambience is reached (Figure 11). Furthermore, this value of 10% of unsatisfied persons can be reduced when compared with 40% of unsatisfied persons with impermeable coverings.

Finally, since the energy needed by a heating, ventilation, and air-conditioning (HVAC) system to reach an adequate indoor ambience is directly related to moist air enthalpy, we could deduce the energy consumption tendency as a function of this thermodynamic variable (represented in Figures 7 and 12). Figure 7 shows a higher enthalpy value during nearly all of the unoccupied periods in office buildings with permeable internal coverings. Consequently, a lower energy consumption of 20% in the HVAC systems was needed to reach more comfortable conditions during the summer season in the first hours of occupation. At the same time, it is interesting to highlight that, during a reduced percentage of time, offices with impermeable internal coverings showed a peak of indoor partial vapor pressure and, as a consequence, a peak of enthalpy and percentage of dissatisfied persons.

Despite these results, these effects are reduced when compared with the values obtained during the summer season. As can be seen in Figure 12 there is a difference in the indoor enthalpy of 10 kJ/kg and the enthalpy of indoor air in offices with impermeable respect permeable internal coverings during the last hours of inoccupation and during the first hours of occupation. This effect implies reduced energy consumption to reach a cool ambience during the summer season, which is really important since such a reduced consumption is becoming one of the key objectives in the majority of today's research related to sustainability [30].

Based on these results, we can deduce that it is possible to model indoor ambiences based on the outdoor conditions by means of NN procedures. At the same time, we can confirm the validity of these predictions owing to its agreement with the results obtained in previous research [1,2] based on the statistical analysis of real sampled data. In this sense, NN predictions let us confirm the effect of permeable internal coverings, such as paper or wood, to control the indoor ambiences toward better local thermal comfort and energy saving, which acted as a mechanical control system. In particular, internal coverings materials behavior depended on the extreme outdoor conditions to which the office was exposed. For instance, the climatic region of this study showed a high relative humidity but no extreme temperature changes. This is the reason why, during the winter season, this effect was reduced and clearly more intense in the summer season.

Finally, this passive method needs to be redesigned and adjusted to the amount of surface and permeability level of covering employed in each case in order to obtain the best possible indoor ambience behavior. Thus, this methodology can be employed for almost all kinds of buildings and weather conditions [31,32]. Furthermore, this methodology would be of interest because climate variations will induce different indoor ambiences in each type of building and, therefore, this and others passive methods must be adjusted for each particular region [33–36]. Other kinds of artificial PCMs [37] are employed as building construction materials rather than the typical permeable internal covering materials, because these PCMs are well modelled in laboratories [37]. However, only limited information is available about the behavior of PCMs in real buildings, which is why this new generation procedure seems promising toward understanding their effect and improving their future design.

6. Conclusions

The present paper shows new and interesting results about a new methodology for internal covering designs to improve these materials effect over indoor ambiences. In this sense, based on the results obtained from previous works, the effect of permeable, semi-permeable, and impermeable internal coverings was analyzed after placing them in the actual final building construction position. The main data was employed to obtain different neural networks that, once trained, were employed to predict the indoor ambience based on input data with the same weather conditions.

This result had a direct implication on the indoor thermal comfort and energy consumption, proposing permeable coverings as the better way to reduce energy peak demands in the first hours of occupation and a better thermal comfort condition, which acted as a mechanical control system. In particular, a reduction of 20% in the expected energy consumption of the HVAC system and a reduction from 40% to 10% of unsatisfied persons was obtained during the summer season when the permeable coverings were employed.

Finally, from these interesting results, we can conclude that, once the effect of internal coverings over thermal comfort and energy consumption has been demonstrated, it is possible to design internal coverings and, consequently, to define the exact amount of internal covering surface and permeability level needed, to reach an adequate behavior for a specific indoor ambience as the main constant to adjust this mechanical control system.

Author Contributions: Conceptualization, J.A.O.; methodology, J.A.O., Á.M.C., and R.B.; software, J.A.O.; validation, D.V., Á.M.C., and R.B.; formal analysis, J.A.O., D.V., Á.M.C., and R.B.; data curation, J.A.O. and D.V.; writing—original draft preparation, J.A.O.; writing—review and editing, J.A.O. and D.V.

Funding: This research was funded by CYPE Ingenieros S.A. in their research project to reduce energy consumption in buildings and its certification, in collaboration with the University of A Coruña (Spain) and the University of Porto (Portugal). (Grant No. 64900).

Acknowledgments: The authors wish to express their deepest gratitude to the Sustainability Specialization Campus of the University of A Coruña for the administrative and technical support.

Conflicts of Interest: The authors declare no conflict of interest.

References

1. Orosa, J.A.; Baaliña, A. Passive climate control in Spanish office buildings for long periods of time. *Build. Environ.* **2008**, *43*, 2005–2012. [CrossRef]
2. Orosa, J.A.; Baaliña, A. Improving PAQ and comfort conditions in Spanish office buildings with passive climate control. *Build. Environ.* **2009**, *44*, 502–508. [CrossRef]
3. Orosa, J.A.; Oliveira, A.C. Energy saving with passive climate control methods in Spanish office buildings. *Energy Build.* **2009**, *41*, 823–828. [CrossRef]
4. Orosa, J.A.; Oliveira, A.C. Reducing energy peak consumption with passive climate control methods. *Energy Build.* **2011**, *43*, 2282–2288. [CrossRef]
5. Orosa, J.A.; Oliveira, A.C.; Ramos, N.M.M. Experimental quantification of the operative time of a passive hvac system using porous covering materials. *J. Porous Media* **2010**, *13*, 637–643. [CrossRef]
6. Hens, H. Indoor Climate in Student Rooms: Measured Values. IEA-EXCO Energy Conservation in Buildings and Community Systems Annex 41 "Moist-Eng" Glasgow Meeting. 2004. Available online: https://www.kuleuven.be/bwf/projects/annex41/protected/data/KUL%20Oct%202004%20Paper%20A41-T3-B-04-6.pdf (accessed on 24 February 2019).
7. Padfield, T. The Role of Absorbent Building Materials in Moderating Changes of Relative Humidity. Ph.D. Thesis, Department of Structural Engineering and Materials, The Technical University of Denmark, Kongens Lyngby, Denmark, October 1998.
8. Hameury, S.; Lundstrom, T. Contribution of indoor exposed massive wood to a good indoor climate: In situ measurement campaign. *Energy Build.* **2004**, *36*, 281–292. [CrossRef]
9. Nicolajsen, A. Thermal transmittance of a cellulose loose-fill insulation material. *Build. Environ.* **2005**, *40*, 907–914. [CrossRef]
10. Plathner, P.; Littler, J.; Stephen, R. Dynamic water vapour sorption: Measurement and modelling. In Proceedings of the Indoor Air Quality 99, Edinburgh, Scotland, 9–13 August 1999.
11. Talukdar, P.; Osanyintola, O.F.; Olutimayin, S.O.; Simonson, C.J. An experimental data set for benchmarking 1-D, transient heat and moisture transfer models of hygroscopic building materials. Part II: Experimental, numerical and analytical data. *Int. J. Heat Mass Transf.* **2007**, *50*, 4915–4926. [CrossRef]
12. Osanyintola, O.F.; Simonson, C.J. Moisture buffering capacity of hygroscopic building materials: Experimental facilities and energy impact. *Energy Build.* **2006**, *38*, 1270–1282. [CrossRef]
13. Simonson, C.J.; Tuomo, O. Moisture performance of buildings envelopes with no plastic vapour retarders in cold climates. In Proceedings of the Healthy Buildings 2000, Espoo, Finland, 6–10 August 2000; SIY Indoor Air Information Oy: Helsinki, Finland, 2000; Volume 3.
14. Johra, H.; Heiselberg, P. Influence of internal thermal mass on the indoor thermal dynamics and integration of phase change materials in furniture for building energy storage: A review. *Renew. Sustain. Energy Rev.* **2017**, *69*, 19–32. [CrossRef]
15. Talukdar, U.P.; Das, A.; Alagirusamy, R. Effect of structural parameters on thermal protective performance and comfort characteristic of fabrics. *J. Text. Inst.* **2016**, *108*, 1430–1441. [CrossRef]
16. Wahid, M.A.; Hosseini, S.E.; Hussen, H.M.; Akeiber, H.J.; Saud, S.N.; Mohammad, A.T. An overview of phase change materials for construction architecture thermal management in hot and dry climate region. *Appl. Therm. Eng.* **2017**, *112*, 1240–1259. [CrossRef]
17. Fedorczak-Cisak, M.; Furtak, M.; Gintowt, J.; Kowalska-Koczwara, A.; Pachla, F.; Stypuła, K.; Tatara, T. Thermal and vibration comfort analysis of a nearly zero-energy building in Poland. *Sustainability* **2018**, *10*, 3774. [CrossRef]
18. Álvarez, J.D.; Costa-Castelló, R.; Castilla, M.D.M. Repetitive control to improve users' thermal comfort and energy efficiency in buildings. *Energies* **2018**, *11*, 976. [CrossRef]
19. Weng, K.; Meng, F.; Mourshed, M. Model-based optimal control of window openings for thermal comfort. *Proceedings* **2018**, *2*, 1134. [CrossRef]
20. SIAM. Environmental Information System of Galicia. 2016. Available online: http://archivo.cesga.es/component/option,com_proyectos/task,view/Itemid,0/catid,46/id,45/lang,en/ (accessed on 24 February 2019).

21. Gemini Data Loggers. 2016. Available online: http://www.geminidataloggers.com (accessed on 24 February 2019).

22. Toftum, J.; Jorgensen, A.S.; Fanger, P.O. Upper limits for indoor air humidity to avoid uncomfortably humid skin. *Energy Build.* **1998**, *28*, 1–13. [CrossRef]

23. Toftum, J.; Jorgensen, A.S.; Fanger, P.O. Upper limits of air humidity for preventing warm respiratory discomfort. *Energy Build.* **1998**, *28*, 15–23. [CrossRef]

24. Simonson, C.J.; Salonvaara, M.; Ojanen, T. The effect of structures on indoor humidity-possibility to improve comfort and perceived air quality. *Indoor Air* **2002**, *12*, 243–251. [CrossRef]

25. Fang, L.; Clausen, G.; Fanger, P.O. Impact of temperature and humidity on perception of indoor air quality during immediate and longer whole-body exposures. *Indoor Air* **1998**, *8*, 276–284. [CrossRef]

26. Matlab Mathworks. Available online: https://es.mathworks.com (accessed on 24 February 2019).

27. Specht, D.F. A general regression neural network. *IEEE Trans. Neural Netw.* **1991**, *2*, 568–576. [CrossRef] [PubMed]

28. Kartal, S.; Oral, M.; Ozyildirim, B.M. Pattern layer reduction for a generalized regression neural network by using a self–organizing map. *Int. J. Appl. Math. Comput. Sci.* **2018**, *28*, 411–424. [CrossRef]

29. MathWorks. Divide Data for Optimal Neural Network Training. 2019. Available online: https://www.mathworks.com/help/deeplearning/ug/divide-data-for-optimal-neural-network-training.html;jsessionid=33c5d9e13c9aa6f19e618b26f0d8 (accessed on 26 April 2019).

30. Costa, A.M.; Bouzón, R.; Vergara, D.; Orosa, J.A. Eco-friendly pressure drop dehumidifier: An experimental and numerical analysis. *Sustainability* **2019**, *11*, 2170. [CrossRef]

31. Csoknyai, T.; Hrabovszky-Horváth, S.; Georgiev, Z.; Jovanovic-Popovic, M.; Stankovic, B.; Villatoro, O.; Szendrő, G. Building stock characteristics and energy performance of residential buildings in Eastern-European countries. *Energy Build.* **2016**, *132*, 39–52. [CrossRef]

32. Vasco, D.A.; Muñoz-Mejías, M.; Pino-Sepúlveda, R.; Ortega-Aguilera, R.; García-Herrera, C. Thermal simulation of a social dwelling in Chile: Effect of the thermal zone and the temperature-dependant thermophysical properties of light envelope materials. *Appl. Thermal Eng.* **2017**, *112*, 771–783. [CrossRef]

33. Glass, S.V.; Kochkin, V.; Drumheller, S.C.; Barta, L. Moisture performance of energy-efficient and conventional wood-frame wall assemblies in a mixed-humid climate. *Buildings* **2015**, *5*, 759–782. [CrossRef]

34. Bhikhoo, N.; Hashemi, A.; Cruickshank, H. Improving thermal comfort of low-income housing in thailand through passive design strategies. *Sustainability* **2017**, *9*, 1440. [CrossRef]

35. Boostani, H.; Hancer, P. A model for external walls selection in hot and humid climates. *Sustainability* **2019**, *11*, 100. [CrossRef]

36. Orosa, J.A.; Vergara, D.; Costa, Á.M.; Bouzón, R. A novel method for nZEB internal coverings design based on neural networks. *Coatings* **2019**, *9*, 288. [CrossRef]

37. Medved, I.; Trník, A.; Vozár, L. Modeling of heat capacity peaks and enthalpy jumps of phase-change materials used for thermal energy storage. *Int. J. Heat Mass Transf.* **2017**, *107*, 123–132. [CrossRef]

*applied
sciences*

MDPI

Article

Energy Consumption Optimization Model of Multi-Type Bus Operating Organization Based on Time-Space Network

Yuhuan Liu [1], Enjian Yao [2,*] and Shasha Liu [1]

1 School of Traffic and Transportation, Beijing Jiaotong University, Beijing 100044, China
2 Key Laboratory of Transport Industry of Big Data Application Technologies for Comprehensive Transport, Beijing Jiaotong University, Beijing 100044, China
* Correspondence: enjyao@bjtu.edu.cn

Received: 13 July 2019; Accepted: 12 August 2019; Published: 15 August 2019

Featured Application: The findings obtained from this study have implications for application of the pure electric bus.

Abstract: As a new type of green bus, the pure electric bus has obvious advantages in energy consumption and emission reduction compared with the traditional fuel bus. However, the pure electric bus has a mileage range constraint and the amount of charging infrastructure cannot meet the demand, which makes the scheduling of the electric bus driving plans more complicated. Meanwhile, many routes are operated with mixing pure electric buses and traditional fuel buses. As mentioned above, we focus on the operating organization problem with the multi-type bus (pure electric buses and traditional fuel buses), aiming to provide guidance for future application of electric buses. We take minimizing the energy consumption of vehicles, the waiting and traveling time of passengers as the objectives, while considering the constraints of vehicle full load limitation, minimal departure interval, mileage range and charging time window. The energy consumption based multi-type bus operating organization model was formulated, along with the heuristic algorithm to solve it. Then, a case study in Beijing was performed. The results showed that, the optimal mixing ratio of electric bus and fuel bus vary according to the variation of passenger flow. In general, each fuel bus could be replaced by two pure electric buses. Moreover, in the transition process of energy structure in public transport, the vehicle scale keeps increasing. The parking yard capacity and the amount of charging facilities are supposed to be further expanded.

Keywords: pure electric buses; multi-type bus operating organization; time-space network; energy consumption; public transport

---

1. Introduction

In China, with the rising amount of motor vehicles, the energy consumption is aggravating. At the same time, the automobile exhausted emissions bring severe air pollution, and the energy and environmental problems caused by transportation vehicles are becoming increasingly serious. Driven by the oil crisis and environmental pressure, the government and enterprises are actively promoting the development of pure electric buses. According to statistics, the cost of energy consumption accounts for 30% to 40% of the operating costs of bus operation enterprises. In addition, under the background of the national policy of energy conservation and emission reduction, many bus operators are confronted the targets of the total energy consumption control. Once the target is not completed, they will face penalties. Pure electric buses have the advantages of high energy efficiency and low pollution and they are rapidly replacing traditional fuel buses, accounting for a gradually increasing proportion in the bus fleet. However, pure electric buses have an insufficient driving mileage range, long charging time and the amount of charging infrastructure is not enough, which adds the complexity of bus operation organization. Under the background of application of the pure electric buses, it is the main support to realize the large-scale pure electric bus operation by strengthening the study on the scheduling driving plan based on the constraints of the pure electric bus mileage range and charging time window, ensuring the vehicles to complete the daily operation tasks with maximum reduction on energy consumption under the condition of pure electric buses and fuel buses mixing operation.

In the field of the bus operating organization considering driving ranges constraint, Bodin L et al. [1] described the problem, and focused on the constraint of the mileage limit of a vehicle without returning to the yard to replenish energy after the departure. Freling. R et al. [2] discussed the driving ranges constraint as well. This vehicle operating organization method is based on a single parking yard, solving the problem by considering the battery life constraint in the single trip of entering and leaving parking yard, without taking into account the situation that the bus drives back to the yard halfway to charge and continue to conduct the tasks, which makes the model not suitable for the problem considering fuel consumption constraints. Haghani et al. [3,4] studied the problem of the regional bus operating organization based on the driving ranges battery life constraint. The paper classified the buses into three shifts, morning, noon and evening, and divided them into the yard matching shifts and the route matching shifts. The model reduced the size of the problem by about 40% without reducing any feasible solutions. The application showed that the research could effectively increase the vehicle operating time and cut the operation cost of bus operators. However, it adopted the post-check strategy, which limited the optimization of the solution to some extent. Ali Haghani et al. [5] constructed a mathematical model for the multi-depot scheduling with path time constraints, and used an exact algorithm and two heuristic algorithms to solve the model. Amar Oukil [6], Guy Desaulners [7] and M.A. Forbes [8] also studied the multi-depot vehicle scheduling problem, but they used the column generation algorithm and exact algorithm, respectively to solve the problem. Ali Haghani [4] compared one multi-depot vehicle scheduling model with two single-depot vehicle scheduling models. Natalia Kliewer [9] and Pablo C. Guedes [10] modeled the multi-depot bus scheduling problem based on the spatio-temporal network.

In terms of solving the operating organization optimization model, hyper-heuristic and hybrid algorithms have attracted much attention in recent years. Pepin et al. [11] compared a variety of hybrid heuristic methods for vehicle scheduling, finding that the heuristic method based on column generation had the best solution quality yet with a long solution time, while the heuristic method based on large-scale neighborhood search was fast and of good quality. Among the soft computing method applied to the vehicle scheduling, VAMPIRES [12,13] was one of the most successful early heuristic methods, having the similar main ideas to 2-opt algorithms. It had successfully solved hundreds of actual public transport vehicle scheduling problems, and later was replaced by the the BOOST object-oriented software system. Wren and Kwan [14] reported the application of the system in a British bus company. In the past ten years, more researches focused on the realistic constraints or characteristics, resulting in the emergence of a series of hyper-heuristic and hybrid approaches. For example, Shen [15] developed a vehicle scheduling method based on the tabu search by applying the 2-opt algorithm. Eliiyi et al. [16] proposed six types of meta-heuristic methods to solve vehicle scheduling problems with multiple vehicle types and continuous driving time constraints.

Since the 1990s, the focus of the problem research has been on solving the exact solution of the problem. Some scholars had proposed the precise branch-bound method and the precise column generation method. In order to reduce the scale of the problem, the time-space network was introduced into the multi-yard vehicle operating organization. Kliewer et al. [17] applied the concept of the time-space network to the multi-station vehicle operating organization problem for the first time and explained the method of cutting empty-drive arcs in the network. In addition, it introduced the method and steps of reducing the network scale, proposed the multi-commodity network flow model of the multi-type multi-yard bus driving plan scheduling problem based on the time-space network, and solved an example problem with a large-scale vehicle operation organization by using the standard mathematical optimization software CPLEX. Naumaim et al. [18] provided a multi-commodity network flow model based on the time-space network and proposed a stochastic programming algorithm to solve the model. It was of great significance to simplify the problem, establish the model and solve the problem. He Di et al. [19] analyzed the connotation of bus regional dispatching and distribution planning problem, constructed a model of bus regional dispatching and distribution planning based on the space-time network, and verified the model and algorithm through an example. Yang Yang et al. [20] transformed the planning problem of the electric bus into a directed network. Bodin L et al. [21] and others put forward the idea of two-stage heuristics to solve the problem of multi-station traffic planning for the first time. Dell Amico et al. [22] took the minimum number of vehicles required as the optimization objective and used the heuristic algorithm to solve the problem in stages. Freling, R et al. [23] considered decision variables describing the connection between the vehicles and assigning vehicles to each station. Laurent [24] solved the problem of the multi-station traffic planning based on the iterative local search algorithm, and analyzed 30 cases.

At present, the research on the bus driving region dispatching is not very mature. Wei Ming et al. [25] established a mixed integer programming model with time windows, aiming at the minimum number of vehicles, vehicle waiting time and empty driving time, taking into account factors such as yard capacity, allowable vehicle refueling and task reliability of each vehicle. On the basis of describing the vehicle scheduling problem with time windows, multiple yard and multiple vehicle types, Yang Yuanfeng [26] established a mathematical model and proposed a simulated annealing genetic algorithm to solve the problem. The strong climbing performance of simulated annealing algorithm could avoid the "premature" of the genetic algorithm and improve the convergence speed of the algorithm. Zou Ying [27] took the regional scheduling of multiple bus lines as the service object, established a bus driving plan model with the objective of minimizing the passengers waiting cost, in-bus cost and the cost of the bus companies and proposed the solution idea of "allocate shifts by-line, optimize to form a network". Li Jun [28] proposed a heuristic algorithm based on dispatching after analyzing the vehicle scheduling problem with time windows. The algorithm defined two kinds of dispatching costs, designed a method to arrange routes in the dispatching process, and conducted case verification. Later, Li Jun [29] proposed a heuristic algorithm based on network optimization, which transformed the vehicle scheduling problem with time windows into several vehicle scheduling problems with a certain start time, then, used the minimum cost and the maximum flow algorithm to solve the vehicle scheduling problem with a certain start time.

To sum up, there are few researches on the optimization theory of regional operating organization currently. From the perspective of the network description, most of them emphasize the fixed allotment relationship between vehicles and vehicle yard. The researches without the fixed allotment relationship between the vehicles and vehicle yard are mainly found in logistics distribution, while fewer studies are in the field of the bus operating organization. The existing researches on the vehicle driving plan model are too simple, which greatly simplify the real operating environment, and the results are difficult to adapt to the complex work. Additionally, the current researches are hardly carried out under the background of the mixing operation with multiple vehicle types such as electric buses and fuel buses, and there are no regional bus driving plan models considering both the mileage range and charging time constraints. Some literature took the battery life into account, but they only discussed the battery life constraint within one shift and did not consider the condition that the vehicle drives back to the yard for charging and continues the tasks. Furthermore, there is little research on bus energy consumption. Therefore, we conduct the research on the regional pure electric bus driving plan with the objectives of energy consumption and bus service level, considering the pure electric bus charging constraints and mileage range constraints, and modify the algorithm to better support the application of pure electric buses in the real operation.

2. The Establishment of Time-Space Network

2.1. Establish the Time-Space Network

The time-space network is composed of a large number of nodes and various arcs. While establishing the network, the simplification of empty-driving arcs should be considered at the generation stage. We do not distinguish the vehicles belonging to different parking yards, and there is no fixed allotment relationship between vehicles and parking yards. As long as the number of vehicles in the yard is unchanged, there is no necessity that the vehicles must return to the original departing parking yard. Therefore, the yards can be regarded as general stations, and used to express the connection between yards and other stations in the same network. The processes of establishing multi-type pure electric buses system time-space network with the characteristic of recharging requirement are as follows:

1. Generate task nodes and task arcs. Generate the task start node and task end node according to the route timetable, and each node has its own time and space attributes. Assuming that there are N task vehicles, S_i is the departure site of number i; e_i is the arrival site of number i; d_i is the departure time of number i; a_i is the arrival time of number i. According to the above process, the start node $t(S_i, d_i)$ can be formed, and the start node of the task can be arranged in chronological order to form a set of task start nodes T; the end node $e(e_i, a_i)$ can be arranged in chronological order to form a set of end-of-task nodes E. Task arc: $(t, e), t \in T, e \in E$. The set of task arcs is A_{task}. Connect the task start-node with the corresponding task end node to generate the task arc.

2. Generate the parking lot node, exit arc, and entry arc. According to the task start node set, the task end node set and the empty-driving time between each station and parking yard, set up the corresponding exit node and entry node (collectively regarded as the parking yard node). Then set up the exit arc connecting the exit node and task start node, and the entry arc connecting the task end node and entry node.

3. Generate the waiting arc. Sort by stations first, then arrange them in time order and generate a collection of all the nodes.

4. Generate the empty-driving arc. Take out the task end node set and generate a DHE set according to the sequence of the task end time from the smallest to the largest. Sort the task start nodes according to the stations firstly, and then generate a T set according to the sequence of time from the smallest to the largest. Initialize the empty-driving arc.

2.2. The Representation of Vehicle Operating States

In the processes of optimizing the energy consumption of the multi-type buses operating organization, it is assumed that when there are no pure electric buses that can operate, the fuel buses will be arranged. Compared with the charging time of electric buses, the refueling time of the fuel buses is negligible, which can be assumed that the fuel buses can run unlimitedly. Assuming that if the energy of a pure electric bus is not enough to run the next task, it cannot charge in the halfway, instead, it needs to return to the yard to charge. Therefore, after the bus completes a certain task, it would have one of the following states. (1) Get into the parking yard through the entry arc and wait for the next task, as shown in Figure 1a; or considering the mileage range of the pure electric bus, if it has reached the maximum mileage, it has to get into the yard through the entry arc to recharge, as shown in Figure 1b. (2) The bus departs from the terminal station either directly or after a period of waiting, conduct the next task departing from this terminal station, which is shown in Figure 1c. (3) The bus arrives at the departure station of another task through an empty-driving arc to operate the next task. There is a certain waiting time between the termination of this task and the departure of the next task, which indicates there may be some empty-driving arcs and a certain number of waiting arcs between running the two tasks, which are shown in Figure 1d.

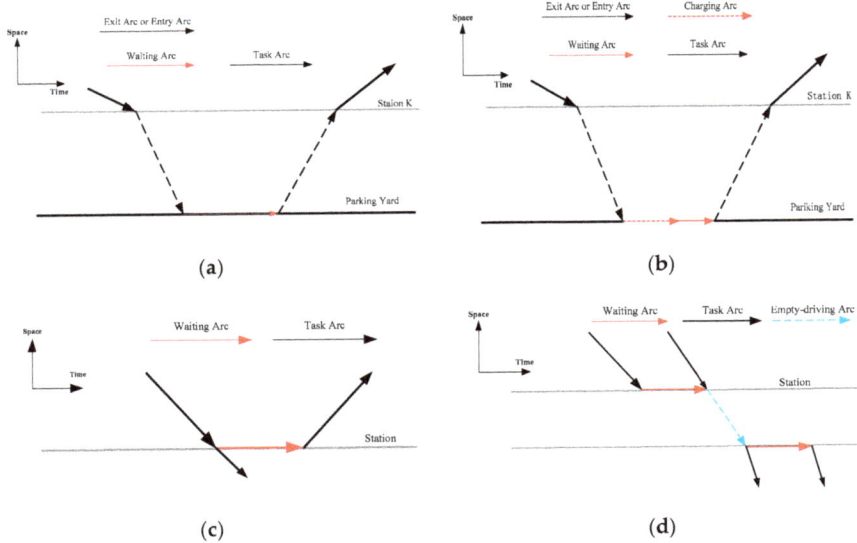

Figure 1. The representation of vehicle operating states in the time-space network. (**a**) The bus returns to the parking yard to wait for next task; (**b**) the pure electric bus gets into the yard to charge; (**c**) the connection of the two tasks at the same site; (**d**) the bus runs two tasks connecting with the empty running arcs.

3. Multi-Type Bus Operating Organization Energy Consumption Optimization Model

3.1. Problem Description

Compared with the operating organization of traditional buses, the multi-type bus operating organization has the constraints of both pure electric buses mileage range and charging time windows. In addition, in the context of "energy conservation and emission reduction", we not only consider the completion of operational tasks, but also take the reduction of energy consumption into consideration. Therefore, the bus operating organization studied in this paper is similar to the vehicle scheduling problem with the constraint of mileage range, charging time and the consideration of energy consumption optimization. At the same time, we study the regional bus driving mode, and the track of vehicles changes from a single route to several routes. The relationship among vehicles, stations, bus routes and task shifts become extremely complicated, making it more difficult to organize the operation of buses.

3.2. Model Establishment

3.2.1. Model Assumption

In the process of replacing fuel buses with electric buses, the condition that fuel buses and electric buses are mixed operated exists. Based on this background, we establish an operating organization energy consumption optimization model with the following assumptions: (1) The bus driving plan is scheduled with the unit of day, and it takes a minute as the smallest unit of time; (2) according to the data collected from the Beijing Bus Group, the maximum mileage for the 12-meter-long electric buses is 133 km and 117 km for the 18-meter-long electric buses. The buses powered by fuel and natural gases have unlimited mileage; (3) the quick charging time for electric buses is 30 min; (4) all the buses can run the task on time, and the model does not consider the condition of delay.

3.2.2. Objective Function

In the process of optimizing the multi-type bus operating plan while controlling the total energy consumption, we consider the service level of the bus, which includes the comfort indicators (waiting time, full load rate), convenience indicators (transfer distance, station coverage rate) and economic indicators (fare). However, for fixed bus lines, the transfer distance and station coverage indicators have been determined, and the ticket fare has no relation to the departure frequency, so the service level indicators considered in the paper mainly contains the waiting time and full load rate. The time cost of passengers waiting for a bus is mainly related to the time of waiting, and the full load rate is associated with the time cost of the passengers' in-bus time, because passengers have different perceptions of the travel time for the same travel distance under different full load rates. Therefore, we establish a multi-objective bus operating organization energy consumption optimization model. The objectives include the total energy consumption, passengers waiting time and passenger's in-bus time. By transforming different objectives into cost, multiple objectives can be converted into a single optimization objective with a unified unit of measurement, which can reduce the complexity and difficulty of solving the model.

The minimum total energy consumption E can be calculated as the sum of the product of vehicle per unit energy consumption and driving mileage for each task, which is shown as Formula (1).

$$minE = \sum_{p \in \Omega} \sum_{k \in K} \sum_{m \in M} e_k L_p \beta_{mp}^k \theta_p \tag{1}$$

Ω is the set of all the feasible bus order chain solutions. P represents the bus order chain. M is the set of buses. m represents the bus. K represents the vehicle type set. e_k is the unit (per kilometer) energy consumption of type k vehicles. L_p is the mileage of the bus order chain P. β_{mp}^k is a variable ranging from zero to one, representing whether the bus order chain P is executed by the bus m of the vehicle type k. If it is, β_{mp}^k is one, otherwise, β_{mp}^k is zero. θ_p is a variable ranging from zero to one, indicating whether the bus order chain P is in the feasible solution. If it is, then θ_p is one, otherwise θ_p is zero. $P = \{1, 2, 3, \ldots, i, n\}$ is a set representing the sequence of the exit arc, task arc, waiting arc, empty-driving arc, and entry are executed by a bus departing from a certain parking yard.

C_1 is the minimum passengers' waiting time. Since the passengers' waiting time cost is mainly decided by the waiting time, and the waiting time is associated with the departure frequency of the bus, the relationship between the departure frequency and the passengers' waiting time can be described as follows. Assuming that the law of the buses and the passenger arriving at the bus stations are subject to the Poisson Distribution and uniform distribution, respectively. The passengers' waiting time on average is the half of the departure interval and the bus lines' departure interval in unit time is equal to the reciprocal of departure frequency. The passengers' waiting time cost is inversely proportional to the departure frequency, and directly proportional to the number of people boarding the bus in the stations, so the calculation of the waiting time is shown in Formula (2).

$$minC_1 = \frac{1}{2f} \sum_n x^n \tag{2}$$

f is the departure frequency of the research bus line in the unit time period, with the unit of times/h. x^n is the number of passengers boarding the bus at station n in a unit time period, with the unit of persons.

C_2 is the minimum passengers in-bus time. The passengers' perception of the in-bus time cost is mainly influenced by the degree of the in-bus crowding (full load rate), which is determined by the departure frequency to some extent. The in-bus time cost of the same travel distance perceived by passengers is various under different crowding degrees, which is mainly affected by the in-bus time perception coefficient. Therefore, we consider that the passenger's in-bus time cost is mainly determined by the cross-sectional passenger flow per unit period, travel time and passenger's in-bus perception coefficient, as shown in Formula (3).

$$\min C_2 = \min \sum_k \sum_n x_{n,n+1} \frac{l_{n,n+1}}{v_k} F_k(x) \tag{3}$$

$$F_k(x) = 1 + \beta_{\min} \left(\frac{x_{n,n+1}}{f \cdot N_k} \right)^{\beta_{\max}}$$

$x_{n,n+1}$ is the cross-sectional passenger flow between station n and station $n + 1$ in a unit time period, with the unit of person. $l_{n,n+1}$ is the operation distance of the research route between station n and station $n + 1$, with the unit of km. v_k is the average velocity of the type k bus. $F_k(x)$ is the passenger perception coefficient of the type k bus when the cross-sectional passenger flow between station n and station $n + 1$ is x. N_k is the specified passenger capacity, with the unit of persons. g represents the time period. $\beta_{\min}$ is the minimum allowable full load rate of the research route, $\beta_{\max}$ is the maximum allowable full load rate of the research route.

Therefore, by converting the multiple objectives into a single optimization objective, the multi-objective operating organization energy consumption optimization model can be represented as Formula (4).

$$\min z = \min(\omega_1 E + \omega_2 C_1 + \omega_3 C_2)$$
$$= \min(\omega_1 \sum_{p \in \Omega} \sum_{k \in K} \sum_{m \in M} e_k L_p \beta_{mp}^k \theta_p + \omega_2 \frac{1}{2f} \sum_n x^n + \omega_3 \sum_k \sum_n x_{n,n+1} \frac{l_{n,n+1}}{v_k} F_k(x)) \tag{4}$$

$\omega_1, \omega_2, \omega_3$ is the cost converting coefficient.

3.2.3. Constraint Conditions

To ensure the original service level of the route, meeting the accurate matching between the transport capacity in the processes of replacing fuel buses with electric buses, while considering the mileage and charging time of electric buses, we establish constraints conditions from the aspect of the bus operating service level including the vehicle full load capacity, departure interval, road capacity, mileage of electric buses and charging time. The specific conditions are as follows:

1. The constraints of the full load rate

$$\beta_{\min} \le \beta_{gh} \le \beta_{\max} \tag{5}$$

$$\beta_{gh} = \frac{q_{gh}}{N_g C} = \frac{q_{gh} d_g}{N_g H_g}$$

In the formula, β_{gh} is the cross-section full load rate of the h section in a g time period of the research route, q_{gh} is the cross-section passenger flow of the h section in a g time period of the research route, N_g is the number of passengers on board of the h section in a g time period of the research route, d_g is the departure interval in a g time period of the research route, with the unit of minute, C is number of vehicles that passed the section h in a g time period, H_g is the duration of the g period, with the unit of minute, N_0 is the standard passenger capacity of the pure electric buses.

2. The constraint of departure interval

If the departure interval is too short, the energy consumption will increase that will result in the resource waste. If the departure interval is too long, the waiting time of passengers will be longer, and the full load rate will increase, then the service level decreases. Therefore, we comprehensively consider the acceptability of enterprises and passengers to the departure interval and refer to the Beijing bus departure interval which is less than 5 min in the peaking period and less than 15 min in the low peak period. We widen the constraints of departure frequency within a certain range and set the following constraint conditions:

$$4 \leq f \leq 60 \tag{6}$$

3. The constraints of mileage and charging time windows

$$\sum_{p \in \Omega} a_{ip} \theta_p = 1 \ i \in A_{task} \tag{7}$$

α_{ip} is a variable either zero or one, indicating whether the bus order chain P contains the arc i. If it does, α_{ip} equals to one, or α_{ip} equals to zero.

$$\sum_{p \in \Omega} s_{dp} \theta_p = \sum_{p \in \Omega} e_{dp} \theta_p \leq capacityd \quad \forall d \in D \tag{8}$$

s_{dp} is a variable either zero or one, representing whether the bus order chain P departs from the parking lot D or not. e_{dp} is a variable either zero or one, representing whether the bus order chain P stops at the parking lot d. D represents the parking yard set. *capacityd* is the capacity of the parking yard $\forall d \in D$. V_d^k is the vehicle set of type $k \in K$ departing from the parking yard $d \in D$.

$$\sum_i l_i a_{ip} \leq DD_k \forall p \in \Omega \quad i \in A_{task} \quad j \in A \tag{9}$$

DD_k represents the maximum mileage of type $k \in K$ bus. l_i is the mileage of arc i, and if the arc i is the waiting arc, $l_i = 0$, otherwise, it equals the distance between the two stations.

$$a_{ip} arrtime_i \leq a_{jp} deptime_j \ i < j, \forall p \in \Omega \tag{10}$$

arrtime$_i$ represents the arrival time of arc i. *deptime*$_i$ represents the departing time of arc i.

$$s_{dp} e_{dp} y_i \theta_p \in \{0, 1\} \ p \in \Omega, d \in D, i \in A_{task} \tag{11}$$

y_i is the variable ranging either zero or one, indicating whether the bus returns to the parking yard to charge after completing the task i (assuming that the bus can only be charged at the yard). If the bus needs to be charged, y_i equals to one, and the corresponding entry time is $a_i + t_{ei} + ST_k$. ST_k is the charging time of the bus, which is 30 min in the paper.

In summary, we establish a bus driving plan optimization model considering the service level with the objective function (4) and constraint conditions from formulas (5) to (11). The constraint condition (7) means every task can only be executed once by one bus. The left-hand side of the constraint condition Formula (8) represents the number of bus order chains departing from parking yard d, and the right-hand side represents the number of bus order chains finally returning to the parking yard, and both of them should not exceed the total number of vehicles in the yard d. Formula (9) means the driving distance should be less than the maximum driving distance if the bus has not been supplied with energy. Formula (10) represents the time connection constraints between the arcs. Constraints condition (11) means the related variables ranging from zero to one.

4. Research on Heuristic Optimization Algorithm

In this paper, the energy consumption optimization problem of multi-type bus operating organizations is abstracted into a set segmentation problem, which can be described as the problem that which vehicle completes task shifts in turn. The integer coding method based on vehicles is adopted for chromosome coding, which can be simply expressed as [1231312...]. The coding represents that task one, four and six are completed by the first bus, task two and seven are completed by the second bus, and task three and five are completed by the third bus.

4.1. The Production of Initial Population in Heuristic Algorithm

To solve the model, the paper sets the heuristic algorithm with the following steps:

1. Arrange the tasks in ascending order according to the task departure time to form T, and the task information includes the task sequence number, route number, departure time, arrival time, departure station, arrival station, travel time, route mileage, the nearest parking yard to the arrival station and the distance between them.
2. Set up the vehicle set in each parking yard, including the parking yard number and its capacity.
3. Assign one bus for the first task (the bus has not executed the task, which means the last task shift is empty), randomly extract an electric bus from the parking yard to execute, and record in the following order: The parking yard which the bus belongs to, the remaining mileage of the bus, the sequence number of the completed shift. Additionally, record the driving path of the current bus: The departure parking yard and the task sequence number.
4. For the rest of task *i*:

Look for whether there are any already used electric buses that can still execute the task i, which should meet the following two requirements: 1. Remaining mileage subtracts possible empty-driving mileage is greater than task mileage adds the mileage returning to the nearest parking yard; 2. the time when the last task was completed adds possible empty-driving time is less than the start time of the task adds 30 min (the charging time).

According to the following rules, relate the bus to the task:

1. Give priority to the buses meeting both requirements;
2. Give priority to the buses meeting the second requirement but not the first. In this case, the bus can return to the parking yard to charge, making it meet the first requirement. Record the nearest charging yard and update the bus information and the driving path information.
3. If all the requirements cannot be satisfied, the fuel buses will be used.

Look for the last task of each vehicle currently, assign the task *i* to the vehicle which has the smallest time difference from task *i*. Determine whether it is necessary to return to the parking yard to link up the tasks and update the vehicle information. All the individuals achieved based on the above heuristic algorithm are all the feasible solutions to the problem.

4.2. Design of Fitness Function

In the process of generating the initial solution, that is, the generation of the running plan of each vehicle, the corresponding task mileage, empty driving mileage, station or waiting time in the yard of the vehicle are recorded. In this paper, ObjV = $\sum_{p\in\Omega} c_p \theta_p$ is taken as the basis of the individual fitness evaluation, and the advantages and disadvantages of individuals are evaluated. The chromosomes that do not meet the constraints are eliminated. Calculate the objective function value of each individual and sort it. The fitness value of each individual is calculated according to its position in the arrangement. The fitness function of individual *i* is as follows:

$$g(i) = 2 - sp + 2(sp - 1)\left[\frac{pos - 1}{N - 1}\right] \tag{12}$$

In the formula, *sp* represents the selected pressure difference which is the difference between fitness values of assigned individuals, with a default value of two; N represents the size of the population; *Pos* represents its position in the arrangement.

4.3. Genetic Manipulation

Through selection, crossover and mutation to achieve the genetic manipulation, the specific process is as follows.

1. Selection. We adopt the roulette strategy RWS as the selection method. Since the probability of each individual being selected is proportional to its fitness function value. The probability of being selected is as follows:

$$p(i) = \frac{g(i)}{\sum_{i=1}^{N} g(i)} \tag{13}$$

On the basis of knowing the probability of each individual being selected, the selected individual is randomly determined. The specific steps are shown in Figure 2.

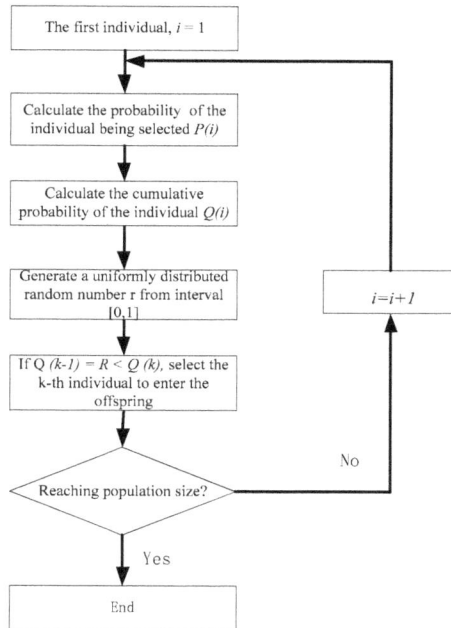

Figure 2. Individual selection step.

2. Crossover. The single point crossover is used. To adapt to the multi-constraint model proposed above, the following ideas are designed in the crossover operation.

    (a) Achieve the population from the selection operation for chromosome pairing;
    (b) According to the crossover probability, determine whether to carry out the crossover operation. If so, move to the next step, otherwise, move to the next species.
    (c) Search for feasible intersections. Decode to achieve the task shifts of each vehicle in the chromosome, randomly select two vehicles, and determine whether there is an intersection. If yes, conduct the crossover. Otherwise, pass it on to the next population until the two parents are inherited.
    (d) Move the chromosomes into the next population, determine whether the population size has been reached. If yes, terminate the crossover operation. Otherwise, repeat the above steps.

3. Variation. The variation ranges from 0.0001 to 0.1. Randomly select a certain task of a certain vehicle in the chromosome. Delete the task and the insert it into other vehicles at random, then determine whether the mileage range constraint can be satisfied. If it is feasible, then insert the task. Otherwise, continue to search for the vehicles that can be inserted. If no suitable insertion

position can be found in the existing vehicles, a new vehicle will be assigned to perform the corresponding task.

4.4. Design of Termination Principle

Before reaching the maximum number of iterations, determine whether the average fitness of successive generations has not changed, or the variation is less than a threshold. If so, the iterative process of the algorithm converges and the algorithm ends. Before reaching the maximum number of iterations, it is judged whether the average fitness of successive generations is unchanged or the change value is less than a minimum threshold. If so, the iteration process of the algorithm converges and the algorithm ends. Otherwise, if the maximum number of iterations has been reached, the new generation population obtained through selection, crossover and mutation will replace the previous generation population.

4.5. Flow Diagram of the Algorithm

The flow chart describing the established driving plan model is shown in Figure 3.

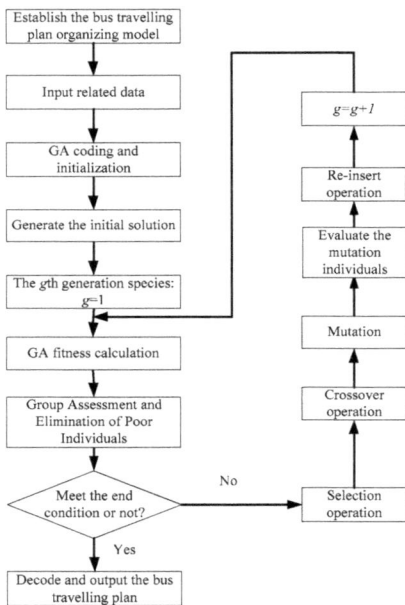

Figure 3. Flow diagram of the algorithm.

1. Input the parameter of the genetic algorithm: Species size, species generation, crossover probability, mutation probability, and the generation gap.
2. Input the task information, parking yard information, empty driving distance, average velocity, vehicle information and maximum mileage of pure electric buses. To make the calculation easier, the task information includes the task sequence number, route number, departure time, arrival time, departure station, arrival station, traveling time and route mileage. The parking yard information includes the parking yard number and the number of parking space in the yard. The vehicle information includes the vehicle number, parking yard number which the vehicle belongs to, remaining mileage, completion time of the last task, the sequence number of the last task, the information that whether the vehicle can execute the current task, the parking lot

number which the vehicle returns to, empty driving mileage, number of charge cycles and the vehicle energy consumption.
3. Input the coefficient of the vehicle energy consumption, the coefficient of passengers waiting cost and coefficient of passengers in-bus cost.

5. Result and Discussion

5.1. Explanation of Basic Information

Taking a certain bus system in Beijing as an example, we study multi-type buses scheduling and analyze an algorithm calculation example. The example contains six bus routes (Route 405, Route 415, Route 538, Route 1, Route 322 and Route 496), 12 timetables, 1040 tasks in total and seven bus parking yards. The parking yard information is shown in Table 1. The number of parking space in Table 1 is one of the constraints which is calculated by Formula (8).

Table 1. Parking yard information.

Number of Parking Yard	Name of Parking Yard	Number of Parking Space
1	Si Hui Junction Station Parking Yard	50
2	Sun He Bus Parking Yard	35
3	Hui Zhong Li Parking Yard	25
4	National Stadium East Parking Yard	25
5	Lao Shan Bus Parking Yard	25
6	Gu Cheng West Bridge Bus Parking Yard	25
7	Kang Jing Nan Li Parking Yard	25

The vehicle information is shown in Table 2. The energy consumption per 100 km is applied in the objective function (1). The specified passenger volume is one of the correlative conditions of the full load rate constraint Formula (5).

Table 2. Vehicle information.

Type of Fuel	Length of Bus(m)	The Sequence Number of Vehicle Type	Energy Consumption per 100 km (Standard Coal)	Specified Passenger Capacity (Person)
Electricity	18	V1	55.76	130
Diesel	16	V2	71.66	140
Natural Gas	16	V3	75.47	164

The bus route information is shown in Table 3. The route mileage is the key index for calculating energy consumption of the objective function (1).

Table 3. Bus route information.

Bus Route	Departure Parking Yard	Terminal Parking Yard	Route Mileage (km)
405	1	2	22.31
415	3	2	17.69
538	4	2	19.76
1	1	5	24.84
322	1	6	18.8
496	1	7	12.78

The empty driving distance between different parking yards is shown in Table 4. When the end of a task site is not the next start of a task site, there is an empty driving distance.

Table 4. Empty driving distance between different parking yards.

	Lao Shan Bus Parking Yard	Si Hui Junction Station Parking Yard	Gu Cheng West Bridge Bus Parking Yard	Sun He Bus Parking Yard	Hui Zhong Li Parking Yard	Kang Jing Nan Li Parking Yard	National Stadium East Parking Yard
Lao Shan Bus Parking Yard	0	37.7	1.9	44.7	29.8	42	30.2
Si Hui Junction Station Parking Yard	23.8	0	23.4	23.6	16.2	11	16.6
Gu Cheng West Bridge Bus Parking Yard	1.4	36.8	0	45.6	24	42.9	24.3
Sun He Bus Parking Yard	40.8	22.3	40.6	0	15.8	16.2	16.8
Hui Zhong Li Parking Yard	31.2	17.1	31.1	19.7	0	17.4	1.9
Kang Jing Nan Li Parking Yard	42.4	10.9	42.2	14.2	17.6	0	18.6
National Stadium East Parking Yard	22.4	19.1	21	24.4	3.5	21.6	3.9

The maximum mileage for the 12-meter-long electric buses is 133 km and 117 km for the 18-meter-long electric buses and the buses powered by fuel and natural gases have unlimited mileage. The quick charging time of pure electric buses is 30 min. The tasks of each route are executed jointly by the buses of seven parking yards. Convert the unit of energy consumption into the standard coal. According to the unit price of standard coal of 71.39 €/ton, $\omega_1 = 3.3$. The average annual salary level in Beijing in 2017 is 11033€. Calculated by 250 working days per year for 8 h per day, the average salary is 5.45 €/h, so $\omega_2 = \omega_3 = 5.45$. The parameters of the example are shown in Table 5.

Table 5. Setting of parameters in the calculation example.

ω_1	ω_2	ω_3
3.3	5.45	5.45

In the genetic algorithm proposed in this paper, the initial species size is 200 individuals, the crossover probability is 0.6, the mutation probability is 0.01, and the retention rate of good genes is 0.1.

5.2. Results Analysis

After calculation, the value of the objective function is 8456 € and a total of 144 buses are required. Among them, the number of vehicles needed for type V1, V2, V3 is 86, 14 and 44, respectively. The number of buses needed to be stopped in the parking yard is 48,36,19,16,8,7,10 respectively for the Si Hui Junction Station Parking Yard, Sun He Bus Parking Yard, Lao Shan Bus Parking Yard, Gu Cheng West Bridge Bus Parking Yard, Kang Jing Nan Li Parking Yard, Hui Zhong Li Parking Yard and National Stadium East Parking Yard. The total mileage of the tasks is 19,741 km and the total number of the charge is 70. The program iteration diagram is shown as Figure 4.

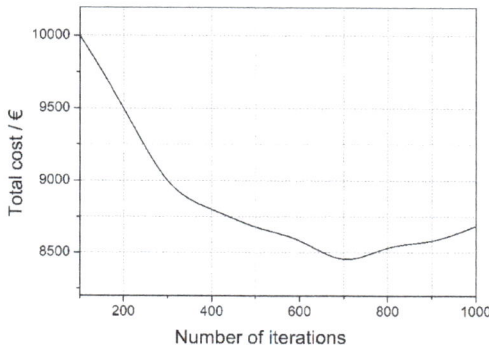

Figure 4. Program iteration diagram.

The bus order chain of each vehicle is shown in Table 6, in which 1041 to 1047 represents the parking yard one to seven, and if they are in the middle of the chain, it means the bus gets into the parking yard to charge.

Table 6. Bus order chain.

Bus Sequence Number	Vehicle Type	Bus Order Chain
1	V1	1041-682-429-759-934-38-882-47-893-1043
2	V1	1043-411-1046-839-771-128-865-1045
3	V3	1044-10-421-321-687-206-1047
4	V1	1046-966-868-1041-236-257-727-203-354-763-1044
5	V1	1042-326-163-1041-870-1043
6	V1	1042-925-1046-545-379-1045
7	V1	1042-511-1043-492-148-1046
8	V1	1047-791-516-157-730-765-582-1043
9	V3	1042-85-234-561-622-697-590-292-552-1041
10	V3	1045-136-653-662-917-838-577-584-1046
...	...	...
142	V2	1043-650-747-372-908-595-949-1044
143	V3	1041-497-996-51-578-41-94-910-1047
144	V1	1041-25-263-422-1046-651-846-931-1035-599-1046

5.3. Comparative Analysis

From the program iteration chart, we can see the variation trend. The optimal scheme is the lowest cost, but in the actual situation, it is impossible to achieve the optimal configuration at once. The capacity of the yard and the charging infrastructure need to be expanded. In the process of gradually replacing fuel vehicles by pure electric buses, we need to pay attention to how costs change and what proportion of fuel vehicles should be replaced by pure electric buses, so we choose different vehicle ratios to discuss. The cost of different plans with different vehicle type proportions is shown in Table 7. Energy consumption cost one refers to the energy consumption cost of pure electric buses, and energy consumption cost two refers to the energy consumption cost of diesel buses and liquefied natural gas buses. The proportion of two types of buses of five different plans and the changes in costs in each case are shown in Table 7.

Table 7. The cost of different vehicle type proportions.

Plans	Number of Buses (Num)			Cost (Euro)					Number of Charge Times
	Total Number of Buses	Number of Electric Buses	Number of Fuel Buses	Energy Consumption Cost1	Energy Consumption Cost 2	Passengers Cost	Total Cost		
Fuel buses increase by 20%	132	59	73	2083	4860	1736	8679		46
Fuel buses increase by 10%	138	70	68	2142	4553	1888	8583		58
The optimal plan	144	83	61	3187	3187	2082	8456		70
Fuel buses decrease by 10%	150	95	55	3808	2539	2185	8533		82
Fuel buses increase by 20%	158	111	47	4080	2197	2306	8583		96
Only electric buses	205	205	0	6188	0	2238	8686		114

It can be concluded from the above table that:

1. From the 1–3 columns of the table, we can see that in the process of pure electric buses replacing fuel vehicles, the total number of vehicles increases gradually. The proportion of the number of reducing fuel vehicles and increasing pure electric buses is about 1:2., because pure electric buses with similar passenger capacity need to return to the parking lot to recharge after reaching the limited mileage. Therefore, in practical application, it is necessary to replace one fuel vehicle with two pure electric vehicles.

2. From the 4–5 columns of the table, we can see that energy consumption cost one increases while energy consumption cost two decreases. However, the energy consumption cost shows a decreasing trend, indicating that the energy consumption of the pure electric bus is much lower than that of the fuel vehicle. On the premise of ensuring the completion of the operation task, the energy consumption of the increased pure electric bus is lower than that of the reduced fuel vehicle. Therefore, considering the energy consumption alone, it is better to keep the higher the proportion of the electric bus.

3. The total cost is shown as concave in column 7 of Table 7, with decreasing energy costs and increasing passengers cost. It indicates that when the number of electric buses increases to a certain proportion, the saving costs on energy are not enough to make up for the rising passengers cost. There is a balance between the two costs in order to achieve the optimal (the lowest total cost). Therefore, it is not acceptable to merely consider energy consumption reduction from the perspective of the total cost. How to choose depends on which aspect the bus enterprises focus on, and at the same time, they should respond to the national policy.

4. As is shown in column 8 of Table 7, increasing charging times is inevitable in the process of increasing the proportion of pure electric buses, which requires the expansion of charging infrastructure.

5. The total number of vehicles in column 1 of Table 7 shows that in the process of replacing fuel buses by pure electric buses, the size of buses keeps increasing, so the capacity of the parking yard and charging equipment need to be expanded.

The current discussion on energy consumption saving is mostly concentrated in reducing the number of cars and formulating policies on reducing car use, but the studies on public transit are less. Based on the characteristics of energy-intensive for the traditional buses, blindly advocate public transit may save energy but it is not the best solution. The application of the electric bus in a reasonable ratio can achieve the reduction on public transport energy consumption. The current situation is that the environmental-friendly electric buses are gradually replacing the fuel vehicles. Although the replacement is not entire, the scale is gradually expanding. Therefore, the study on the schedule of the bus driving plan to reduce the energy consumption is particularly important. The variation of the single bus type operation is not reasonable, which may easily lead to the waste of capacity during flat peak periods and the shortage of capacity during peak periods. Different types of buses can carry different passenger loads, so different bus type matching ratios should be considered according to the passenger flow.

6. Conclusions

This paper makes an in-depth analysis on the relationship between the bus driving energy consumption and bus dispatching and bus type matching ratio under the background that pure electric buses gradually replaces traditional fuel buses, and many routes are operated with mixing pure electric buses and traditional fuel buses. There are mainly two steps to solve the problem. The first is to establish an optimization model of the multi-type bus operation energy consumption based on the time-space network and reduce the scale of the problem by cutting the empty driving arc. In the second, the genetic algorithm is applied to optimize the multi-objective function to obtain the optimal driving scheme. The proportion of electric vehicles replacing fuel vehicles is analyzed with examples under the situation of gradual reduction of fuel vehicles. The optimal vehicle scheduling scheme and vehicle type ratio are obtained. In addition, the energy consumption cost and passenger cost are calculated in each case, and the suggestion of expanding the parking yards is given.

Currently, the Chinese government vigorously promotes the public transport with low energy consumption. The ratio of electric buses in the bus fleets increases, and mixing operation with multiple bus types makes the bus operating organization more complex. The paper innovatively takes the energy consumption as the objective and organizes the bus operation. The difference from the traditional fuel bus operating organization lies in the constraints of mileage range and charging time. In the multi-type energy consumption optimization model, the solution scale of the problem is reduced by establishing the time-space network and cutting the empty driving arcs. Due to the constraint of the charging time window of pure electric buses, two pure electric buses need to be added to replace one fuel bus, and the parking yard capacity needs to be expanded correspondingly. The bus type matching ratio is different for the situation considering the energy consumption cost alone and the situation considering the total cost. The decision depends on the preference of the decision makers. However, under the dual pressure of the environment and energy consumption, the growth of pure electric buses is a trend, and it also needs a stronger policy support from the government.

Author Contributions: Conceptualization, Y.L. and E.Y.; Methodology, Y.L. and E.Y.; Validation, E.Y.; Formal analysis, Y.L., S.L.; Writing—Original draft preparation, Y.L.; Writing—Review and editing, Y.L., E.Y., S.L.

Funding: This research was funded by National Key R&D Program of China, grant number 2018YFB1601300.

Conflicts of Interest: The authors declare no conflict of interest.

References

1. Bodin, L.; Golden, B.; Assad, A.; Ball, M. Routing and scheduling of vehicles and crews: The state of the art. *Comput. Oper. Res.* **1983**, *10*, 63–211. [CrossRef]
2. Freling, R.; Paixão, J.M.P. *Vehicle Scheduling with Time Constraint*; Springer: Berlin/Heidelberg, Germany, 1995.
3. Banihashemi, M.; Haghani, A. Optimization model for large-scale bus transit scheduling problems. *Transp. Res. Rec.* **2000**, *10*, 23–30. [CrossRef]
4. Haghani, A.; Banihashemi, M.; Chiang, K.H. A comparative analysis of bus transit vehicle scheduling models. *Transp. Res. Part B* **2003**, *37*, 301–322. [CrossRef]
5. Haghani, A.; Banihashemi, M. Heuristic approaches for solving large-scale bus transit vehicle scheduling problem with route time constraints. *Transp. Res. Part A* **2002**, *36*, 309–333. [CrossRef]
6. Oukil, A.; Amor, H.B.; Desrosiers, J.; El Gueddari, H. Stabilized column generation for highly degenerate multiple-depot vehicle scheduling problems. *Comput. Oper. Res.* **2007**, *34*, 817–834. [CrossRef]
7. Desaulniers, G.; Lavigne, J.; Soumis, F. Multi-depot vehicle scheduling problems with time windows and waiting costs. *Eur. J. Oper. Res.* **1998**, *111*, 479–494. [CrossRef]
8. Forbes, M.A.; Holt, J.N.; Watts, A.M. Exact Solution of Locomotive Scheduling Problems. *J. Oper. Res. Soc.* **1991**, *42*, 825–831. [CrossRef]
9. Kliewer, N.; Mellouli, T.; Suhl, L. A time—Space network based exact optimization model for multi-depot bus scheduling. *Eur. J. Oper. Res.* **2006**, *175*, 1616–1627. [CrossRef]
10. Guedes, P.C.; Borenstein, D. Column generation based heuristic framework for the multiple-depot vehicle type scheduling problem. *Comput. Ind. Eng.* **2015**, *90*, 361–370. [CrossRef]
11. Pepin, A.S.; Desaulniers, G.; Hertz, A.; Huisman, D. A Comparison of Five Heuristics for the Multiple Depot Vehicle Scheduling Problem. *J. Sched.* **2009**, *12*, 17–30. [CrossRef]
12. Wren, A.; Holliday, A. Computer Scheduling of Vehicles from One or More Depots to a Number of Delivery Points. *Oper. Res. Q.* **1972**, *23*, 333–344. [CrossRef]
13. Smith, B.M.; Wren, A. VAMPIRES and TASC: Two Successfully Applied Bus Scheduling Programs. *Comput. Sched. Public Transp.* **1981**, 97–124.
14. Wren, A.; Kwan, R.S.K. Installing an Urban Transport Scheduling System. *J. Sched.* **1999**, *2*, 3–17. [CrossRef]
15. Shen, Y.; Ni, Y. A Public Transport Scheduling System Based On Tabu Search. *Urban Public Transp.* **2006**, *9*, 34–39. [CrossRef]

16. Eliiyi, D.T.; Ornek, A.; Karakiitiik, S.S. A Vehicle Scheduling Problem with Fixed Trips and Time Limitations. *Int. J. Prod. Econ.* **2009**, *117*, 150–161. [CrossRef]

17. Kliewer, N.; Mellouli, T.; Suhl, L. A new solution model for multi-depot multi-vehicle-type vehicle scheduling in (sub) urban public transport. In Proceedings of the 13th Mini-EURO Conference and the 9th Meeting of the EURO Working Group on Transportation, Politechnic of Bari, Bari, Italy, 10–13 June 2002.

18. Naumaim, M.; Leena, S.; Stefan, K. A Stchastic programming approach for robust vehicle scheduling in public bus transport. *Procedia Soc. Behav. Sci.* **2011**, *20*, 826–835. [CrossRef]

19. He, D. *Research on Regional Bus Scheduling Problem under APTS*; Southwest Jiaotong University: Chengdu, China, 2009.

20. Yang, Y.; Guan, W.; Ma, J.H. Research on Scheduling Optimization of electric bus scheduling based on column generation algorithm. *Transp. Syst. Eng. Inf.* **2016**, *16*, 198–204.

21. Bodin, L.; Golden, B. Classification in vehicle routing and scheduling. *Networks* **1981**, *11*, 97–108. [CrossRef]

22. Dell'Amico, M.; Fischetti, M.; Toth, P. Heuristic Algorithms for the Multiple Depot Vehicle Scheduling Problem. *Manag. Sci.* **1999**, *39*, 115–125. [CrossRef]

23. Freling, R.; Wagelmans, A.P.M.; Paixão, J.M.P. An Overview of Models and Techniques for Integrating Vehicle and Crew Scheduling. In *Computer-Aided Transit Scheduling*; Springer: Berlin/Heidelberg, Germany, 1999; pp. 21–28.

24. Laurent, B.; Hao, J.K. Iterated local search for the multiple depot vehicle scheduling problem. *Comput. Ind. Eng.* **2009**, *57*, 277–286. [CrossRef]

25. Wei, M.; Jin, W.Z.; Sun, B. Bi-Levei Programming Model for Scheduling and Procurement Scheme of Regional Bus. *J. South China Univ. Technol.* **2011**, *39*, 118–123. [CrossRef]

26. Yang, Y.F. *Research and Application of Multi-Depot Vehicle Scheduling Problem Based on Simulated Annealing Genetic Algorithms*; Suzhou University: Suzhou, China, 2006.

27. Zou, Y. Research on the Method of Bus Area Dispatching Planning. *Transp. Syst. Eng. Inf.* **2007**, *7*, 78–82.

28. Li, J. Assignment Heuristic Algorithms System for Vehicle Scheduling Problem. *Eng. Theory Pract.* **1999**, *19*, 27–33.

29. Li, J. Network Heuristic Algorithms for Vehicle Scheduling Problem with Time Window. *Syst. Eng.* **1999**, *17*, 66–71.

applied sciences

MDPI

Article

Robust Planning of Energy and Environment Systems through Introducing Traffic Sector with Cost Minimization and Emissions Abatement under Multiple Uncertainties

Cong Chen [1,*], **Xueting Zeng** [2], **Guohe Huang** [3,*], **Lei Yu** [4] and **Yongping Li** [5]

[1] Donlinks School of Economics and Management, University of Science and Technology Beijing, Beijing 100083, China
[2] School of Labor Economics, Capital University of Economics and Business, Beijing 100070, China; zxt1231@sina.com
[3] Professor and Canada Research Chair, Environmental Systems Engineering Program, Faculty of Engineering and Applied Science, University of Regina, Regina, SK S4S 0A2, Canada
[4] School of Water Conservancy & Environment, Zhengzhou University, Zhengzhou 450001, China; yulei1060220069@sina.com
[5] Environment and Energy Systems Engineering Research Center, School of Environment, Beijing Normal University, Beijing 100875, China; yongping.li33@gmail.com
* Correspondence: chencong@ustb.edu.cn (C.C.); huang@iseis.org (G.H.); Tel.: 010-62332207 (C.C.); 306-585-4095 (G.H.)

Received: 4 December 2018; Accepted: 13 February 2019; Published: 5 March 2019

Abstract: Motor vehicles have been identified as a growing contributor to air pollution, such that analyzing the traffic policies on energy and environment systems (EES) has become a main concern for governments. This study developed a dual robust stochastic fuzzy optimization—energy and environmental systems (DRSFO-EES) model for sustainable planning EES, while considering the traffic sector through integrating two-stage stochastic programming, robust two-stage stochastic optimization, fuzzy possibilistic programming, and robust fuzzy possibilistic programming methods into a framework, which can be used to effectively tackle fuzzy and stochastic uncertainties as well as their combinations, capture the associated risks from fuzzy and stochastic uncertainties, and thoroughly analyze the trade-offs between system costs and reliability. The proposed model can: (i) generate robust optimized solutions for energy allocation, coking processing, oil refining, heat processing, electricity generation, electricity power expansion, electricity importation, energy production, as well as emission mitigation under multiple uncertainties; (ii) explore the impacts of different vehicle policies on vehicular emission mitigation; (iii) identify the study of regional atmospheric pollution contributions of different energy activities. The proposed DRSFO-EES model was applied to the EES of the Beijing-Tianjin-Hebei (BTH) region in China. Results generated from the proposed model disclose that: (i) limitation of the number of light-duty passenger vehicles and heavy-duty trucks can effectively reduce vehicular emissions; (ii) an electric cars' policy is enhanced by increasing the ratio of its power generated from renewable sources; and (iii) the air-pollutant emissions in the BTH region are expected to peak around 2030, because the energy mix of the study region would be transformed from one dominated by coal to one with a cleaner pattern. The DRSFO-EES model can not only provide scientific support for the sustainable managing of EES by cost-effective ways, but also analyze the desired policies for mitigating pollutant emissions impacts with a risk adverse attitude under multiple uncertainties.

Keywords: dual robust optimization; risk aversion; energy and environmental systems; vehicular emissions; multiple uncertainties

Appl. Sci. **2019**, *9*, 928

1. Introduction

1.1. Background

China has experienced severe environmental pollution in recent years, which pose a critical threat to public health and sustainable development [1,2]. Energy-related activities are the dominant sources of air pollution [3], with the amounts of carbon dioxide (CO_2) and air pollutant emitted from electricity generation plants accounting for approximately 40% and 30% of the total CO_2 and air pollutant emissions, respectively [2,4]. Further, motor vehicles have been identified as growing contributors to air pollution due to the rapid growth of transportation, accounting for approximately 20–67% of carbon monoxide (CO) emissions, 12–36% of oxynitride (NOx) emissions, and 12–39% of hydrocarbon compound (HC) emissions [5,6]. The scale of emissions of most of China's regions has exceeded the capacity of self-purification and air-pollutants' diffusion from the atmosphere. There is currently a severe conflict between increasing energy demand, excessive vehicle population, and "high coal" energy mix on the one hand, and the imperative of mitigating air pollution on the other hand [7]. To tackle the above-mentioned problems, several policies and measures have been implemented with regard to the development of renewable energy resources: adjustment of the energy structure; encouragement of the use of electric cars (EVs); improvement of energy conversion efficiencies; and enhancement of vehicular emission standards. However, it remains unclear how much pollution reduction can be achieved by these control measures and policies. This situation has forced local managers to propose ambitious schemes for planning energy and environment systems (EES), and to deeply analyze the impacts of different emission mitigation policies and measures on these EES [8,9]. However, EES are complicated by many systemic uncertainties regarding the relevant environmental, economic, energy, and social factors. For instance, electricity demands are often shown as stochastic uncertainties that vary over time based on extant policies and highly variable conditions [10]. Moreover, many economic data and energy demands often exhibit ambiguity [11,12]. Failure to consider these uncertainties may result in less robust decision support [8,13]. Therefore, EES planning as well as considering uncertainty information are required to help confront such problems of EES, and to ensure sustainable economic development and environmental protection [14,15].

1.2. Literature Review

Numerous non-deterministic programming approaches have been used to handle uncertainties in EES. Table 1 lists some previous studies on non-deterministic programming problems.

Table 1. Previous studies related to the subject.

Ref. No.	Non-Deterministic Programming				Research Area		Considering Traffic Sector	
	TSP	RTSO	FPP	RFPP	Energy Systems	Others	Yes	No
[5]	√				√			√
[17]	√				√			√
[18]	√				√		√	
[19]	√				√		√	
[20]	√				√			√
[21]		√			√			
[22]	√	√			√			√
[23]			√		√			√
[24]	√	√				√		√
[25]			√	√	√			√
[26]		√				√		√
[27]			√	√		√		√
[28]	√		√			√		√
[29]			√			√		√
[30]			√			√		√
[31]			√		√			√
[32]			√	√		√		√

Note: TSP, two-stage stochastic programming; RTSO. robust two-stage stochastic optimization; FPP, fuzzy possibilistic programming; and RFPP, robust fuzzy possibilistic programming.

Among them, two-stage stochastic programming (TSP) has been widely used to tackle uncertainties expressed as a probability distribution [3,16–18]. For instance, Gong et al. [16] proposed a two-stage programming method to optimize electric power systems considering air pollutant emissions and CO_2 mitigation. Mavromatidis et al. [19] developed a two-stage integer linear program model to optimize distributed energy systems, enable cost-optimized design decisions regarding technology selection and sizing before the determination of uncertain parameters. Mohan et al. [20] presented a two-stage stochastic method for managing the energy reserve of a microgrid system, with an emphasis on the different levels and sources of uncertainties. However, the TSP is unable to regard the variability of stochastic recourse values because it is based on the assumption that the manager adopts a risk-neutral attitude. Thus, TSP may become infeasible when managers are risk-averse under the conditions of high variability [21].

The robust two-stage stochastic optimization (RTSO) method is an attractive method for tackling the above shortcomings of TSP. It is specifically used to penalize the costs of the second-stage that are greater than the expected values and capture the associated risk of stochastic uncertainties [22–24]. In the last few decades, the RTSO method has been extensively employed in many research areas, such as supply chain systems, electric power systems, solid waste management, and water resource allocation [3,23,25]. For example, Govindan and Cheng [26] developed a stochastic robust programming method for improving retail supply chain planning through supply chain coordination, risk reduction, vendor selection, and sustainability assessment. Xu et al. [24] proposed a robust TSP method for tackling water resource allocation problems, enabling the handling of uncertainties expressed as stochastic, and analysis of policy scenarios regarding economic penalties for the violation of predefined policies.

However, in the real world, many economic parameters and energy demands often exhibit ambiguities, which can be shown as fuzzy sets [27,28]. Fuzzy possibilistic programming (FPP) theory can effectively address the fuzzy uncertainties of goals and constraints [29]. For instance, Vahdani et al. [30] employed an FPP method for closed-loop recycling collection networks, in which some uncertainty information (e.g., distance, capacity, demand, costs, as well as returned products quantity) were tackled by FPP. Lu et al. [31] proposed an interval FPP method for managing China's energy systems with CO_2 emissions constraints, which could address the uncertainties presented in terms of fuzzy-boundary intervals in both the objective and constraints.

However, an FPP algorithm is unable to ensure the minimization of objective function under all conditions because minimizing of the expected objective value is used as the objective function. This can result in significant deviations of the optimized decision schemes, even in the event of system optimization failure. Robust fuzzy possibilistic programming (RFPP) was developed by Pishvaee et al. [32] to overcome the drawbacks of FPP methods and involves the extension of robust optimization from stochastic algorithms to fuzzy algorithms. RFPP considers three sections in objective function: (i) the minimization of the weight sum of the expected objective values; (ii) the difference between two possible extreme objective values; and (iii) the penalty for constraint violation as the objective function [27,33]. It has, however, been limitedly applied to EES planning.

Generally, although previous research works can effectively deal with EES issues under multiple uncertainties, several gaps still need to be remedied. Firstly, few of the previous studies considered the traffic sector, which has been identified as a growing contributor to air pollution. Currently, a series of traffic policies have been adopted to alleviate the air pollution caused by motor vehicles. However, it remains unclear how much pollution reduction can be achieved by these control measures and policies. Secondly, most of these studies are incapable of considering the system risks from the stochastic and fuzzy uncertainties during the optimization process, which may lead to significant deviations in the optimized decision schemes, even in the event of system optimization failure. Thirdly, the RFPP method is commonly used to plan water resource allocation, and solid management, and is scarcely applied to plan EES systems.

1.3. Objective

The objective of this study was the development of a dual robust stochastic fuzzy optimization—energy and environmental systems (DRSFO-EES) model for planning EES while considering the traffic sector. This study is the first attempt at planning an EES while considering the traffic sector by integrating TSP, FPP, RTSO, and RFPP into a single framework. The proposed model can effectively tackle stochastic and fuzzy uncertainties as well as their combinations, capture associated risks from fuzzy and stochastic uncertainties, and thoroughly analyze trade-offs between system costs and reliability. The proposed DRSFO-EES model was applied to the Beijing-Tianjin-Hebei (BTH) region in China, which experiences severe smog and haze associated with high concentrations of air pollutants. The following is a detailed enumeration of the capabilities of the proposed model: (i) exploration of the impacts of different vehicle policies (such as regarding EVs usage, EV's power source, and vehicular emission standards) on vehicular emission mitigation via scenario analysis; (ii) generation of robust optimized solutions for energy allocation, coking processing, oil refining, heat processing, electricity generation, electricity power expansion, electricity importation, energy production, as well as emission mitigation under multiple uncertainties; (iii) identification of the study regional atmospheric pollution contributions of different energy activities such as coke processing, heat processing, oil refining, electricity generation, and motor vehicle operation.

2. Methodology

The robust two-stage stochastic optimization (RTSO) method brings risk aversion into stochastic programming methods, and finds robust schemes for system management [33,34]. The RTSO method can deal with the stochastic uncertainties of real-world management problems, analyze the policy scenarios associated with economic penalties when the predefined policies of the first-stage are violated, capture the variability of the second-stage costs that are greater than the expected values, and evaluate trade-offs between system economy and risk [23,24]. An RTSO method is formulated as follows:

$$\text{Min} f = C_{T_1} X + \sum_{h=1}^{s} p_h D_{T_2} Y_h + \rho_1 \sum_{h=1}^{s} p_h V \tag{1}$$

subject to

$$A_r X \le B_r, r \in M; M = 1, 2, \ldots, m_1 \tag{2}$$

$$A_r X + A'_{rh} Y_h \ge w'_{ih}, i \in M; i = 1, 2, \ldots, m_2; h = 1, 2, \ldots, s \tag{3}$$

$$x_j \ge 0, x_j \in X; j = 1, 2, \ldots, n_1 \tag{4}$$

$$y_{jh} \ge 0, y_{jh} \in Y; j = 1, 2, \ldots, n_2; h = 1, 2, \ldots, v \tag{5}$$

$$V = D_{T_2} Y_h - \sum_{h=1}^{s} p_h D_{T_2} Y_h + 2\theta_h, h = 1, 2, \ldots, s \tag{6}$$

$$v_{jh} \ge 0, v_{jh} \in V; j = 1, 2, \ldots, n_2; h = 1, 2, \ldots, s \tag{7}$$

where, X and Y_h denote the decision variables of the first-stage and the second-stage, respectively; p_h are occurrence probability of scenario h, $\sum_{h=1}^{H} p_h = 1$; C represent coefficients of X and D_h are coefficients of Y_h; w'_{ih} are random variables with probability levels p_h; A_r is the fixed coefficient of X; A_{rh}' are coefficients of Y_h. B_r represent the boundary vectors of the right-hand side of constraints. θ_h denote slack variables used for achieving looser constraints. ρ_1 denotes a goal programming weight of stochastic uncertainties; the managers can regulate the variability of the stochastic recourse cost through adjusting the ρ_1 level [24,35]. When $\rho_1 = 0$, the RTSO model becomes a conventional TSP, which indicates that the managers adopt risk-neutral attitudes and the variability of the stochastic uncertain recourse costs is not considered. However, when $\rho_1 = 1$, the managers adopt risk- aversive attitudes and the variability of the second-stage cost is considered. V denotes the deviation of an expected value from the given scenario's cost [3]. Besides, constraints (6) and (7) can define the positive variability of the recourse costs.

However, RTSO is inefficient in addressing the uncertainties expressed by fuzzy sets. Thus, fuzzy possibilistic programming (FPP) is joined to RTSO as a hybrid robust stochastic-fuzzy optimization (RSFO) model as follows:

$$\text{Min } \tilde{f} = \tilde{C}_{T_1} X + \sum_{h=1}^{s} p_h \tilde{D}_{T_2} Y_h + \rho_1 \sum_{h=1}^{s} p_h V \tag{8}$$

subject to

$$Cr\{\tilde{B} | A_r X \leq \tilde{B}_r\} \geq \alpha, r \in M; M = 1, 2, \ldots, m_1 \tag{9}$$

$$A_r X + A'_{rh} Y_h \geq w'_{ih}, i \in M; i = 1, 2, \ldots, m_2; h = 1, 2, \ldots, s \tag{10}$$

$$x_j \geq 0, x_j \in X; j = 1, 2, \ldots, n_1 \tag{11}$$

$$y_{jh} \geq 0, y_{jh} \in Y; j = 1, 2, \ldots, n_2; h = 1, 2, \ldots, v \tag{12}$$

$$V = D_{T_2} Y_h - \sum_{h=1}^{s} p_h D_{T_2} Y_h + 2\theta_h, h = 1, 2, \ldots, s \tag{13}$$

$$v_{jh} \geq 0, v_{jh} \in V; j = 1, 2, \ldots, n_2; h = 1, 2, \ldots, s \tag{14}$$

where $Cr\{\cdot\}$ represents the credibility measure of a fuzzy event in $\{\cdot\}$; $\tilde{C}_{T_1}$ and $\tilde{D}_{T_2}$ are cost coefficients expressed as a triangular fuzzy number; $\tilde{B}_r(B_r^1, B_r^2, B_r^3)$ is the boundary vectors of the right-hand side of constraints, which expresses as triangular fuzzy sets with its membership functions $\mu(\tilde{B}_r)$. α denotes the predetermined confidence-level. $Cr\{\tilde{B} | A_r X \leq \tilde{B}_r\} \geq \alpha$ denotes the credibility of satisfying $A_r X \leq \tilde{B}_r$ is higher than or equal to confidence-level α. According to reference [29], the detailed solution procedures of model (2) can be summarized as: firstly, transforming objective function (8) into its expected value form; secondly, converting constraints (9) into their crisp equivalents. A series of solutions can be obtained under difference confidence-levels [25].

However, an FPP algorithm cannot ensure the minimization of objective function under all conditions because the expected objective value is used as the objective function. This may result in significant deviations of the optimized decision solutions, even in the event of system optimization failure. Robust fuzzy possibilistic programming (RFPP) was developed by Pishvaee et al. [32] to overcome the above-mentioned drawbacks of FPP methods. Additionally, the RFPP method is scarcely applied to EES systems, which are commonly used to plan water resource allocation, and solid management [25,27].

Therefore, this study developed a dual robust stochastic fuzzy optimization (DRSFO) method through integrating TSP, RTSP, FPP, and RFPP methods in a single framework, which can be used to effectively tackle fuzzy and stochastic uncertainties as well as their combinations, capture the associated risks from fuzzy and stochastic uncertainties, as well as thoroughly analyze the trade-offs between system costs and reliability. In detail, it is formulated as follows:

$$
\begin{aligned}
\text{Min } \tilde{f} &= \overline{f} + \lambda_1 \cdot (f_{\max} - \overline{f}) + \lambda_2 \cdot (\overline{f} - f_{\min}) + \rho_2 \cdot [B^R(\alpha) - B_r{}^1] \\
&= 0.25 \cdot (C_{T_1}^1 + 2 \cdot C_{T_1}^2 + C_{T_1}^3) \cdot X + \lambda_1 \cdot [C_{T_1}^3 - 0.25 \cdot (C_{T_1}^1 + 2 \cdot C_{T_1}^2 + C_{T_1}^3)] \cdot X \\
&+ \lambda_2 \cdot [0.25 \cdot (C_{T_1}^1 + 2 \cdot C_{T_1}^2 + C_{T_1}^3) - C_{T_1}^1] \cdot X \\
&+ 0.25 \cdot (D_{T_2}^1 + 2 \cdot D_{T_2}^2 + D_{T_2}^3) \cdot \sum_{h=1}^{s} p_h \cdot Y_h \\
&+ \lambda_1 \cdot [D_{T_2}^3 - 0.25 \cdot (D_{T_2}^1 + 2 \cdot D_{T_2}^2 + D_{T_2}^3)] \cdot \sum_{h=1}^{s} p_h \cdot Y_h \\
&+ \lambda_2 \cdot [0.25 \cdot (D_{T_2}^1 + 2 \cdot D_{T_2}^2 + D_{T_2}^3) - D_{T_2}^1] \cdot \sum_{h=1}^{s} p_h \cdot Y_h \\
&+ \rho_1 \cdot \sum_{h=1}^{s} p_h \, V + \rho_2 \cdot [B^R(\alpha) - B_r{}^1]
\end{aligned}
\tag{15}
$$

subject to:

$$
Cr\{\tilde{B} | A_r X \le \tilde{B}_r\} \ge \alpha, r \in M; \; M = 1, 2, \ldots, m_1 \tag{16}
$$

$$
A_r X + A'_{rh} Y_h \ge w'_{ih}, \; i \in M; \; i = 1, 2, \ldots, m_2; \; h = 1, 2, \ldots, s \tag{17}
$$

$$
x_j \ge 0, \; x_j \in X; j = 1, 2, \ldots, n_1 \tag{18}
$$

$$
y_{jh} \ge 0, \; y_{jh} \in Y; j = 1, 2, \ldots, n_2; \; h = 1, 2, \ldots, v \tag{19}
$$

$$
V = D_{T_2} Y_h - \sum_{h=1}^{s} p_h D_{T_2} Y + 2\theta_h, \; h = 1, 2, \ldots, s \tag{20}
$$

$$
v_{jh} \ge 0, v_{jh} \in V; \; j = 1, 2, \ldots, n_2; \; h = 1, 2, \ldots, s \tag{21}
$$

where $\overline{f}$ denotes the expected value of objective function (8); $f_{\max}$ and $f_{\min}$ are the maximum value and minimum value of objective function (8); $B^R(\alpha)$ is the maximum values of all potential values, i.e., $B^R(\alpha) = \sup\{B | B = \mu^{-1}(\alpha)\}$, μ^{-1} is the inverse of μ. The first section in the objective function (i.e., $\overline{f}$) is used to denote the system cost, while minimizing the weight sum among the expected objective values; the second section [i.e., $\lambda_1 \cdot (f_{\max} - \overline{f}) + \lambda_2 \cdot (\overline{f} - f_{\min})$] represents the differences between two extreme possible values, which can enhance the robustness of optimization solutions, where λ_1 and λ_2 represent their respective weight coefficients; and the last section (i.e., $\rho_2 \cdot [B^R(\alpha) - B_r{}^1]$) denotes the difference between the extreme values of fuzzy uncertainty information, which can control the feasibility robustness of the generated solution. ρ_2 is a goal programming weight for fuzzy uncertainties [25]. The evaluation of the trade-offs between system cost and risk can be obtained under different coefficients $\lambda_1, \lambda_2, \rho_1, \rho_2,$ and α. The constraint (16) can be tackled by a conventional FPP algorithm.

3. Applications

3.1. Statement of Problem

The BTH region is regarded as China's economic, political, cultural, and technological innovation center, representing the most dynamic urban cluster in the country, including 11 prefecture-level cities of Hebei Province, as well as two megacities (Beijing and Tianjin) [36,37]. The region has experienced rapid economic development, with a high annual growth rate of 6.64% for gross domestic product (GDP) over the last five years [1–3]. The BTH region's energy demands have dramatically increased along with its rapid urbanization, industrialization, and economic development. According to the National Bureau of Statistics of the People's Republic of China, the region's electricity demand reached 50.92 million kWh in 2016, representing an increase of 13.31% compared with five years earlier. The domestic energy structure is heavily dependent on coal resources, with coal-based power accounting for 85.04% of the region's electricity power generation, while renewable energy contributes only 8%.

Fossil fuels in general are the major contributor to air pollution. For example, approximately 50%, 70%, 90%, and 65% of the total CO_2 emissions, nitrogen oxide (NO_x), sulfur dioxide (SO_2), and particulate matter (PM) are contributed by coal resources [4,38]. In addition, the BTH region is one of the high-traffic regions of China, with a vehicle population of 20.67 million in 2016, representing an increase of 63% from 2011. Motor vehicles are recognized as a major contributor of regulated pollutants such as CO, NO_x, HC, and PM. Based on several references, vehicular emissions contributed approximately 20–67% of CO emission, 12–36% of NO_x emission, and 12–39% of HC emissions [5].

The BTH region experiences severe air pollution due to increasing energy consumption and vehicle population, as well as a "high coal" energy mix characteristic. Cities in the region generally occupy more than half of the top-ten spots for the most polluted cities in China [39]. The scale of emissions in the BTH region has exceeded the capacity of self-purification and air-pollutants' diffusion from the atmosphere [25,33]. This situation has forced local managers to propose ambitious schemes for planning the EES of the BTH region.

To tackle the above-mentioned problems, several policies and measures have been implemented with regard to the development of renewable energy resources: adjustment of the energy structure; encouragement of the use of EVs; improvement of energy conversion efficiencies; and enhancement of vehicular emission standards. For example, several strategies have been proposed for reducing vehicular emissions, such as the deployment of EVs and the implementation of the China VI vehicular emission standard. However, it remains unclear how much pollution reduction can be achieved by these control measures and policies. As noted earlier, this paper proposed the DRSFO-EES model for the assessment of the impacts of different emission mitigation policies and measures on EES; development of robust optimization solutions for EES planning; determination of the regional atmospheric pollution contributions of different energy sectors such as energy processing, electricity generation, and vehicular traffic.

3.2. Schematic Overview of This Study

The proposed DRSFO-EES model was applied to EES planning for the BTH region. As shown in Figure 1, the considered EES included several energy-related activities such as heat processing, coke processing, oil refining, electricity conversion, EVs expansion, and energy import. The electricity demands of the BTH region are supplied by local power plants and import from adjacent power grids. The major local electricity conversion technologies include natural gas-based, coal-based, wind and solar power. The vehicle population consists of heavy-duty passenger vehicles (HDVs), light-duty passenger vehicles (LDVs), light duty trucks (LDTs), heavy duty trucks (HDTs), EVs, and "others". The DRSFO-EES model was used in this study to determine the air pollutants and greenhouse gases (NO_x, SO_2, PM, CO, HC, and CO_2) emitted from the different energy related-activities in the region. The EES considered a number of uncertainties (such as electricity demand, and many economic parameters expressed as random variables and fuzzy sets) that would affect the optimization scheme.

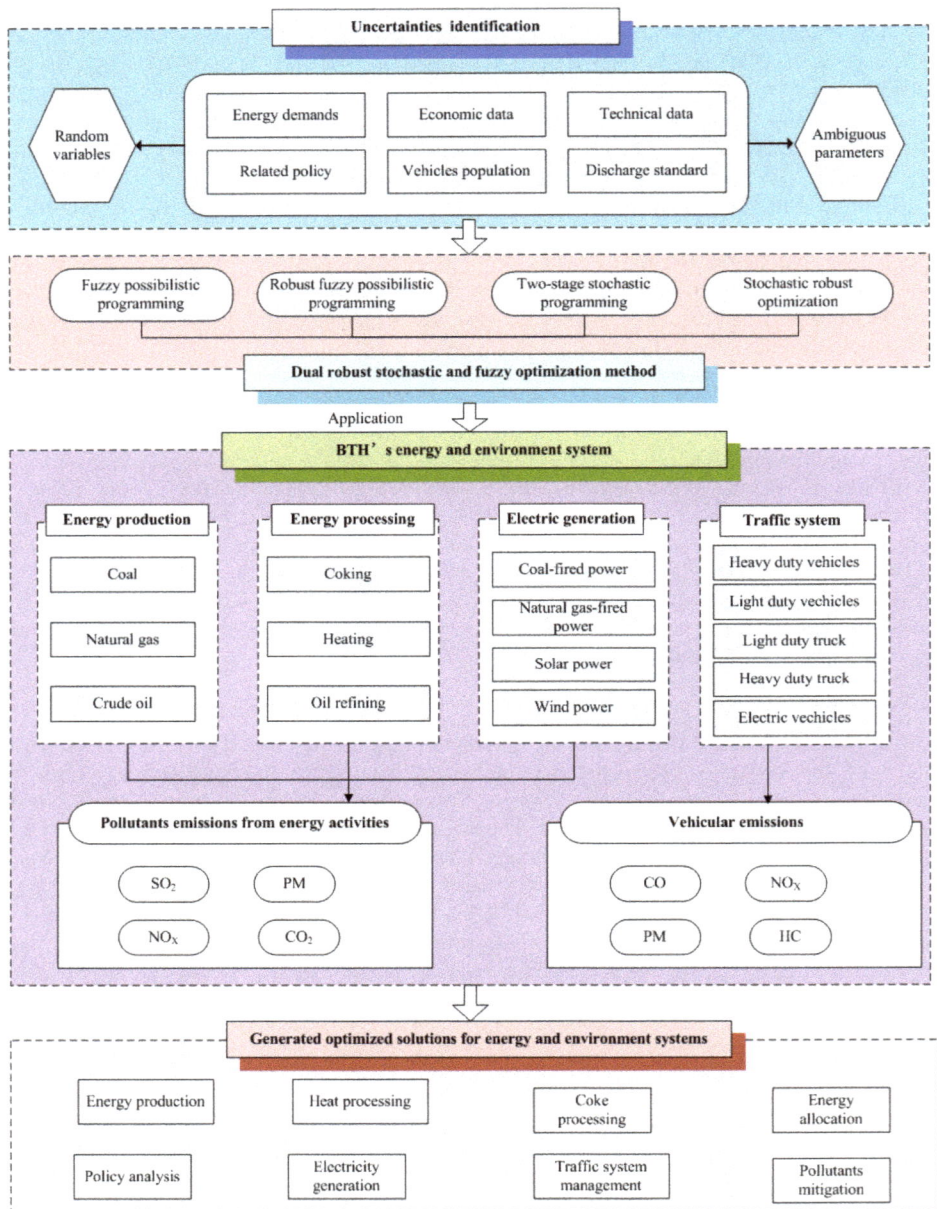

Figure 1. Schematic overview of this study.

3.3. Development of DRSFO-EES Model

Minimizing the system cost is the objective of DRSFO-EES, which includes costs for coke processing, heat processing, input for coal-based power, oil refining, input for natural gas-based power, imported electricity, first-stage of electricity generation, second-stage of electricity generation, electricity expansion, energy production, subsidy for solar power generation, subsidy for wind power generation, EVs charging piles, EVs charging stations, pollutants treatment, and risk recourse for the stochastic and fuzzy uncertainties. The proposed DRSFO-EES can be solved by Lingo 11.0 software. The schematic diagram of DRAOM-EWNS model is presented in Figure 2.

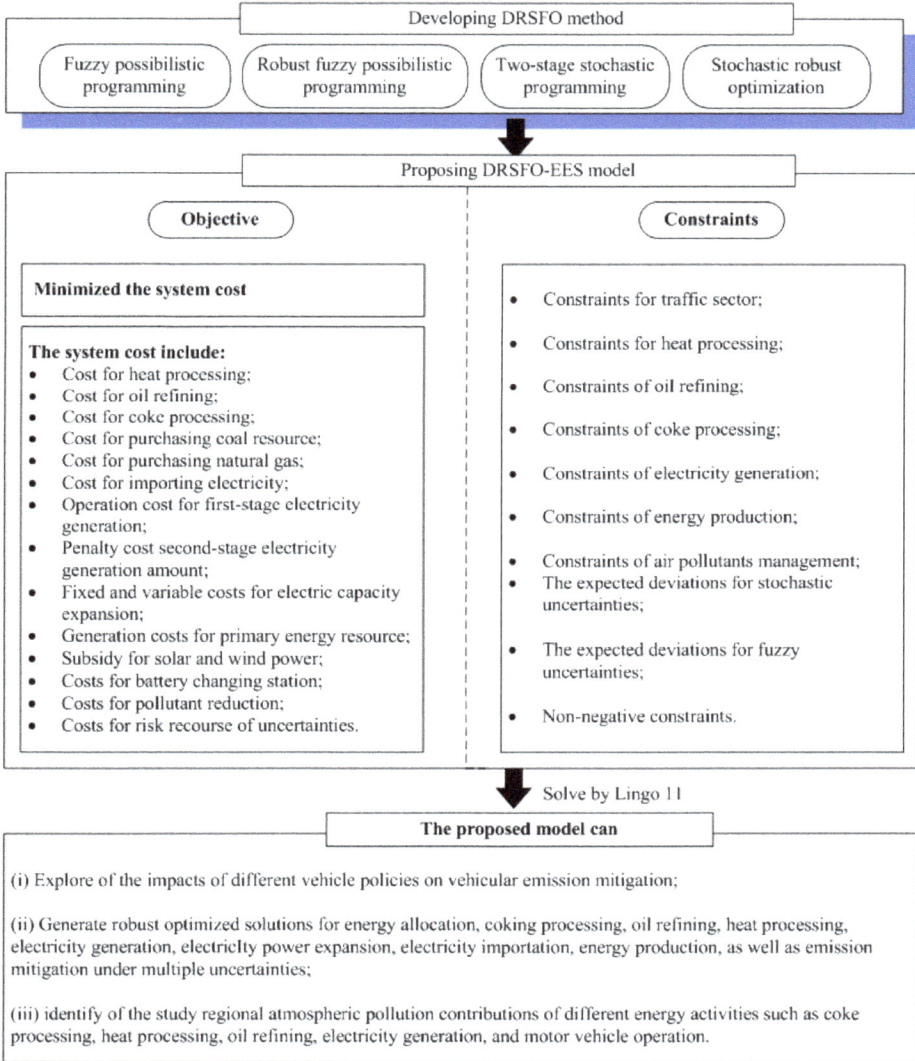

Figure 2. The schematic diagram of dual robust stochastic fuzzy optimization—energy and environmental systems (DRSFO-EES).

In the DRSFO-EES, *t* denotes the planning periods, period 1 (2020), period 2 (2025), and period 3 (2030); *i* denotes the primary energy production type, $i = 1$ is coal, $i = 2$ is natural gas, and $i = 3$ is crude oil; *k* denotes the electric conversion technology, $k = 1$ is coal-based power, $k = 2$ is natural gas-based power, $k = 3$ is solar power, and $k = 4$ is wind power; *h* expresses the electricity demand-level, $h = 1$

(low demand level), $h = 2$ (medium demand level), $h = 3$ (high demand level), $\sum\limits_{h=1}^{3} p_h = 1$; g are the vehicles types, $g = 1$ is HDVs, $g = 2$ is LDVs, $g = 3$ is TDTs, $g = 4$ is HDTs, and $g = 5$ is others. The proposed DRSFO-EES is formulated as follows:

Objective:

$$
\begin{aligned}
\text{Min } \tilde{f} &= \overline{f} + \lambda_1 (f_{\max} - \overline{f}) + \lambda_2 (\overline{f} - f_{\min}) + VF_t \\
&= f_1 + f_2 + f_3 + f_4 + f_5 + f_6 + f_7 \\
&\quad + f_8 + f_9 + f_{10} - f_{11} - f_{12} + f_{13} + f_{14} \\
&\quad + f_{15} + f_{16}
\end{aligned}
\tag{22}
$$

where $\tilde{f}$ represents the system cost; $\tilde{f}_1$ denotes the heat processing cost; $\tilde{f}_2$ denotes oil refining cost; $\tilde{f}_3$ denotes coke processing cost; $\tilde{f}_4$ denotes the cost of coal inputs for coal-based power; $\tilde{f}_5$ denotes the cost of natural gas inputs for natural gas-based power; $\tilde{f}_6$ denotes importing electricity cost; $\tilde{f}_7$ denotes the cost of first-stage electricity generation; $\tilde{f}_8$ denotes the cost of second-stage electricity generation; $\tilde{f}_9$ denotes electricity expansion cost; $\tilde{f}_{10}$ denotes the energy production cost; $\tilde{f}_{11}$ denotes the subsidy for solar power; $\tilde{f}_{12}$ denotes the subsidy for wind power; $\tilde{f}_{13}$ denotes the cost of charging pile; $\tilde{f}_{14}$ denotes the cost of changing station; $\tilde{f}_{15}$ denotes air-pollutants removal cost; $\tilde{f}_{16}$ denotes the risk recourse costs of stochastic and fuzzy uncertainties. VF_t represents the positive deviation between maximum values and worst value of fuzzy parameters. λ_1 and λ_2 represent their respective weight coefficients, where it is assumed λ_1 and λ_2 are fixed (i.e., $\lambda_1 = \lambda_2 = 1$).

(1) Cost of heat processing. This cost is used for heat processing, and it is calculated based on the unit-price and amount of heat processing.

$$
\begin{aligned}
\tilde{f}_1 &= \sum_{t=1}^{3} [0.25(HPP_t{}^1 + 2HPP_t{}^2 + HPP_t{}^3) \cdot HGA_t \\
&\quad + \lambda_1 \cdot \sum_{t=1}^{3} [HPP_t{}^3 - 0.25(HPP_t{}^1 + 2HPP_t{}^2 + HPP_t{}^3)] \, HGA_t \\
&\quad + \lambda_2 \cdot \sum_{t=1}^{3} [0.25(HPP_t{}^1 + 2HPP_t{}^2 + HPP_t{}^3) - HPP_t{}^1] \, HGA_t
\end{aligned}
\tag{23}
$$

where $H\tilde{P}P_t(HPP_{it}{}^1, HPP_{it}{}^2, HPP_{it}{}^3)$ is the cost for unit of heat processing, which is expressed as a triangular fuzzy number; HGA_t is the heat processing amount.

(2) Cost for oil refining. This cost is calculated in terms of the unit-price and the amount of oil refining.

$$
\begin{aligned}
\tilde{f}_2 &= \sum_{t=1}^{3} [0.25(PVO_t{}^1 + 2PVO_t{}^2 + PVO_t{}^3) \cdot OFOIL_t \\
&\quad + \lambda_1 \cdot \sum_{t=1}^{3} [PVO_t{}^3 - 0.25(PVO_t{}^1 + 2PVO_t{}^2 + PVO_t{}^3)] OFOIL_t \\
&\quad + \lambda_2 \cdot \sum_{t=1}^{3} [0.25(PVO_t{}^1 + 2PVO_t{}^2 + PVO_t{}^3) - PVO_t{}^1] \, OFOIL_t
\end{aligned}
\tag{24}
$$

where $P\tilde{V}O_t \, (PVO_t{}^1, PVO_t{}^2, PVO_t{}^3)$ is the cost for unit of oil refining; $OFOIL_t$ is the crude oil consumption of oil refining. (3) Cost for coke processing. This cost is calculated in terms of the unit-price and the amount of coke processing.

$$
\begin{aligned}
\tilde{f}_3 &= \sum_{t=1}^{3} [0.25(PWO_t{}^1 + 2PWO_t{}^2 + PWO_t{}^3) \cdot CKPA_t \\
&\quad + \sum_{t=1}^{3} [PWO_t{}^3 - 0.25(PWO_t{}^1 + 2PWO_t{}^2 + PWO_t{}^3)] \, CKPA_t \\
&\quad + \sum_{t=1}^{3} [0.25(PWO_t{}^1 + 2PWO_t{}^2 + PWO_t{}^3) - PWO_t{}^1] \, CKPA_t
\end{aligned}
\tag{25}
$$

where $\widetilde{PWO}_t(PWO_t{}^1, PWO_t{}^2, PWO_t{}^3)$ is the cost for unit of coke processing; and $CKPA_t$ is the coke processing amount.

(4) Cost for purchasing coal resources. This cost is used for purchasing coal resources, and it is calculated based on the unit-price and the amount of coal resource.

$$
\begin{aligned}
\tilde{f}_4 = & \sum_{t=1}^{3} [0.25(NSC_t{}^1 + 2NSC_t{}^2 + NSC_t{}^3)] \cdot ECOALM_t \\
& + \lambda_1 \cdot \sum_{t=1}^{3} [NSC_t{}^3 - 0.25(NSC_t{}^1 + 2NSC_t{}^2 + NSC_t{}^3)] ECOALM_t \\
& + \lambda_2 \cdot \sum_{t=1}^{3} [0.25(NSC_t{}^1 + 2NSC_t{}^2 + NSC_t{}^3) - NSC_t{}^1] ECOALM_t
\end{aligned}
\tag{26}
$$

where $\widetilde{NSC}_t(NSC_t{}^1, NSC_t{}^2, NSC_t{}^3)$ is the price for unit of coal resource; and $ECOALM_t$ is the coal consumption of coal-based power.

(5) Cost for purchasing natural gas resources. This cost is used for purchasing natural gas resources, and it is calculated based on the unit-price and the amount of natural gas resource.

$$
\begin{aligned}
\tilde{f}_5 = & \sum_{t=1}^{3} [0.25(NSN_t{}^1 + 2NSN_t{}^2 + NSN_t{}^3)] \cdot ENGM_t \\
& + \lambda_1 \cdot \sum_{t=1}^{3} [NSN_t{}^3 - 0.25(NSN_t{}^1 + 2NSN_t{}^2 + NSN_t{}^3)] ENGM_t \\
& + \lambda_2 \cdot \sum_{t=1}^{3} [0.25(NSN_t{}^1 + 2NSN_t{}^2 + NSN_t{}^3) - NSN_t{}^1] ENGM_t
\end{aligned}
\tag{27}
$$

where $\widetilde{NSN}_t(NSN_t{}^1, NSN_t{}^2, NSN_t{}^3)$ is the price for unit of natural gas resource; and $ENGM_t$ is the natural gas consumption of natural gas-based power.

(6) Cost for importing electricity. This cost is calculated in terms of the unit-price and the amount of importing electricity.

$$
\begin{aligned}
\tilde{f}_6 = & \sum_{t=1}^{3} \sum_{h=1}^{3} [0.25(NE_t{}^1 + 2NE_t{}^2 + NE_t{}^3) \cdot ED_{th} \\
& + \lambda_1 \cdot \sum_{t=1}^{3} \sum_{h=1}^{3} [NE_t{}^3 - 0.25(NE_t{}^1 + 2NE_t{}^2 + NE_t{}^3)] ED_{th} \\
& + \lambda_2 \cdot \sum_{t=1}^{3} \sum_{h=1}^{3} [0.25(NE_t{}^1 + 2NE_t{}^2 + NE_t{}^3) - NE_t{}^1] ED_{th}
\end{aligned}
\tag{28}
$$

where $\widetilde{NE}_t(NE_t{}^1, NE_t{}^2, NE_t{}^3)$ is the price for unit of imported electricity; ED_{th} is the amount of imported electricity.

(7) Operation cost for first-stage electricity generation. The cost represents the operation cost of first-stage electricity generation facilities (i.e., coal-based power, gas-based power, wind power and solar power) during the planning periods. It is calculated in terms of the operation costs and the amount of electricity generation for each of the electricity generation facilities. Furthermore, the first-stage is given by $W_{kt}^{\pm} = W_{kt}^- + rr_{kt} \cdot \Delta W$, where rr_{kt} denotes the decision variables, $\Delta W = W_{kt}^+ - W_{kt}^-$.

$$
\begin{aligned}
\tilde{f}_7 = & \sum_{k=1}^{4} \sum_{t=1}^{3} [0.25(PV_{kt}{}^1 + 2 \cdot PV_{kt}{}^2 + PV_{kt}{}^3)] \cdot (W_{kt}{}^- + rr_{kt} \cdot \Delta W_{kt}) \\
& + \lambda_1 \sum_{k=1}^{4} \sum_{t=1}^{3} [PV_{it}{}^3 - 0.25(PV_{kt}{}^1 + 2 \cdot PV_{kt}{}^2 + PV_{kt}{}^3)] \cdot (W_{kt}{}^- + rr_{kt} \cdot \Delta W_{kt}) \\
& + \lambda_2 \sum_{k=1}^{4} \sum_{t=1}^{3} [0.25(PV_{kt}{}^1 + 2 \cdot PV_{kt}{}^2 + PV_{kt}{}^3) - PV_{kt}{}^1] \cdot (W_{kt}{}^- + rr_{kt} \cdot \Delta W_{kt})
\end{aligned}
\tag{29}
$$

where $P\widetilde{V}_{kt}(PV_{kt}^1, PV_{kt}^2, PV_{kt}^3)$ denotes the operation cost of different power conversion technologies; $W_{kt}^{\pm} = W_{kt}^- + rr_{kt}\cdot\Delta W_{kt}$ is the first-stage electricity generation amount, which is determined by decision variables rr_{kt} ($\Delta W = W_{kt}^+ - W_{kt}^-$ and $rr_{kt} \in [0, 1]$).

(8) Penalty cost for second-stage electricity generation amounts (i.e., shortage electricity amount of first-stage). It is calculated in terms of the penalty costs and the amount of second-stage electricity generation by each electricity generation technology.

$$\widetilde{f}_8 = \sum_{k=1}^{4}\sum_{t=1}^{3}\sum_{h=1}^{3}[0.25(PP_{kt}^1+2\cdot PP_{kt}^2+ PP_{kt}^3)]\cdot p_{th}\cdot Y_{kth}$$
$$+ \lambda_1\cdot\sum_{k=1}^{4}\sum_{t=1}^{3}\sum_{h=1}^{3}[PP_{kt}^3 - 0.25(PP_{kt}^1+2\cdot PP_{kt}^2+ PP_{kt}^3)]\cdot p_{th}\cdot Y_{kth} \qquad (30)$$
$$+ \lambda_2\cdot\sum_{k=1}^{4}\sum_{t=1}^{3}\sum_{h=1}^{3}[0.25(PP_{kt}^1+2\cdot PP_{kt}^2+ PP_{kt}^3) - PP_{kt}^1]\cdot p_{th}\cdot Y_{kth}$$

where $P\widetilde{P}_{kt}(PP_{kt}^1, PP_{kt}^2, PP_{kt}^3)$ is the operating cost for the second-stage electricity generation amount (Y_{kth}). Y_{kth} is the second-stage electricity generation amount.

(9) Fixed and variable costs for electric capacity expansion. These costs include the fixed and variable electric capacity expansion costs of four electricity generation technologies in the planning horizon.

$$\widetilde{f}_9 = \sum_{i=1}^{4}\sum_{t=1}^{3}\sum_{h=1}^{3} p_{th}[0.25(A_{kt}^1+2\cdot A_{kt}^2+ A_{kt}^3)]\cdot Q_{kth}$$
$$+\lambda_1\cdot\sum_{i=1}^{4}\sum_{t=1}^{3}\sum_{h=1}^{3} p_{th}[A_{kt}^3 - 0.25(A_{kt}^1+2\cdot A_{kt}^2+ A_{kt}^3)]\cdot Q_{kth}$$
$$+\lambda_2\cdot\sum_{i=1}^{4}\sum_{t=1}^{3}\sum_{h=1}^{3} p_{th}[0.25(A_{kt}^1+2\cdot A_{kt}^2+ A_{kt}^3) - A_{kt}^1]\cdot Q_{kth}$$
$$+\sum_{i=1}^{4}\sum_{t=1}^{3}\sum_{h=1}^{3} p_{th}[0.25(B_{kt}^1+2\cdot B_{kt}^2+ B_{kt}^3)]\cdot Z_{kth} \qquad (31)$$
$$+\lambda_1\cdot\sum_{i=1}^{4}\sum_{t=1}^{3}\sum_{h=1}^{3} p_{th}[B_{kt}^3 - 0.25(B_{kt}^1+2\cdot B_{kt}^2+ B_{kt}^3)]\cdot Z_{kth}$$
$$+\lambda_2\cdot\sum_{i=1}^{4}\sum_{t=1}^{3}\sum_{h=1}^{3} p_{th}[0.25(B_{kt}^1+2\cdot B_{kt}^2+ B_{kt}^3) - B_{kt}^1]\cdot Z_{kth}$$

where $\widetilde{A}_{kt}(A_{kt}^1, A_{kt}^2, A_{kt}^3)$ denote fixed-charge cost for different electric capacity expansion. Q_{kth} are binary variables for identifying whether or not the capacity expansion needs to be undertaken by power conversion technology k. $\widetilde{B}_{kt}(B_{kt}^1, B_{kt}^2, B_{kt}^3)$ denote the variable cost of capacity expansion. Z_{kth} denotes the continuous variable of the capacity expansion amount.

(10) Generation costs for primary energy resources. This cost is calculated in terms of the unit-price of primary energy (i.e., coal, natural gas, and crude oil) and the amount of primary generation.

$$\widetilde{f}_{10} = \sum_{i=1}^{3}\sum_{t=1}^{3} 0.25(EPP_{it}^1 +2\cdot EPP_{it}^2 + EPP_{it}^3)\cdot EPA_{it}$$
$$+\lambda_1\cdot\sum_{i=1}^{3}\sum_{t=1}^{3} [EPP_{it}^3 - 0.25\cdot(EPP_{it}^1+2\cdot EPP_{it}^2 + EPP_{it}^3)]\cdot EPA_{it} \qquad (32)$$
$$+\lambda_2\cdot\sum_{i=1}^{3}\sum_{t=1}^{3} [0.25\cdot(EPP_{it}^1+2\cdot EPP_{it}^2 + EPP_{it}^3) - EPP_{it}^1]\cdot EPA_{it}$$

where $E\widetilde{P}P_{it}(EPP_{it}^1, EPP_{it}^2, EPP_{it}^3)$ is the production cost per unit of energy resource; and EPA_{it} is the production amount.

(11) Subsidy for solar power. The government provides a subsidy for units of electricity generated by solar power, and it is calculated in terms of the unit-subsidy of solar power and the amount of electricity generation amount of solar power.

$$
\begin{aligned}
\widetilde{f}_{11} = &\sum_{t=1}^{3} [0.25(SP_t{}^1 + 2SP_t{}^2 + SP_t{}^3) \cdot (W_{3t}{}^- + rr_{3t} \cdot \Delta W_{3t}) \\
&+ \sum_{t=1}^{3} [SP_t{}^3 - 0.25(SP_t{}^1 + 2SP_t{}^2 + SP_t{}^3)] (W_{3t}{}^- + rr_{3t} \cdot \Delta W_{3t}) \\
&+ \sum_{t=1}^{3} [0.25(SP_t{}^1 + 2SP_t{}^2 + SP_t{}^3) - SP_t{}^1] (W_{3t}{}^- + rr_{3t} \cdot \Delta W_{3t})
\end{aligned}
\tag{33}
$$

where $S\widetilde{P}_t(SP_t{}^1, SP_t{}^2, SP_t{}^3)$ is the solar subsidy provided by government for unit of electricity generated by solar power.

(12) Subsidy for wind power. The government provides a subsidy for units of electricity generated by wind power, and it is calculated in terms of the unit-subsidy of wind power and the amount of electricity generation amount of wind power.

$$
\begin{aligned}
\widetilde{f}_{12} = &\sum_{t=1}^{3} [0.25(WP_t{}^1 + 2WP_t{}^2 + WP_t{}^3) \cdot (W_{4t}{}^- + rr_{4t} \cdot \Delta W_{4t}) \\
&+ \sum_{t=1}^{3} [WP_t{}^3 - 0.25(WP_t{}^1 + 2WP_t{}^2 + WP_t{}^3)] (W_{4t}{}^- + rr_{4t} \cdot \Delta W_{4t}) \\
&+ \sum_{t=1}^{3} [0.25(WP_t{}^1 + 2WP_t{}^2 + WP_t{}^3) - WP_t{}^1] (W_{4t}{}^- + rr_{4t} \cdot \Delta W_{4t})
\end{aligned}
\tag{34}
$$

where $W\widetilde{P}_t(WP_t{}^1, WP_t{}^2, WP_t{}^3)$ is the wind subsidy provided by government for unit of electricity generated by wind power.

(13) Costs for battery charging piles. It is calculated in terms of the average charging pile amount for unit of electric vehicle, the EVs population, as well as the investment cost per EV charging pile.

$$
f_{13} = \sum_{t=1}^{3} EVA_t REC_t FCD_t
\tag{35}
$$

where EVA_t is the EVs population; REC_t is the average charging pile amount for unit of electric vehicle; FCD_t is the investment cost per EV charging pile.

(14) Costs for battery changing station. It is calculated in terms of the average changing station amount for unit of electric vehicle, the EVs population, as well as the investment cost per EV changing station.

$$
f_{14} = \sum_{t=1}^{3} EVA_t REV_t FVD_t
\tag{36}
$$

where REV_t is the average changing station amount for unit of electric vehicles; FVD_t is the investment cost per EV changing station.

(15) Costs for pollutant reduction. These costs include reduction costs for three pollutants, i.e., SO_2, NO_X, and PM emissions.

$$
\widetilde{f}_{15} = \widetilde{f}_{15s} + \widetilde{f}_{15N} + \widetilde{f}_{15PM}
\tag{37}
$$

where $\widetilde{f}_{15s}$ is the cost of desulfurization; $SO2K_t$ is the SO_2 emissions from coking processing; $SO2CE_t$ is the SO_2 emissions from coal-based power; $SO2NE_t$ is the SO_2 emissions from natural gas-based power; $SO2H_t$ is the SO_2 emissions from heating processing; $SO2O_t$ is the SO_2 emissions from oil refining; $T\widetilde{P}S_t(TPS_t{}^1, TPS_t{}^2, TPS_t{}^3)$ is the desulfurization cost of per unit of SO_2 emission; η_s is the desulfurization efficiency.

(15a) Costs for SO_2 emissions reduction. The cost is calculated in terms of the total SO_2 emissions, the desulfurization cost per unit of SO_2 emission, and the desulfurization efficiency.

$$
\begin{aligned}
\widetilde{f}_{15S} = {} & \{ \sum_{t=1}^{3} (SO2K_t + SO2CE_t + SO2NE_t + SO2H_t + SO2O_t) \\
& \cdot 0.25 \cdot (TPS_t{}^1 + TPS_t{}^2 + TPS_t{}^3) \cdot \eta_s \} \\
& + \lambda_1 \cdot \{ \sum_{t=1}^{3} (SO2K_t + SO2CE_t + SO2NE_t + SO2H_t + SO2O_t) \\
& \cdot [TPS_t{}^3 - 0.25 \cdot (TPS_t{}^1 + TPS_t{}^2 + TPS_t{}^3)] \cdot \eta_s \} \\
& + \lambda_2 \cdot \{ \sum_{t=1}^{3} (SO2K_t + SO2CE_t + SO2NE_t + SO2H_t + SO2O_t) \\
& \cdot [0.25 \cdot (TPS_t{}^1 + TPS_t{}^2 + TPS_t{}^3) - TPS_t{}^1] \cdot \eta_s \}
\end{aligned}
\tag{38}
$$

where $\widetilde{f}_{15s}$ is the cost of desulfurization; $SO2K_t$ is the SO_2 emissions from coking processing; $SO2CE_t$ is the SO_2 emissions from coal-based power; $SO2NE_t$ is the SO_2 emissions from natural gas-based power; $SO2H_t$ is the SO_2 emissions from heating processing; $SO2O_t$ is the SO_2 emissions from oil refining; $T\widetilde{P}S_t(TPS_t{}^1, TPS_t{}^2, TPS_t{}^3)$ is the desulfurization cost per unit of SO_2 emission; η_s is the desulfurization efficiency.

(15b) Costs for NO_X emissions reduction. The cost is calculated in terms of the total NOx emissions, the desulfurization cost per unit of NOx emission, and the desulfurization efficiency.

$$
\begin{aligned}
\widetilde{f}_{15N} = {} & \{ \sum_{t=1}^{3} (NOXK_t + NOXCE_t + NOXNE_t + NOXH_t + NOXO_t) \\
& \cdot 0.25 \cdot (TPN_t{}^1 + TPN_t{}^2 + TPN_t{}^3) \cdot \eta_N \} \\
& + \lambda_1 \cdot \{ \sum_{t=1}^{3} (NOXK_t + NOXCE_t + NOXNE_t + NOXH_t + NOXO_t) \\
& \cdot [TPN_t{}^3 - 0.25 \cdot (TPN_t{}^1 + TPN_t{}^2 + TPN_t{}^3)] \cdot \eta_N \} \\
& + \lambda_2 \cdot \{ \sum_{t=1}^{3} (NOXK_t + NOXCE_t + NOXNE_t + NOXH_t + NOXO_t) \\
& \cdot [0.25 \cdot (TPN_t{}^1 + TPN_t{}^2 + TPN_t{}^3) - TPN_t{}^1] \cdot \eta_N \}
\end{aligned}
\tag{39}
$$

where $\widetilde{f}_{15N}$ is the cost of denitration; $NOXK_t$ is the NO_x emissions from coking processing; $NOXCE_t$ is the NO_x emissions from coal-based power; $NOXNE_t$ is the NO_x emissions from natural gas-based power; $NOXH_t$ is the NO_x emissions from heat processing; $NOXO_t$ is the NO_x emissions from oil refining; $T\widetilde{P}N_t(TPN_t{}^1, TPN_t{}^2, TPN_t{}^3)$ is the denitration cost per unit of NO_x emission; η_N is the denitration efficiency.

(15c) Costs for PM emissions reduction. The cost is calculated in terms of the total PM emissions, the desulfurization cost per unit of PM emission, and the desulfurization efficiency.

$$
\begin{aligned}
\widetilde{f}_{15PM} = {} & \{ \sum_{t=1}^{3} (PMK_t + PMCE_t + PMNE_t + PMH_t + PMO_t) \\
& \cdot 0.25 \cdot (TPPM_t{}^1 + TPPM_t{}^2 + TPPM_t{}^3) \cdot \eta_{PM} \} \\
& + \lambda_1 \cdot \{ \sum_{t=1}^{3} (PMK_t + PMCE_t + PMNE_t + PMH_t + PMO_t) \\
& \cdot [TPPM_t{}^3 - 0.25 \cdot (TPPM_t{}^1 + TPPM_t{}^2 + TPPM_t{}^3)] \cdot \eta_{PM} \} \\
& + \lambda_2 \cdot \{ \sum_{t=1}^{3} (PMK_t + PMCE_t + PMNE_t + PMH_t + PMO_t) \\
& \cdot [0.25 \cdot (TPPM_t{}^1 + TPPM_t{}^2 + TPPM_t{}^3) - TPPM_t{}^1] \cdot \eta_{PM} \}
\end{aligned}
\tag{40}
$$

where $\widetilde{f}_{15PM}$ is the PM removal cost; PMK_t is the PM emissions from coking processing; $PMCE_t$ is the PM emissions from coal-based power; $PMNE_t$ is the PM emissions from natural gas-based power; PMH_t is the PM emissions from heating processing; PMO_t is the NO_X emissions from oil refining;

$T\widetilde{P}PM_t(TPPM_t{}^1, TPPM_t{}^2, TPPM_t{}^3)$ is the treatment cost per unit of PM emission; η_{PM} is the PM removal efficiency.

(16) Costs for risk recourse of stochastic and fuzzy uncertainties.

$$\widetilde{f}_{16} = V\widetilde{T}S + V\widetilde{T}F = \rho_1 \cdot \sum_{i=1}^{I} \sum_{t=1}^{T} \sum_{h=1}^{H} p_{th}(V_{ith} + VC_{ith}) + \rho_2 \cdot \sum_{t=1}^{3} VF_t \qquad (41)$$

where $V\widetilde{T}S$ denotes the risk recourse cost for stochastic uncertainties; $V\widetilde{T}F$ denotes the risk recourse cost of fuzzy uncertainties, with capturing the difference between the extreme values of fuzzy parameters. The nonnegative factors ρ_1 and ρ_2 represent robust levels (its range from 0 to 1), which can help managers make trade-offs between system economy and reliability.

Constraints:

The constraints of the proposed DRSFO-EES model include traffic sector, heat processing, coke processing, oil refining, electricity generation, energy production, air-pollutants treatment, and risk recourse cost for stochastic uncertainties and risk recourse for fuzzy uncertainties.

(1) Constraints for traffic sector.

This constraint represents that the optimized vehicles population must be not less than the lower bounds of vehicle population. $Cr\{M\widetilde{A}L_{gt}|MA_{gt} \geq M\widetilde{A}L_{gt}\} \geq \alpha$ means the credibility of the $MA_{gt} \geq M\widetilde{A}L_{gt}$ is higher than or equal to confidence-level α.

$$Cr\{M\widetilde{A}L_{gt}|MA_{gt} \geq M\widetilde{A}L_{gt}\} \geq \alpha \qquad (42)$$

where MA_{gt} is the optimized solutions of vehicles population; $M\widetilde{A}L_{gt}(MAL_{gt}{}^1, MAL_{gt}{}^2, MAL_{gt}{}^3)$ is the lower bounds of vehicle population, which is expressed as a triangular fuzzy number.

Constraints for vehicle-emissions, including CO, NOx, HC, and PM emissions. These constraints represent that the vehicle-emissions are calculated in terms of the annual average mileage, the proportion of electric vehicles and the emission factors of CO, NOx, HC and PM emissions.

$$TRCO_{gt|g=1,3,4,5} = MA_{gt|g=1,3,4,5} \cdot MQ_{gt|g=1,3,4,5} \cdot TRCOE_{gt|g=1,3,4,5} \qquad (43)$$

$$TRCO_{2t} = (MA_{gt} - MA_{gt} \cdot REV_t) \cdot MQ_{gt} \cdot TRCOE_{gt} \qquad (44)$$

$$TRNOX_{gt|g=1,3,4,5} = MA_{gt|g=1,3,4,5} \cdot MQ_{gt|g=1,3,4,5} \cdot TRNOE_{gt|g=1,3,4,5} \qquad (45)$$

$$TRNOX_{2t} = (MA_{gt} - MA_{gt} \cdot REV_t) \cdot MQ_{gt} \cdot TRNOE_{gt} \qquad (46)$$

$$TRHC_{gt|g=1,3,4,5} = MA_{gt|g=1,3,4,5} \cdot MQ_{gt|g=1,3,4,5} \cdot TRHC_{gt|g=1,3,4,5} \qquad (47)$$

$$TRHC_{2t} = (MA_{gt} - MA_{gt} \cdot REV_t) \cdot MQ_{gt} \cdot TRHC_{gt} \qquad (48)$$

$$TRPM_{gt|g=1,3,4,5} = MA_{gt|g=1,3,4,5} \cdot MQ_{gt|g=1,3,4,5} \cdot TRPM_{gt|g=1,3,4,5} \qquad (49)$$

$$TRPM_{2t} = (MA_{gt} - MA_{gt} \cdot REV_t) \cdot MQ_{gt} \cdot TRPM_{gt} \qquad (50)$$

where $TRCO_{gt}$ is CO emissions from vehicle g in period t; MQ_{gt} is the annual average mileage; REV_t is the proportion of electric vehicles; $TRCOE_{gt}$ the emission factor of CO; $TRNOE_{gt}$ the emission factor of NOx; $TRHC_{gt}$ the emission factor of HC; $TRPM_{gt}$ the emission factor of PM.

Constraints of the electricity demands of EVs. This constraint is calculated in terms of the electricity consumption amounts per hundred kilometers of EVs and the amount of EVs.

$$EVEC_t = MA_{2t} \cdot REV_t \cdot MQ_{2t} \cdot ECPEV_t \qquad (51)$$

where $EVEC_t$ is the electricity consumption amounts of EVs; $ECPEV_t$ is the electricity consumption amounts per hundred kilometers of EVs.

(2)   Constraints of heat processing. Constraint (52) depicts that the heat generation amounts must not be less than the heat demands, and $Cr\{D\widetilde{M}H_t|HGA_t \geq D\widetilde{M}H_t\} \geq \alpha$ means the credibility of the $HGA_t \geq D\widetilde{M}H_t$ is higher than or equal to confidence-level α. Constraints (53) and (54) represent the consumption amounts of natural gas and coal for heat processing.

$$Cr\{D\widetilde{M}H_t|HGA_t \geq D\widetilde{M}H_t\} \geq \alpha \tag{52}$$

$$CFHGC_t = CFH_t \cdot HGA_t \cdot HCP_t \tag{53}$$

$$CFHGN_t = CFH_t \cdot HGA_t \cdot HNP_t \tag{54}$$

where $D\widetilde{M}H_t(DMH_t{}^1, DMH_t{}^2, DMH_t{}^3)$ is the heat demand; CFH_t denotes the energy consumption for the unit of kerosene processing. $CFHGC_t$ denotes the coal consumption of heat processing; $CFHGN_t$ is the natural gas consumption of heat processing; HCP_t denotes the ratio of coal consumption of heat processing; HNP_t denotes the ratio of natural gas consumption of heat processing.

(3)   Constraints of oil refining. The constraint depicts the input amounts of crude oil for oil refining.

$$\begin{aligned} OFOIL_t = {} & ZCY_t \cdot ZCYE_t + ZQY_t \cdot ZQYE_t + ZMY_t \cdot ZMYE_t \\ & + ZRLY_t \cdot ZRLYE_t + ZSNY_t \cdot ZSNYE_t \\ & + ZSJY_t \cdot ZSJYE_t + ZLPG_t \cdot ZLPGE_t + ZQT_t \end{aligned} \tag{55}$$

where ZCY_t is the diesel processing amount; $ZCYE_t$ denotes the crude oil consumption for unit of diesel processing; ZQY_t is the processing amount of gasoline; $ZQYE_t$ denote the crude oil consumption for unit of gasoline processing; ZMY_t is the processing amount of kerosene; $ZMYE_t$ denotes the crude oil consumption for unit of kerosene processing; $ZRLY_t$ is the fuel oil processing amount; $ZRLYE_t$ is the crude consumption for unit of fuel oil processing; $ZSNY_t$ is the processing amount for naphtha; $ZSNYE_t$ denotes the crude oil consumption for unit of naphtha processing; $ZSJY_t$ is the processing amount of asphaltic pyrobitumen; $ZSJYE_t$ denotes the crude oil consumption for unit of asphaltic pyrobitumen processing; $ZLPG_t$ is the processing amount of LPG; $ZLPGE_t$ denotes the crude consumption for unit of LPG processing; ZQT_t denotes the crude oil consumption of other oil products.

(4)   Constraints of coke processing. Constraint (56) means that the amount of coke processing is not less than the coke demands, and $Cr\{D\widetilde{MC}J_t|CKPA_t \geq D\widetilde{MC}J_t\} \geq \alpha$ means the credibility of the $CKPA_t \geq D\widetilde{MC}J_t$ is higher than or equal to confidence-level α. Constraint (57) depicts the coal consumption amount for coke processing.

$$Cr\{D\widetilde{MC}J_t|CKPA_t \geq D\widetilde{MC}J_t\} \geq \alpha \tag{56}$$

$$CFCJ_t = CKPA_t / CTCKE_t \tag{57}$$

where $D\widetilde{MC}J_t(DMCJ_t{}^1, DMCJ_t{}^2, DMCJ_t{}^3)$ is the coke demand expressed as a triangular fuzzy number. $CTCKE_t$ denotes the coal consumption for unit of coke processing; $CFCJ_t$ is the coal consumption of coke processing.

(5)   Constraints of electricity generation.
    Constraints for mass balance of coal and natural gas resources. These constraints are established to calculate the consumption amounts of coal and natural gas for electricity generation.

$$[(W_{1t}^- + rr_{1t} \cdot WC_{1t}) + Y_{1th} + EVEC_t] \cdot FE_{1t} = ECOALM_t \tag{58}$$

$$[(W_{2t}^- + rr_{2t} \cdot WC_{2t}) + Y_{2th}] \cdot FE_{2t} = ENGM_t \tag{59}$$

where FE_{it} is the coal ($i = 1$) and natural gas ($i = 2$) consumption for a unit of electricity generation.

Constraints of electricity demand and supply balance. Constraint (60) is established to ensure the electricity demand be satisfied by domestic electricity generation and importation. And $Cr\{\tilde{D}_{th} | \sum_{i=1}^{4} \sum_{t=1}^{3} [(W_{it}^{-} + rr_{it}WC_{it}) + Y_{ith} + EVEC_t] + Ed_{th} \geq \tilde{D}_{th}\} \geq \alpha$ means the credibility of the

$\sum_{i=1}^{4} \sum_{t=1}^{3} [(W_{it}^{-} + rr_{it}WC_{it}) + Y_{ith} + EVEC_t] + Ed_{th} \geq \tilde{D}_{th}$ is higher than or equal to confidence-level α. For constraint (61), the optimized amount of electricity generated in the first-stage is given by $W_{kt}^{\pm} = W_{kt}^{-} + rr_{kt} \cdot \Delta W$, where rr_{kt} denotes the decision variables, and $rr_{kt} \in [0, 1]$.

$$Cr\{\tilde{D}_{th} | \sum_{i=1}^{4} \sum_{t=1}^{3} [(W_{it}^{-} + rr_{it}WC_{it}) + Y_{ith} + EVEC_t] + Ed_{th} \geq \tilde{D}_{th}\} \geq \alpha \tag{60}$$

$$0 \leq rr_{it} \leq 1 \tag{61}$$

where $\tilde{D}_{th}$ is the electricity demand under various electricity demand levels, which show the characteristics of stochastic and fuzzy sets. rr_{it} denotes the decision variable of first-stage electricity generation.

Constraints for electricity capacities. These constraints mainly depict that the amount of generated electricity must not exceed its existing and expanded capacities, which are established to ensure that the available electricity-generation capacity is greater than the generated electricity.

$$\sum_{i=1}^{4} (W_{it}^{-} + rr_{it} \cdot WC_{it}) + Y_{ith} + EVEC_t \leq \sum_{i=1}^{2} [CF_{it} \cdot (RC_{it} + Z_{ith})] \tag{62}$$

$$\sum_{i=1}^{4} (RC_i + Z_{ith}) \geq U_t; \forall t, h \tag{63}$$

$$\sum_{i=1}^{4} (RC_i + Z_{ith}) \leq US_t; \forall t, h \tag{64}$$

$$\sum_{i=1}^{4} Y_{ith} \leq \sum_{i=1}^{4} CF_{it} \cdot Z_{ith} \tag{65}$$

$$0 \leq Y_{ith} \leq (W_{it}^{-} + rr_{it} \cdot WC_{it}); \forall i, t, h \tag{66}$$

$$Q_{ith} = \begin{cases} = 1, & \text{if capacity expansion of is undertaken } \forall i, t, h \\ = 0, & \text{if otherwise} \end{cases} \tag{67}$$

where CF_{kt} is the operating hours of electricity conversion technology k; RC_{kt} is the original capacity; U_t is the lower bound of load demands; US_t is the upper bound of load demands.

(6) Constraints of energy production. The constraint is established to ensure the primary energy production amount must be less than the lower bound and higher than the upper bound.

$$EPAL \leq EPA_{it} \leq EPAU_{it} \tag{68}$$

where $EPAU_{it}$ is the upper bound of amount of primary energy production. $EPAL_{it}$ is the lower bound of amount of primary energy production.

(7) Constraints for air-pollutants management.

SO$_2$ emissions. Constraints (69) to (74) represent the SO$_2$ emissions from energy activities (i.e., coke processing, heat processing, electricity generation, and oil refining). Constraints (74) are used for ensuring that the SO$_2$-emissions are satisfied by the pollutant-emission permits.

$$SO2K_t = CFCJ_t \cdot CS_t \tag{69}$$

$$SO2CE_t = ECOALM_t \cdot CS_t \tag{70}$$

$$SO2NE_t = NEGM_t \cdot CS_t \tag{71}$$

$$SO2H_t = CFHGC_t \cdot CS_t + CFHGN_t \cdot NS_t \tag{72}$$

$$SO2O_t = OFOIL_t \cdot OS_t \tag{73}$$

$$(SO2K_t + SO2CE_t + SO2NE_t + SO2H_t + SO2O_t) \cdot (1 - \eta_s) \leq TS_t \tag{74}$$

where CS_t is the SO$_2$ emission factor of coal; NS_t is the SO$_2$ emission factor of natural gas; OS_t is the SO$_2$ emission factor of crude oil; η_s is the desulfurization efficiency; TS_t is the upper bounds of SO$_2$ emission.

NO$_X$ emissions. Constraints (75) to (79) represent the NOx emissions from energy activities (i.e., coke processing, heat processing, electricity generation, and oil refining). Constraints (80) are used for ensuring that the NOx-emissions are satisfied by the pollutant-emission permits.

$$NOXK_t = CFCJ_t \cdot CN_t \tag{75}$$

$$NOXCE_t = ECOALM_t \cdot CN_t \tag{76}$$

$$NOXNE_t = NEGM_t \cdot NN_t \tag{77}$$

$$NOXH_t = CFHGC_t \cdot CN_t + CFHGN_t \cdot NN_t \tag{78}$$

$$NOXO_t = OFOIL_t \cdot ON_t \tag{79}$$

$$(NOXK_t + NOXCE_t + NOXNE_t + NOXH_t + NOXO_t) \cdot (1 - \eta_N) \leq TN_t \tag{80}$$

where CN_t is the NO$_x$ emission factor of coal; NN_t is the NO$_x$ emission factor of natural gas; ON_t is the NO$_x$ emission factor of crude oil; η_N is the denitration efficiency; TN_t is the upper bounds of NO$_x$ emission.

PM emissions. Constraints (81) to (85) represent the PM emissions from energy activities (i.e., coke processing, heat processing, electricity generation, and oil refining). Constraints (86) are used for ensuring that the PM-emissions are satisfied by the pollutant-emission permits.

$$PMK_t = CFCJ_t \cdot CPM_t \tag{81}$$

$$PMCE_t = ECOALM_t \cdot CPM_t \tag{82}$$

$$PMNE_t = NEGM_t \cdot NPM_t \tag{83}$$

$$PMH_t = CFHGC_t \cdot CPM_t + CFHGN_t \cdot NPM_t \tag{84}$$

$$PMO_t = OFOIL_t \cdot OPM_t \tag{85}$$

$$(PMK_t + PMCE_t + PMNE_t + PMH_t + PMO_t) \cdot (1 - \eta_{PM}) \leq TPM_t \tag{86}$$

where CPM_t is the PM emission factor of coal; NPM_t is the PM emission factor of natural gas; OPM_t is the PM emission factor of crude oil; η_{PM} is the PM removal efficiency; TPM_t is the upper bound of PM emission.

CO_2 emissions. Constraints (81) to (85) represent the CO_2- emissions from energy activities (i.e., coke processing, heat processing, electricity generation, and oil refining).

$$CO2K_t = CFCJ_t \cdot CCO2_t \tag{87}$$

$$CO2CE_t = ECOALM_t \cdot CCO2_t \tag{88}$$

$$CO2NE_t = NEGM_t \cdot NCO2_t \tag{89}$$

$$CO2H_t = CFHGC_t \cdot CCO2_t + CFHGN_t \cdot NCO2_t \tag{90}$$

$$CO2O_t = OFOIL_t \cdot OCO2_t \tag{91}$$

where $CCO2_t$ is the CO_2 emission factor of coal; $NCO2_t$ is the CO_2 emission factor of natural gas; $OCO2_t$ is the CO_2 emission factor of crude oil.

(8) The expected deviations for stochastic uncertainties. These constraints are used for capturing the risk from stochastic uncertainties.

$$V_{ith} = P\tilde{P}_{it} \cdot Y_{ith}^{\pm} - \sum_{h=1}^{H} p_{th} P\tilde{P}_{it} Y_{ith}^{\pm} + 2\theta_h; \forall i, t, h \tag{92}$$

$$VC_{ith} = (\tilde{A}_{it} Q_{ith} + \tilde{B}_{it} Z_{ith}) - \sum_{h=1}^{H} p_{th} (\tilde{A}_{it} Q_{ith} + \tilde{B}_{it} Z_{ith}) + 2\theta_h; \forall i, t, h \tag{93}$$

$$V_{ith}^{\pm} \geq 0, \ VC_{ijcth}^{\pm} \geq 0; \tag{94}$$

where $\theta_h \geq 0$ are slack variables that can achieve looser constraints; V_{ith} and V_{ijcth} are the weighted values of the expected deviations from stochastic uncertainties.

(9) The expected deviations for fuzzy uncertainties. These constraints are used for capturing the risk from fuzzy uncertainties.

$$VF_t = (DMCJ_t^R(\alpha) - DMCJ_t^1) + (DMH_t^R(\alpha) - DMH_t^1)$$
$$+ \sum_{g=1}^{5} (MAL_{gt}^R(\alpha) - MAL_{gt}^1) + \sum_{h=1}^{3} [D_{th}^R(\alpha) - D_{th}^1]; \forall g, t, h \tag{95}$$

$$VF_t \geq 0 \tag{96}$$

where VF_t represents the positive deviation between maximum values and worst value of fuzzy parameters of right-side of constraints. $DMCJ_t^R(\alpha)$, $DMH_t^R(\alpha)$, $MAL_t^R(\alpha)$, and $D_{th}^R(\alpha)$ are the maximum values of all potential values, i.e., $DMCJ_t^R(\alpha) = \sup\{B | B = \mu^{-1}(\alpha)\}$, μ^{-1} is the inverse of μ. $DMCJ_t^1(\alpha)$, $DMH_t^1(\alpha)$, $MAL_t^1(\alpha)$, and $D_{th}^1(\alpha)$ represent the possible worst values of fuzzy numbers $D\tilde{M}CJ_t(\alpha)$, $D\tilde{M}H_t(\alpha)$, $M\tilde{A}L_t(\alpha)$, and $\tilde{D}_{th}(\alpha)$, respectively.

(10) Non-negative constraints. This constraint assures that only positive variables are considered in the solutions, eliminating infeasibility while calculating the solution.

$$MA_{gt}, EPA_{it}, HGA_t, CFHGC_t, CFHGN_t, OFOIL_t,$$
$$CKPA_t, ECOALM_t, ENGM_t, Ed_{th}, rr_{it}, EVEC_t, \tag{97}$$
$$Y_{ith}, Q_{ith}, Z_{ith} \geq 0$$

3.4. Data Acquirement

The data sources of the DRSFO-EES are related references, government reports, and statistical yearbooks. The data regarding the vehicle population of the BTH region in particular was obtained from reference [38]. The annual average vehicle kilometers travelled (VKT) data were obtained from related reference [40–42]. Table 2 details the vehicular emission factors of China-V and China-VI [40,41,43–48]. Table 3 gives the electricity demands and representative technical data [1,23,49–52]. The electricity demands were expressed as stochastic uncertain with three probability levels (20%, 60%, and 20% corresponding to low, medium, and high levels of electricity demand, respectively). The residual capacities and operation times of the different electricity generation technologies were obtained from [49]. The CO_2 and air-pollutant emission factors were acquired from related references and Intergovernmental Panel on Climate Change (IPCC) reports [2,3,53,54].

Table 2. Vehicular emission factors with different emissions standards (mg/km•unit).

		CO	NO_X	HC	PM
HDV	China-V	300	4610	35	100
	China-VI	300	4610	35	100
LDV	China-V	1400	60	230	5
	China-VI	700	35	115	5
LDT	China-V	5800	60	1200	5
	China-VI	2900	35	600	5
HDT	China-V	200	3530	35	100
	China-VI	200	3530	35	100
Other	China-V	2750	150	855	20
	China-VI	2750	150	855	20

Table 3. Electricity demand and technological data.

	Period		
	Period 1	Period 2	Period 3
Electricity demand (10^9 kWh)			
Low demand level	[517.71, 537.71, 557.71]	[553.82, 573.82, 593.82]	[594.67, 614.67, 634.67]
Medium demand level	[547.25, 567.25, 587.25]	[595.73, 615.73, 635.73]	[636.18, 656.18, 676.18]
High demand level	[559.07, 579.07, 599.07]	[646.05, 666.05, 686.05]	[686.87, 706.87, 726.87]
Electricity generation target (10^9 kWh)			
Coal-fired power	[221.07, 279.81]	[213.27, 268.01]	208.93, 269.50]
Gas-fired power	[58.41, 63.41]	[76.91, 80.91]	[91.88, 95.88]
Wind	[40.19, 49.55]	[45.79, 56.48]	[51.01, 61.94]
Solar power	[1.20, 1.70]	[3.20, 3.60]	[5.99, 6.22]
Energy consumption amounts per unit of electricity production			
Coal (ton of SCE/10^3 kWh)	30.50	30.50	30.50
Natural gas (m^3/10^3 kWh)	142.80	142.80	142.80
Coal consumption amounts per unit of coke processing (ton of SCE/ton)			
	1.35	1.35	1.35
Energy consumption amounts for unit of heat processing (ton of SCE/10^9 kJ)			
	36.00	36.00	36.00

4. Results analysis

4.1. Analysis of Vehicular Emissions of BTH Region

Several policies on vehicular emission mitigation, such as the development of EVs, EV power sources and vehicular emission standards, have been proposed by the government of China. In this study, five scenarios labeled S1–S5 were designed to analyze the potentials of different emission mitigation strategies and policies for reducing vehicular emissions (NOx, HC, CO, and PM) in the

BTH region. Table 4 gives the parameter settings of the different scenarios. Following is a description of each scenario:

S1: Without consideration of EVs, and with the local governments implementing the China V vehicular emission standard.

S2: With consideration of EVs, which account for 1.5% of the LDVs population. EV power sources are 100% based on coal-based power, and the China V vehicular emission standard is implemented.

S3: With consideration of EVs, which account for 1.5% of the LDVs population. EV power sources are based 50% on coal-based and 50% on renewable power. It is assumed that coal-based power is local coal-based power, and renewable power included wind, solar power and imported electricity. The China V vehicular emission standard is implemented.

S4: With consideration of EVs, which account for 1.5% of the LDVs population and with power sources based 100% on renewable energy. The China V vehicular emission standard is implemented.

S5: With consideration of EVs, which account for 1.5% of the LDVs population and with power sources based 100% on renewable energy. The China VI vehicular emission standard (gasoline standard) is implemented.

Table 4. Parameter settings of scenarios.

	Vehicular Emissions Standards	The Proportion of EVs	Power Sources for EVs
S1	China V	0%	—
S2	China V	1.50%	100% coal-fired power based
S3	China V	1.50%	50% coal-fired power based, 50% renewable energy based
S4	China V	1.50%	100% renewable energy based
S5	China VI	1.50%	100% renewable energy based

4.1.1. Contributions of Different Vehicle Categories to Vehicular Emissions

Numerous results were obtained by the DRSFO-EES model, and the situation of $\alpha = 0.5$, $\rho = 0.2$, and scenario S2 is used here as an example to illustrate the optimized solutions for the EES. The rapid development of traffic systems has led to large amounts of vehicular emissions such as CO, HC, NO$_x$, and PM. Figure 3 presents the contributions of the different vehicle categories to vehicular emissions in the BTH region over the planning periods. As indicated in Figure 3a–d, due to the sharp growth of LDVs, the total CO, NO$_x$, HC, and PM emissions are expected to increase by 23.28%, 17.07%, 17.19%, and 22.22% over the planning periods, respectively. The results indicated that LDVs are the major contributors of CO and HC emissions, to which they contribute 65.46–73.72% and 60.15–69.15%, respectively, over the planning horizon. It is therefore necessary for local governments to limit the number of LDVs through appropriate measures and policies such as improvement of the public transit system and encouragement of the use of EVs.

Further, HDTs are the major contributor of NO$_x$ and PM, to which they are expected to contribute 67.07–63.08% and 58.58%–54.98%, respectively, between 2020 and 2030. Limiting the development of HDTs would therefore be effective for reducing vehicular NO$_x$ and PM emissions. The results also indicated that the traffic sector is set to be one of the major emitters for NO$_x$ in the BTH region. For example, the NO$_x$ emission from vehicles will account for 58.65%, 60.13%, and 61.90% of the total NO$_x$ emission (the total NO$_x$ emission includes contributions from oil refining, coke processing, heat processing, natural gas-based power, and coal-based power) in 2020, 2025, and 2030, respectively.

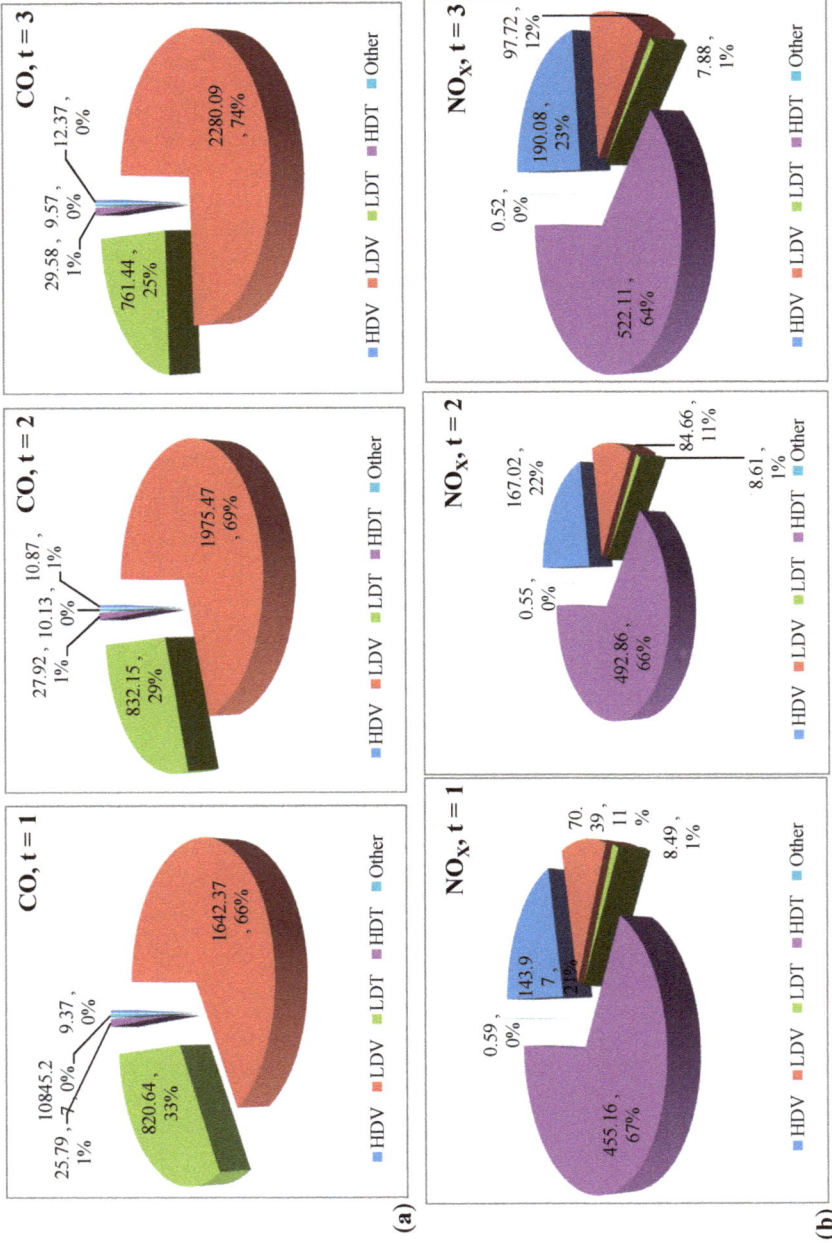

Figure 3. *Cont.*

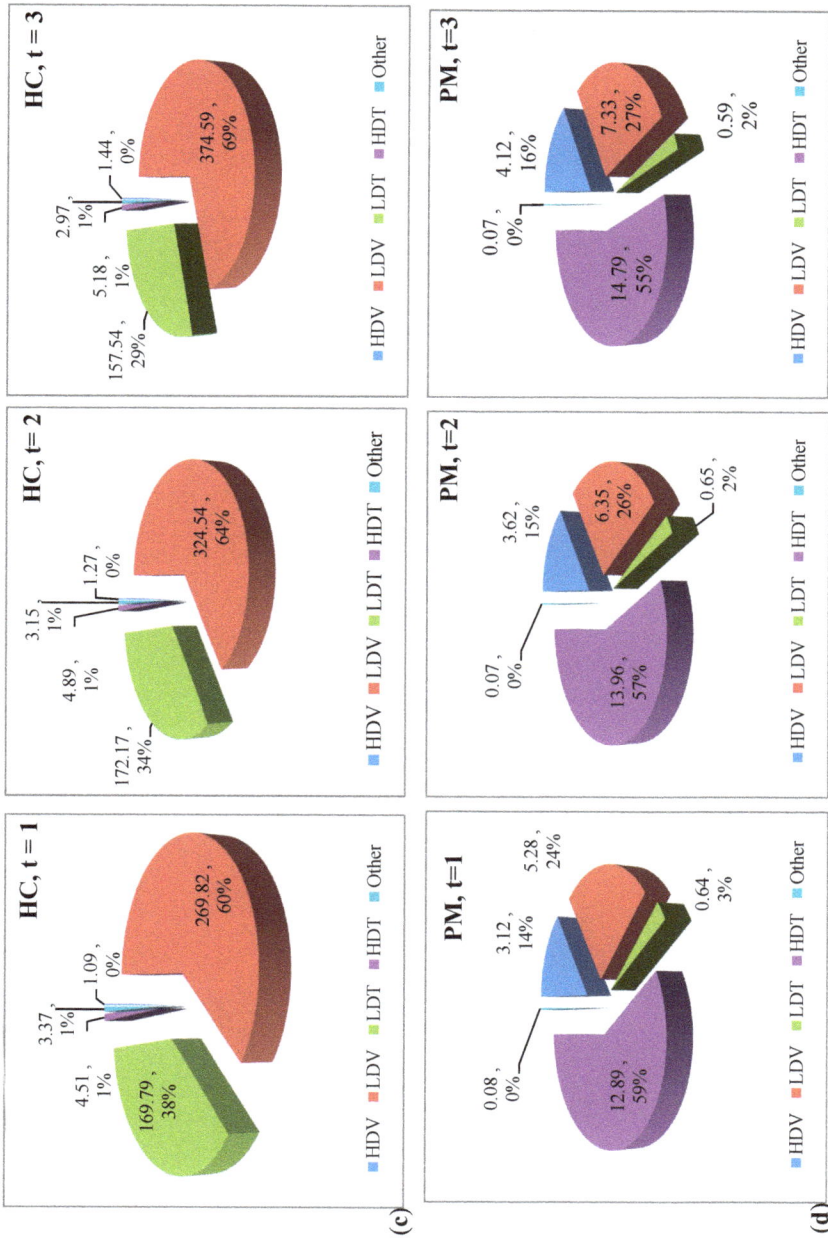

Figure 3. (**a**) CO emissions from traffic system. (**b**) NOX emissions from traffic system. (**c**) HC emissions from traffic system. (**d**) PM emissions from traffic system.

4.1.2. Analysis of the Impact of Traffic Policies on Vehicle Pollutants

Figure 4 shows the vehicular emissions for different scenarios and can be used to analyze the pollutant emission mitigation potential of different vehicle policies. As per the results, implementation of the stringent China VI emission standard is effective for mitigating vehicular emissions, especially CO and HC emissions. The standard is specifically projected to afford CO, HC, NO_x, and PM emission reductions of 49.08%, 49.00%, 4.88%, and 0.00%, respectively, in 2020. As also indicated in Figure 4, the adoption of an EVs policy could reduce CO, NO_x, HC, and PM emissions from vehicles by 25.01×10^3, 1.07×10^3, 4.11×10^3, and 0.08×10^3 ton, respectively, by 2020. However, approximately 2.6×10^9, 3.2×10^9, and 3.7×10^9 kWh of electricity would have to be added by 2020, 2025, and 2030, respectively, to meet the power requirements of EVs. This implies the emission of additional air pollutants (SO_2, NO_x, PM, and CO_2) through the increased electricity generation. Further analysis is thus required to explore the comprehensive impact of the adoption of EVs on pollutant emissions.

The sources of electricity for EVs generally include coal-based power and renewable energy. In this study, it is assumed that electricity from coal-based power was locally generated and renewable power means solar, wind power and imported electricity. Based on different power sources of EVs, there would be different impacts on the environment. Figure 5 shows the additional CO_2 and air pollutant (PM, SO_2, and NO_x) emissions that result from the additional electricity generation required by the adoption of EVs. As can be seen, an EVs policy promises to effectively mitigate CO and HC emissions in all the scenarios. However, it increases SO_2 and CO_2 emissions. Moreover, no reduction in NO_x and PM emissions would be achieved by the EVs policy if the power required by the EVs were entirely generated by coal-based power. Actually, in such a case, the total NO_x emission would increase by 0.72×10^3 ton by 2020, 0.87×10^3 ton by 2025, and 1.32×10^3 ton by 2030. Conversely, if the power for EVs was generated 50% by coal-based and 50% from renewable sources, as in scenario S3, the EVs policy would reduce NO_x emission by 0.17×10^3 ton by 2020, 0.21×10^3 ton by 2025, and 0.24×10^3 ton by 2030. If the EVs power was entirely generated from renewable sources, as in scenario S4, the policy would remarkably reduce NO_x emission by 1.07×10^3 ton by 2020, 1.29×10^3 ton by 2025, and 1.49×10^3 by 2030. These results indicate that the vehicle-emissions could be reduced directly; but through the extra electricity generated from local region, leading to additional air-pollutants. Thus, an EVs policy should be enhanced by increasing the ratio of the needed power generated from renewable sources.

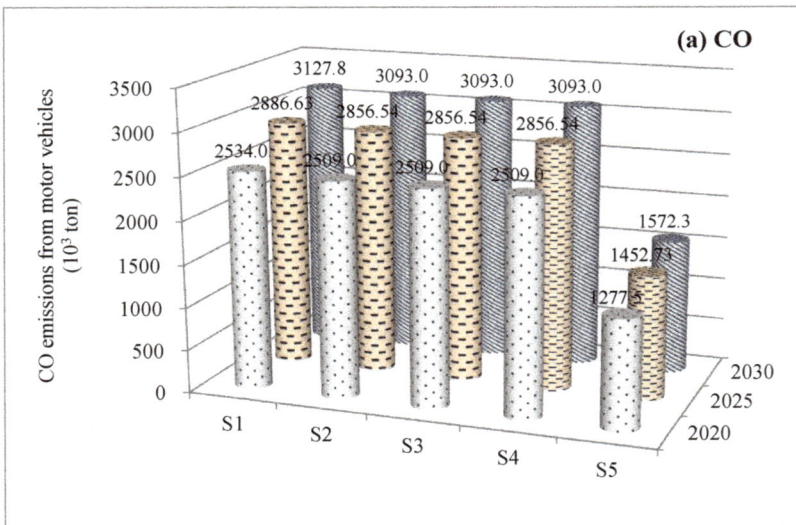

Figure 4. *Cont.*

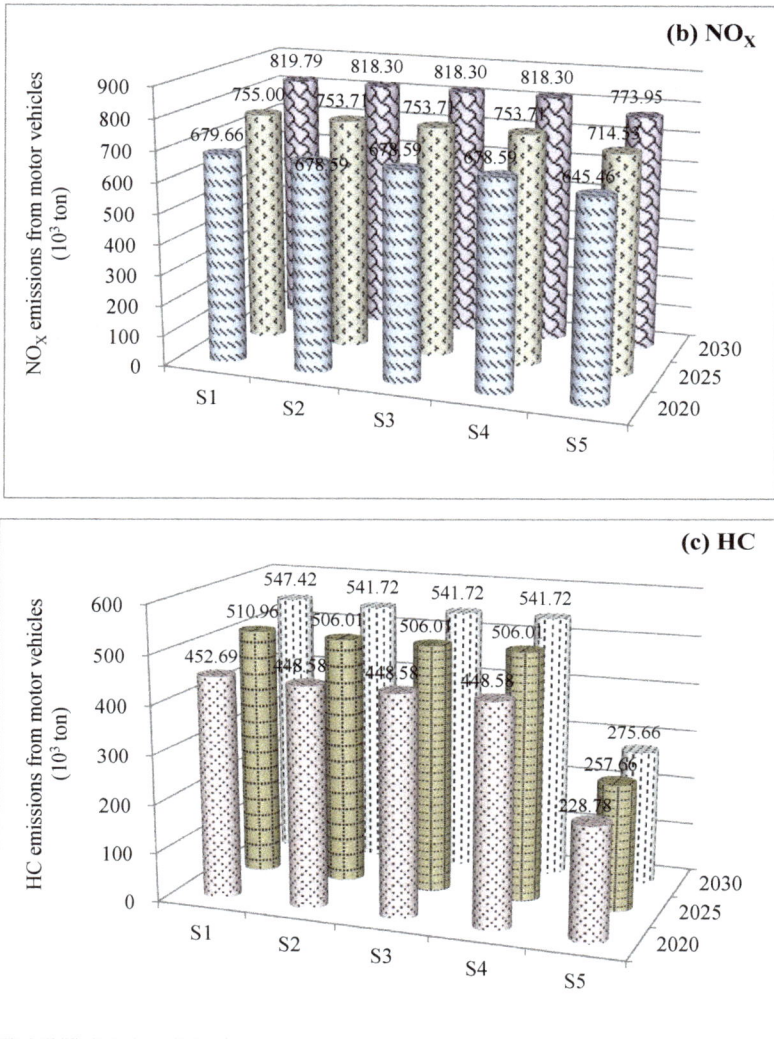

(b) NO$_X$

(c) HC

Figure 4. *Cont.*

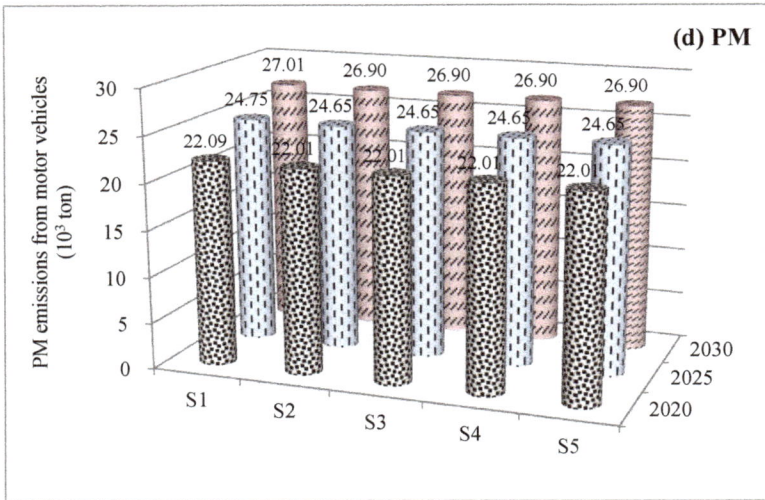

Figure 4. Vehicular emissions from traffic systems under different scenarios.

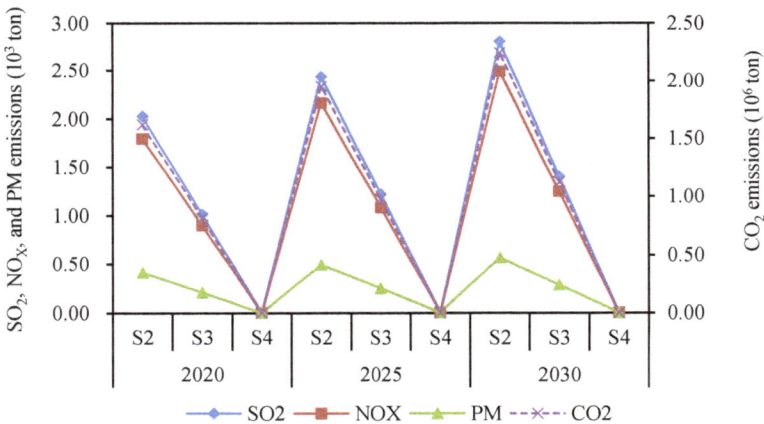

Figure 5. Additional emissions caused by electric vehicles under different scenarios.

4.2. Optimized Robust Solutions for Energy and Environment Systems

4.2.1. Optimized Schemes of Energy Allocation

Figure 6 shows the optimized energy consumption of energy processing (heat processing, coke processing, and oil refining) and electricity generation (natural gas-based power and coal-based power) between period 1 and period 3. The use of natural gas is expected to substantially increase toward achieving a sustainably developed society. For example, the natural gas inputs to heat processing and natural gas-based power generation are projected to respectively increase by 47.16% (from 3.15×10^9 m^3 in 2020 to 4.63×10^9 m^3 in 2030) and 53.86% (from peak m 9.51×10^9 m^3 to 14.63×10^9 m^3) between 2020 and 2030. Crude oil input to oil refining, coal input to coke processing, and coal input to heating processing are expected gradually increase by 10.88%, 10.95%, and 3.86%, respectively, in periods 1, 2 and 3. Conversely, the coal consumption of coal-based power generation is expected to decrease by 2.69% between 2020 and 2030, specifically from 79.83 × 10⁶ ton to 77.68 × 10⁶ ton between period 1 and period 3. An appropriate energy mix can thus be effectively used to reduce coal consumption, and hence mitigate pollutant emissions.

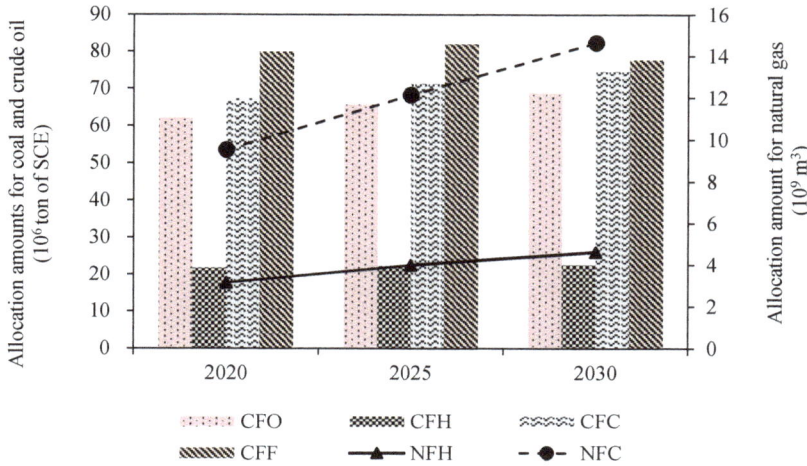

Figure 6. Optimized solutions of energy input amounts for energy processing and electricity generation between 2020 and 2030. CFO represents crude oil input to oil refining; CFH represents coal input to heat processing; NFH represents natural gas input to heat processing; CFC represents coal input to coke processing; CFF represents coal input to coal-fired power; NFC represents natural gas input to coal-fired power.

4.2.2. Optimized schemes of electricity supply

Figure 7a shows the optimized electricity generation scheme between 2020 and 2030. The major electricity power conversion technologies in the BTH region include gas-based, coal-based, wind, and solar power. Optimized electricity generation is defined by the following equation: $OEG_{kth\ opt} = W^{\pm}_{kt\ opt} + Y_{kth\ opt}$, where $W^{\pm}_{ktopt}$ and Y_{ktopt} are the amounts of electricity generated in the first and second stages. Coal-based power is expected to play the dominant role, contributing what would be 259.06 × 10⁹ kWh, 265.41 × 10⁹ kWh, and 250.97 × 10⁹ kWh in periods 1, 2 and 3 (h = 3), respectively. With the implementation of a series of energy mix policies for pollutant emission mitigation, such as the "Paris Agreement" and "Chinese Action Plan of Air Pollution Prevention and Control", the ratio of coal-based power generation would decrease from 68.79% in 2020 to 55.48% in 2030. The utilization of natural gas-based and renewable energy would rapidly develop from 2020 to 2030. Natural gas-based power, wind power, solar power and generation are projected to increase by 53.86%, 87.96%, and 266.34% over the planning horizon, contributing 22.65%, 20.44%, and 1.44%, respectively to the total power generation by 2030.

Further, the amount of locally generated electricity is insufficient for the needs of the BTH region, with approximately 36% of the consumed electricity imported from other regions. Figure 7b describes imported electricity under different electricity demand levels between 2020 and 2030. Imported electricity is expected to increase from 202.47 × 10⁹ to 254.47 × 10⁹ kWh over the planning horizon under a high electricity demand level. This indicates that imported electricity is expected to play an increasingly important role from period 1 to period 3, especially at a high demand level.

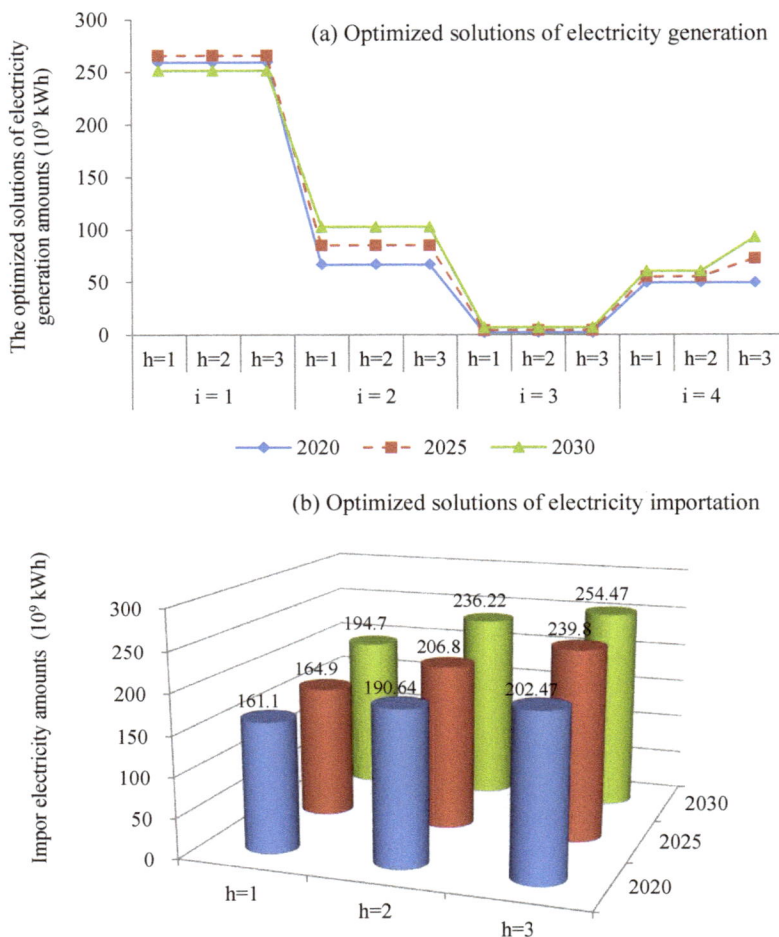

Figure 7. Optimized solutions for electricity supply of the Beijing-Tianjin-Hebei (BTH) region between 2020 and 2030.

4.2.3. Optimized Schemes of Energy Processing

Figure 8 shows the amounts of processed secondary energy (i.e., coke, heat, diesel, gasoline, fuel oil, liquefied petroleum gas, kerosene, naphtha, and tar) during periods 1 to 3. With the rapid development of society over the entire considered time, the energy processing amounts for heat, gasoline, coke, kerosene, naphtha, fuel oil, tar, and liquefied petroleum gas (LPG) are expected to respectively increase by 10.37%, 29.98%, 10.59%, 19.97%, 1.11%, 7.48%, 6.90%, and 8.84%, between 2020 and 2030.

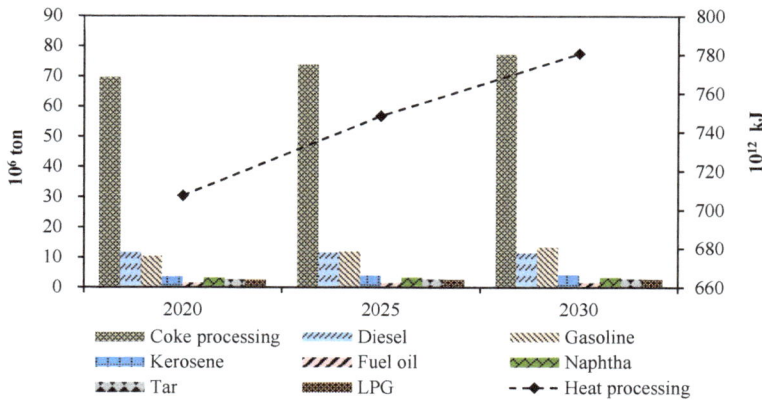

Figure 8. Optimized solutions for energy processing between 2020 and 2030.

4.2.4. Pollutant Emissions from Energy Activities

Figure 9 shows the CO_2 and air pollutant (NO_x, SO_2, and PM) emissions from natural gas-based power generation, coal-based power generation, heat processing, coke processing, and oil refining) over the planning horizon. As can be observed, CO_2, NO_x, SO_2, and PM emissions are expected to gradually increase by 9.37%, 5.25%, 4.43%, and, 3.59%, respectively, between 2020 and 2030. The air-pollutants emissions would approach their peak values around 2030. For example, the SO_2 emission would increase by 4.14% between 2020 and 2025, and by 0.27% between 2025 and 2030, indicative of peaking around 2030. Coal-based power generation would be the major contributor of NO_x, SO_2, and PM emissions, respectively accounting for 36.66–33.90%, 41.89–39.03%, and 47.33–44.46% of these emissions between 2020 and 2030, with the specific contributions progressively decreasing. Coke processing would contribute 30.86–32.53%, 35.26%–37.46%, and 39.84%–42.67% of NO_x, SO_2, and PM emissions, respectively.

As indicated in Figure 9d, oil refining, coal-based power generation, and coke processing would be the major sources of CO_2 emissions between 2020 and 2030 in BTH region, accounting for 33.83–34.30%, 28.08–24.99%, and 23.64–23.98%, respectively. In conclusion, coal-based power generation and coke processing would be the major enablers of SO_2, NO_x, and PM emissions reduction, while oil refining, coal-based power generation, and coke processing would be the major contributors of CO_2 emission.

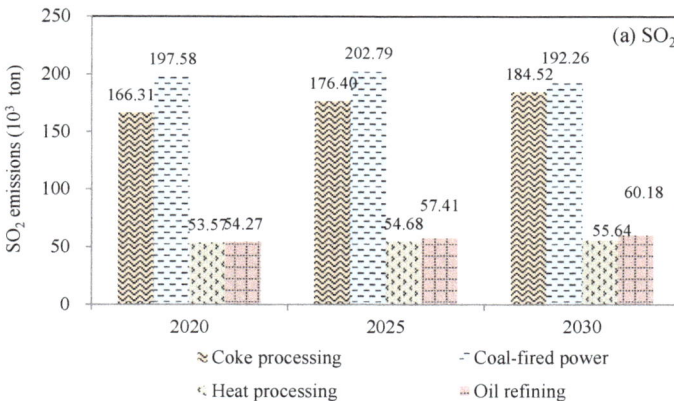

Figure 9. *Cont.*

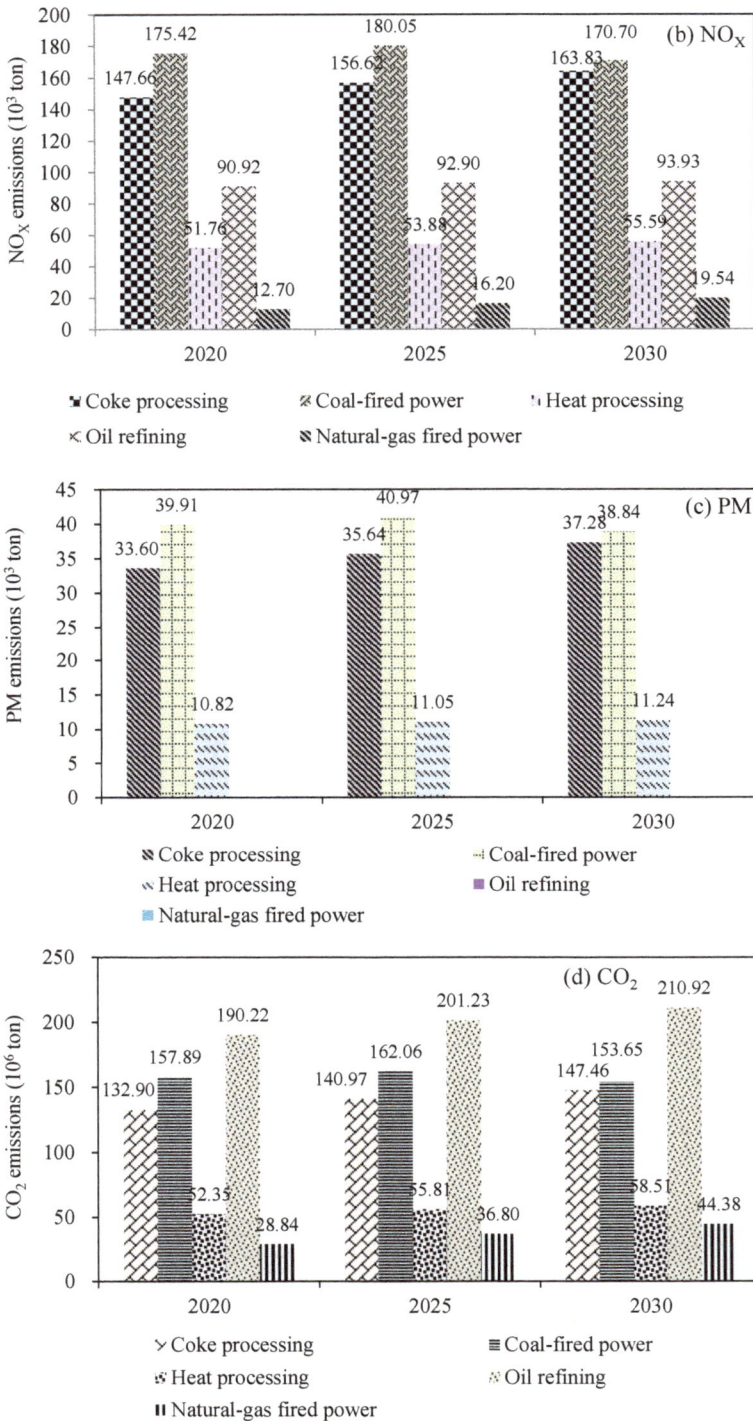

Figure 9. Pollutants and CO₂ emissions from energy activities between 2020 and 2030.

4.2.5. Analysis of system cost

The system costs are composed of 16 main components, namely, the costs of heat processing, coke processing, oil refining, coal input to coal-based power generation, natural gas input to natural gas-based power generation, imported electricity, first-stage of electricity generation, second-stage of electricity generation, electricity expansion, subsidy for solar power generation, subsidy for wind power generation, energy production, EVs charging piles, EVs charging stations, pollutants treatment, and risk recourse for the stochastic and fuzzy uncertainties. As shown in Figure 10, the costs of electricity generation, imported electricity, energy production, and coal input for coal-based power generation account for 30.51%, 16.90%, 16.67%, and 11.98% of the total system costs, respectively. Solar and wind power would be subsidized by the local government because their current price and technology limitations make them uncompetitive in the market. Government subsidies for wind and solar power would be RMB ¥ 40.99×10^9, and RMB ¥ 6.52×10^9, respectively.

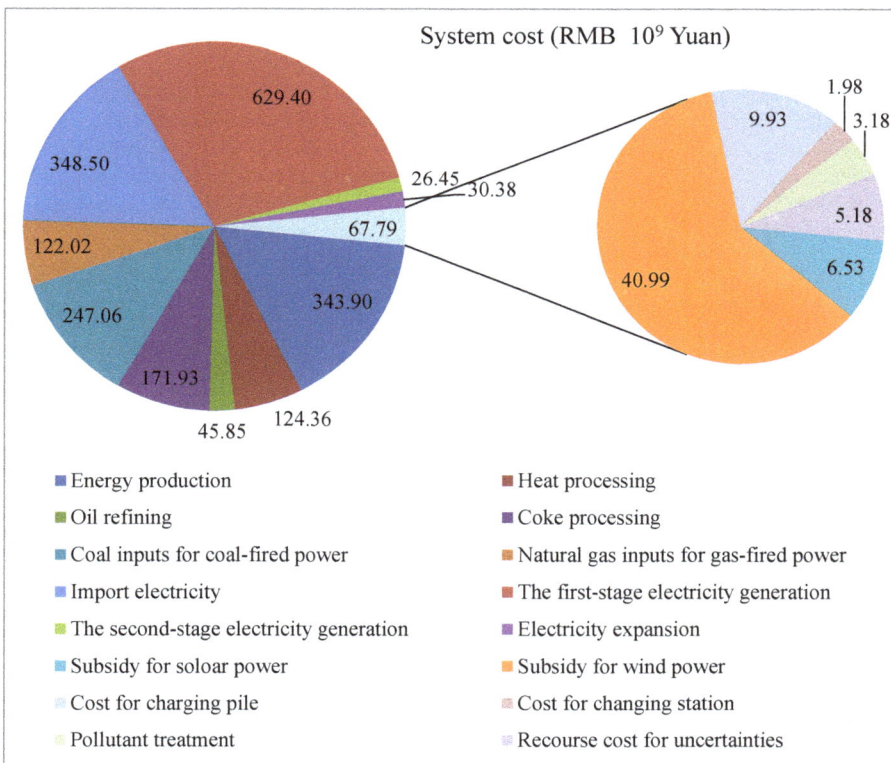

Figure 10. System cost over the planning periods.

5. Discussion

5.1. Analysis of Stochastic Uncertainties

In EES in the real-world, many parameters are expressed as random distribution (i.e., electricity demand, electricity generation, and electricity expansion). In this study, the TSP method was used to tackle the stochastic uncertainties. The decision variables of the electricity generation were divided into two subsets, namely, $W_{ktopt}^{\pm}$ and Y_{ktopt}, which are respectively the amounts of electricity generated in the first-stage (which must be predetermined) and the second-stage (obtained after the determination of the random variables) [55]. Generally, shortages may occur if the electricity demand-levels are continuously high, and the second-stage electricity generation amount Y_{ktopt} would be undertaken to avoid insufficient electricity supply.

In the TSP method, the final optimized electricity generation schemes were equal to $W_{kt}^{\pm} + Y_{kth}^{\pm}$. In detail, the optimized amount of electricity generated in the first-stage is given by $W_{kt}^{\pm} = W_{kt}^{-} + rr_{kt} \cdot \Delta W$, where rr_{kt} denotes the decision variables, $\Delta W = W_{kt}^{+} - W_{kt}^{-}$, and $rr_{kt} \in [0, 1]$. Take period 1 for example, $\lambda_{11opt} = 0.55$, $\lambda_{21opt} = 1$, $\lambda_{31opt} = 1.00$, and $\lambda_{41opt} = 1.00$, indicating that the optimized electricity generation amounts for coal-based power, natural gas-based power, solar power and wind power are supposed to be $(221.08 + 0.52 \times 58.74) \times 10^9$, $(58.41 + 1.00 \times 5.00) \times 10^9$ kWh, $(1.78 + 1.00 \times 0.50) \times 10^9$ kWh, and $(49.54 + 1.00 \times 9.36) \times 10^9$ kWh, respectively. Generally, the variation of rr_{kt} represents diverse policies of electricity generation under stochastic electricity demands. When $rr_{kt} = 0$, the cost would be relatively low, although a higher penalty may have to be paid when the generated electricity does not meet the demand. On the contrary, when $rr_{it} = 1$, the cost would be higher, but accompanied by a lower risk of violating the target, and hence of incurring a lower penalty.

5.2. Analysis of Fuzzy Uncertainties

In the real world, energy price and energy demands often exhibit vagueness and ambiguity because of the subjectivity of human judgment [25]. According to FPP theory, the minimum, medium, and maximum values of these parameters are sufficient for expressing a triangular fuzzy parameter. And the proposed DRSFO-EES model can be used to effectively address the uncertainties expressed as triangular fuzzy parameter of the objectives and constraints. For FPP, the confidence level α is an indication of the manager's violation risk attitude towards imprecise information [56,57]. In the present application of the DRSFO-EES model, four confidence levels ($\alpha = 0.5, 0.6, 0.8,$ and 0.9) were used to examine the impacts of different confidence levels on the EES. Generally, a higher confidence level implies a higher likelihood of satisfying the fuzzy confidence constraints, resulting in less uncertainty about the imprecise constraints. For instance, a confidence level of 0.8 indicates that the credibility of the constraint (e.g., $Cr\{M\widetilde{A}L_{gt} | MA_{gt} \geq M\widetilde{A}L_{gt}\} \geq \alpha$) is greater than or equal to 0.8. However, a higher confidence level increases the system costs. Contrarily, a lower confidence level implies a more aggressive attitude of the decision maker regarding the expected total system costs, and increases the uncertainty of the fuzzy constraints, resulting in a higher risk of violating the energy demand. Table 5 gives the optimized solution under different confidence levels α during the planning periods. As can be seen, a higher α increases the coefficients of the right-side constraints, further necessitating electricity import and increasing pollutant emission and the system costs. Vehicle ownership, heat processing, and coke processing also increase with α, all accompanied by pollutant emissions.

Table 5. (a) Optimized solutions under different α levels in 2020.

	α = 0.5	α = 0.6	α = 0.8	α = 0.9
Traffic system and its relative pollutants				
Vehicle ownership (10^6)	22.11	22.34	22.80	23.03
CO emissions (10^3 ton)	2534.02	2561.58	2616.70	2644.26
NO_X emissions (10^3 ton)	679.66	685.44	696.98	702.75
HC emissions (10^3 ton)	452.69	457.67	467.62	472.59
PM emissions (10^3 ton)	22.09	22.28	22.67	22.86
Heat processing (10^{12} kJ)				
	707.31	708.31	710.31	711.31
Coke processing (10^6 ton)				
	69.61	70.11	71.11	71.61
Import electricity amounts (10^9 kWh, h = 1)				
	161.10	163.10	167.10	169.10
Air pollutants and CO_2 from energy processing and electricity generation (10^3 ton)				
SO_2 emissions	667.86	669.95	674.12	676.20
NO_X emissions	476.67	479.17	484.17	486.67
PM emissions	83.92	84.18	84.69	84.95
CO_2 emissions	560593.92	564482.45	572259.52	576148.05

(b) Optimized solutions under different α levels in 2025.

	α = 0.5	α = 0.6	α = 0.8	α = 0.9
Traffic system and its relative pollutants				
Vehicle ownership (10^6)	26.16	26.39	26.84	27.07
CO emissions (10^3 ton)	2886.63	2914.19	2969.31	2996.87
NO_X emissions (10^3 ton)	754996.10	760.77	772.31	778.09
HC emissions (10^3 ton)	510.96	515.93	525.88	530.86
PM emissions (10^3 ton)	24.75	24.94	25.33	25.52
Heat processing (10^{12} kJ)				
	748.38	749.38	751.38	752.38
Coke processing (10^6 ton)				
	73.84	74.34	75.34	75.84
Import electricity amounts (10^9 kWh, h = 1)				
	164.89	166.89	170.89	172.89
Air pollutants and CO_2 from energy processing and electricity generation (10^3 ton)				
SO_2 emissions	689.20	691.28	695.45	697.53
NO_X emissions	497.49	499.95	504.85	507.30
PM emissions	87.16	87.41	87.93	88.18
CO_2 emissions	594920.14	598809.23	606587.42	610476.51

(c) Optimized solutions under different α levels in 2030.

	α = 0.5	α = 0.6	α = 0.8	α = 0.9
Traffic system and its relative pollutants				
Vehicle ownership (10^6)	29.69	29.92	30.38	30.61
CO emissions (10^3 ton)	3127.77	3155.33	3210.45	3238.01
NO_X emissions (10^3 ton)	819.79	825.56	837.11	842.88
HC emissions (10^3 ton)	547.42	552.40	562.35	567.32
PM emissions (10^3 ton)	27.01	27.21	27.59	27.78
Heat processing (10^{12} kJ)				
	780.64	781.64	783.64	784.64
Coke processing (10^6 ton)				
	77.24	77.74	78.74	79.24
Import electricity amounts (10^9 kWh, h = 1)				
	194.72	195.44	196.88	197.60
Air pollutants and CO_2 from energy processing and electricity generation (10^3 ton)				
SO_2 emissions	688.19	690.27	694.44	696.52
NO_X emissions	501.10	503.75	509.05	511.70
PM emissions	86.79	87.05	87.56	87.81
CO_2 emissions	612669.67	617113.59	626001.42	630445.34

5.3. Risk Analysis

A robust optimization method can be effectively used to determine the associated risk from stochastic and fuzzy uncertainties. In this study, two risk recourse actions were adopted to make the model robust, which were used to capture the risks from stochastic and fuzzy uncertainties, respectively. Figure 11 presents the weighted values of the expected deviations from the stochastic uncertainties ($V\widetilde{T}S = \rho \cdot \sum\limits_{i=1}^{I} \sum\limits_{t=1}^{T} \sum\limits_{h=1}^{H} p_{th}(V_{ith} + VC_{ijcth}]$) and fuzzy uncertainties ($V\widetilde{T}F = \rho \cdot \sum\limits_{t=1}^{3} VF_t$) under different robustness levels of 0.2, 0.6, 0.8, and 1.0, respectively. ρ is a goal programming weight, through varying the ρ level, the decision makers can then control the variability of the recourse cost. Generally, a lower ρ corresponds to a lower weight value of the expected deviations and system costs, indicating an aggressive attitude of the manager regarding the system costs. However, this might be associated with a higher risk level because of the expected deviations from the uncertainties. On the contrary, a plan with a higherρwould better resist a deviation from the uncertainties of the EES. A decision with a higher robust level would thus correspond to a lower risk of system failure and higher system reliability. As results, the weighted values of the expected deviations from the stochastic and fuzzy uncertainties increased with increasing robustness level ρ. For instance, $V\widetilde{T}S$ would be RMB 5.17×10^9 when $\rho = 0.2$, RMB 15.54×10^9 when $\rho = 0.6$, RMB 20.71×10^9 when $\rho = 0.8$, and 25.89×10^9 when $\rho = 1$. This analysis demonstrates the trade-offs between system costs and reliability. The results enable the manager to plan with a reasonable consideration of both system costs and reliability.

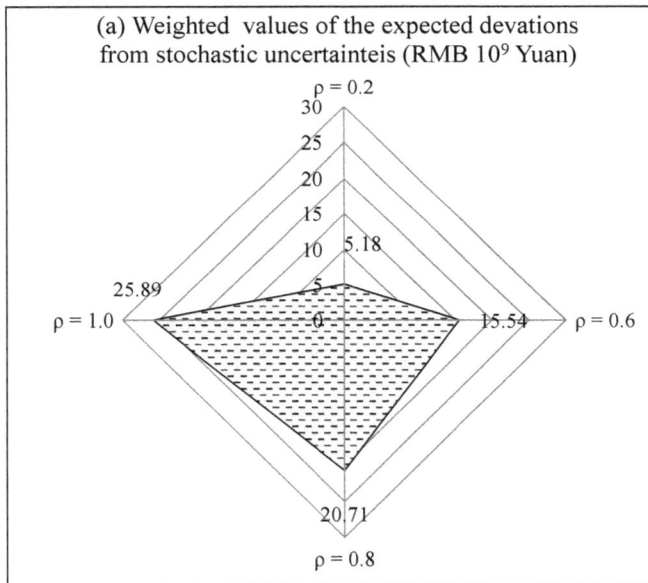

Figure 11. *Cont.*

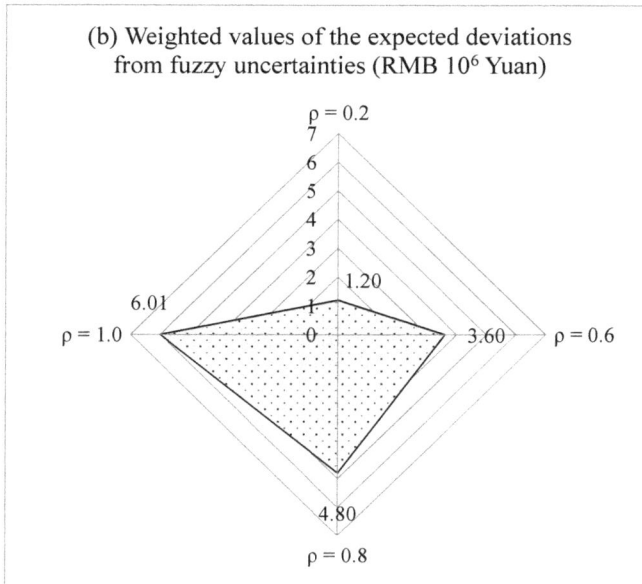

Figure 11. Weighted values of the expected deviations under different robust levels.

6. Conclusions

In this study, a DRSFO-EES model was developed for planning an EES while considering the traffic sector, which integrates the TSP, FPP, RTSO, and RFPP methods in a single framework for the effective handling of EES uncertainties expressed as fuzzy sets and stochastic uncertainties as well as their combinations, capturing of the associated risks from the stochastic and fuzzy uncertainties, and analyzing the trade-offs between system costs and reliability. Four confidence levels ($\alpha = 0.5$, $\alpha = 0.6$, $\alpha = 0.8$, and $\alpha = 0.9$) and four robust levels ($\rho = 0.2$, $\rho = 0.6$, $\rho = 0.8$, and $\rho = 1.0$) were used for examining the impacts of uncertainties on the objective function, constraints and optimized solutions of the DRSFO-EES model.

The proposed model was applied to the EES of the BTH region in China. Following is a summary of the findings and the identified policy implications:

(1) Limiting the numbers of LDVs and HDTs could effectively reduce vehicular emissions. LDVs are expected to be the major contributors of CO and HC emissions, and HDTs are expected to be the major contributors of NO_x and PM emissions.

(2) A EVs policy would be enhanced by increasing the ratio of power generated for EVs from renewable sources. The emission reduction effect of an EVs policy would thus be limited, especially with regard to NO_x and PM emissions, if the EVs power source was entirely coal-based.

(3) Optimizing the energy mix and developing the renewable energy can effectively reduce air-pollutant and CO_2 emissions. Air-pollutant amounts of NO_x, SO_2, and PM emissions in the BTH region are expected to peak around 2030, because the energy mix of the study region would be transformed from one dominated by coal to one with a cleaner pattern, with vigorous development of the utilization of natural gas and renewable energy.

(4) Enhancement of the energy utilization efficiencies of coal-based power generation, oil refining, and coke processing would effectively reduce CO_2 and air-pollutant emissions. Coal-based power generation and coke processing are expected to be the major contributors of air-pollutant

missions, while oil refining, coal-based power generation and coke processing would be the chief sources of CO_2 emissions.

Although the DRSFO method was the first attempt for planning an EES while considering the traffic sector of the BTH region, results indicated that DRSFO-EES could: (i) explore the impacts of different vehicle policies (i.e., EVs deployment, EVs power source for EVs, and vehicular emission standards) on vehicular emissions; (ii) generate robust optimized solutions for energy allocation, oil refining, coking processing, heat processing, electricity generation and expansion, electricity importation, as well as emission mitigation under multiple uncertainties; (iii) identify the atmospheric pollution contributions of different energy activities such as coke processing, electricity generation, heat processing, oil refining, and motor vehicle operation. The proposed model could help to balance the contradiction between increasing energy demands and an increasing vehicle population, "high coal" energy systems, and the pressures of emission mitigation. Moreover, the proposed model could be applied at both city and regional scales, which would support policymakers adjusting current energy and environmental strategies in sustainable and robust ways.

However, the DRSFO-EES also has potential limitations and extensions should be addressed in future study. Firstly, the developed model, based on historical data of annual electricity demand to predict future electricity demands, does not consider the specific parameters such as hourly or seasonal electricity load curves, which may result in significant deviations from the optimized decision schemes, even in the event of electricity shortage. Thus, further study is required for considering the hourly and seasonal electricity load curves; secondly, the TSP, FPP, RTSO, and RFPP methods were combined into a single framework to formulate the DRSFO-EES model, leading to relatively high computational requirements. As a result, simplifying the calculation procedure could be required in the further study work; thirdly, the DRSFO-EES model mainly focused on economic objectives, whilst scarcely considering the trade-off between economic and environmental objectives. Therefore, further study should make improvements in the handling of multi-objective problems and better balance the tension between energy and environmental systems.

Author Contributions: C.C. proposed and calculated the DRSFO-EES model through Lingo 11, as well as analyzed the results and wrote this original manuscript; X.T.Z. and contributed in conceptualization of idea; G.H.H. contributed in supervision and revised the manuscript; Y.L. has collected the data, analyzed related data; Y.P.L. contributed in editing the manuscript.

Funding: This research was supported by the national natural science fund projects (No. 41701621, 71673022); Beijing Municipal Social Science Foundation (No. 18LJC006,17LJB004); Self-designed project of Heilongjiang Institute of Water Conservancy Science (ZN201806). The authors are grateful to the editors and the anonymous reviewers for their insightful comments and suggestions.

Conflicts of Interest: The authors declare no conflict of interest.

References

1. Chen, C.; Long, H.L.; Zeng, X.T. Planning a sustainable urban electric power system with considering effects of new energy resources and clean production levels under uncertainty: A case study of Tianjin, China. *J. Clean. Prod.* **2018**, *17*, 67–81. [CrossRef]
2. Chen, C.; Qi, M.Z.; Kong, X.M.; Huang, G.H.; Li, Y.P. Air pollutant and CO_2 emissions mitigation in urban energy systems through a fuzzy possibilistic programming method under uncertainty. *J. Clean. Prod.* **2018**, *192*, 115–137. [CrossRef]
3. Chen, C.; Zhu, Y.; Zeng, X.T.; Huang, G.H.; Li, Y.P. Analyzing the carbon mitigation potential of tradable green certificates based on a TGC-FFSRO model: A case study in the Beijing-Tianjin-Hebei region, China. *Sci. Total Environ.* **2018**, *630*, 469–486. [CrossRef] [PubMed]
4. Ministry of Environment Protection of the People's Republic of China. *China's Environmental Statistics Yearbook*; China Statistics Press: Beijing, China, 2017.
5. Guo, X.R.; Fu, L.W.; Ji, M.; Lang, J.L.; Chen, D.S. Scenario analysis to vehicular emission reduction in Beijing-Tianjin-Hebei (BTH) region, China. *Environ. Pollut.* **2016**, *216*, 470–479. [CrossRef] [PubMed]

6.   Yu, L.; Li, Y.P. A flexible-possibilistic stochastic programming method for planning municipal-scale energy system through introducing renewable energies and electric vehicles. *J. Clean. Prod.* **2019**, *207*, 772–787. [CrossRef]

7.   Odetayo, B.; MacCormack, J.; Rosehart, W.D.; Zareipour, H.; Seifi, A.R. Integrated planning of natural gas and electric power systems. *Electr. Power Energy Syst.* **2018**, *103*, 593–602. [CrossRef]

8.   Chen, J.P.; Huang, G.H.; Baetz, B.W.; Lin, Q.G.; Dong, C.; Cai, Y.P. Integrated inexact energy systems planning under climate change: A case study of Yukon Territory, Canada. *Appl. Energy* **2018**, *229*, 493–504. [CrossRef]

9.   Sahabmanesh, A.; Saboohi, Y. Model of sustainable development of energy system, case of Hamedan. *Energy Policy* **2017**, *104*, 66–79. [CrossRef]

10.   Peker, M.; Kocaman, S.A.; Kara, B.Y. A two-stage stochastic programming approach for reliability constrained power system expansion planning. *Electr. Power Energy Syst.* **2018**, *103*, 458–469. [CrossRef]

11.   Markel, E.; Sims, C.; English, B.C. Policy uncertainty and the optimal investment decisions of second-generation biofuel producers. *Energy Econ.* **2018**, *76*, 89–100. [CrossRef]

12.   Yu, L.; Li, Y.P.; Huang, G.H. Planning municipal-scale mixed energy system for stimulating renewable energy under multiple uncertainties—The City of Qingdao in Shandong Province, China. *Energy* **2019**, *166*, 1120–1133. [CrossRef]

13.   Sarkar, M.; Sarkar, B.; Iqbal, M.W. Effect of Energy and Failure Rate in a Multi-Item Smart Production System. *Energies* **2018**, *11*, 2958. [CrossRef]

14.   Wen, W.; Zhou, P.; Zhang, F.Q. Carbon emissions abatement: Emissions trading vs consumer awareness. *Energy Econ.* **2018**, *76*, 34–47. [CrossRef]

15.   Zeng, X.T.; Zhu, Y.; Chen, C.; Tong, Y.F.; Li, Y.P.; Huang, G.H.; Nie, S.; Wang, X.Q. A production-emission nexus based stochastic-fuzzy model for identification of urban industry-environment policy under uncertainty. *J. Clean. Prod.* **2017**, *154*, 61–82. [CrossRef]

16.   Gong, J.W.; Li, Y.P.; Suo, C. Full-infinite interval two-stage credibility constrained programming for electric power system management by considering carbon emissions trading. *Electr. Power Energy Syst.* **2019**, *105*, 440–453. [CrossRef]

17.   Ervural, B.C.; Zaim, S.; Delen, D. A two-stage analytical approach to assess sustainable energy efficiency. *Energy* **2018**, *164*, 822–836. [CrossRef]

18.   Simic, V. End-of life vehicles allocation management under multiple uncertainties: An interval-parameter two-stage stochastic full-infinite programming approach. *Resour. Conserv. Recycl.* **2016**, *114*, 1–17. [CrossRef]

19.   Marvromatidis, G.; Orehounig, K.; Garmeliet, J. Design of distributed energy systems under uncertainty: A two-stage stochastic programming approach. *Appl. Energy* **2018**, *222*, 932–950. [CrossRef]

20.   Mohan, V.; Singh, J.G.; Ongsakul, W. An efficient two-stage stochastic optimal energy and reserve management in a microgrid. *Appl. Energy* **2015**, *160*, 28–38. [CrossRef]

21.   Bai, D.; Carpenter, T.J.; Mulvey, J.M. Making a case for robust models. *Manag. Sci.* **1997**, *43*, 895–907. [CrossRef]

22.   Chen, C.; Li, Y.P.; Huang, G.H.; Li, Y.F. A robust optimization method for planning regional-scale electric power systems and managing carbon dioxide. *Electr. Power Energy Syst.* **2012**, *40*, 70–84. [CrossRef]

23.   Chen, C.; Li, Y.P.; Huang, G.H. Interval-fuzzy municipal-scale energy model for identification of optimal strategies for energy management—A case study of Tianjin, China. *Renew. Energy.* **2016**, *86*, 1161–1177. [CrossRef]

24.   Xu, Y.; Huang, G.H.; Qin, X.S. Inexact two-stage stochastic robust optimization model for water Resources management under uncertainty. *Environ. Eng. Sci.* **2009**, *26*, 1765–1774. [CrossRef]

25.   Xu, Y.; Huang, G.H. Development of an improved fuzzy robust chance-constrained programming model for air quality management. *Environ. Model Assess.* **2015**, *20*, 533–548. [CrossRef]

26.   Govindan, K.; Cheng, T.C.E. Advances in stochastic programming and robust optimization for supply chain planning. *Comput. Oper. Res.* **2018**, *100*, 262–269. [CrossRef]

27.   Xu, Y.; Huang, G.H.; Li, J.J. An enhanced fuzzy robust optimization model for regional solid waste management under uncertainty. *Eng. Optim.* **2016**, *48*, 1869–1886. [CrossRef]

28.   Promentilla, M.A.B.; Janairo, J.J.B.; Yu, D.E.C.; Pausta, C.M.J.; Beltran, A.B.; Huelgas-Orbecideo, A.P.; Tapia, J.F.D.; Aviso, K.B.; Tan, R.R. A stochastic fuzzy multi-criteria decision-making model for optimal selection of clean technologies. *J. Clean. Prod.* **2018**, *183*, 1289–1299. [CrossRef]

29. Jiménez, M.; Arenas, M.; Bilbao, A.; Rodriguez, M.V. Linear programming with fuzzy parameters: An interactive method resolution. *Eur. J. Oper. Res.* **2007**, *177*, 1599–1609. [CrossRef]

30. Vahdani, B.; Razmi, J.; Tavakkoli-Moghaddam, R. Fuzzy possibilistic modeling for closed loop recycling collection networks. *Environ. Model. Assess.* **2012**, *17*, 623–637. [CrossRef]

31. Lu, W.T.; Dai, C.; Fu, Z.H.; Liang, Z.Y.; Guo, H.C. An interval-fuzzy possibilistic programming model to optimize China energy management system with CO_2 emission constraint. *Energy* **2018**, *142*, 1023–1039. [CrossRef]

32. Pishvaee, M.S.; Razmi, J.; Torabi, S.A. Robust possibilistic programming for socially responsible supply chain network design: A new approach. *Fuzzy Sets Syst.* **2012**, *1*, 1–20. [CrossRef]

33. Xu, Y.; Huang, G.H.; Shao, L.G. A stochastic fuzzy chance-constrained programming model for energy–environment system planning and management in the City of Beijing. *Int. J. Green Energy* **2017**, *14*, 171–183. [CrossRef]

34. Mulvey, J.M.; Vanderbei, R.J. Robust optimization of large-scale systems. *Oper. Res.* **1995**, *43*, 264–281. [CrossRef]

35. Ahmed, S.; Sahinidis, N.V. Robust process planning under uncertainty. *Ind. Eng. Chem. Res.* **1998**, *37*, 1883–1892. [CrossRef]

36. Sun, L.; Pan, B.L.; Gu, A.; Lu, H.; Wang, W. Energy–water nexus analysis in the Beijing–Tianjin–Hebei region: Case of electricity sector. *Renew. Sustain. Energy Rev.* **2018**, *93*, 27–34. [CrossRef]

37. Zhang, Z.Z.; Wang, W.X.; Cheng, M.M.; Liu, S.J.; Xu, J.; He, Y.J.; Meng, F. The contribution of residential coal combustion to $PM_{2.5}$ pollution over China's Beijing-Tianjin-Hebei region in winter. *Atmos. Environ.* **2017**, *159*, 147–161. [CrossRef]

38. National Bureau of Statistics of the People's Republic of China. *Chinese Statistical Yearbook*; China Statistics Press: Beijing, China, 2017.

39. Chen, L.; Guo, B.; Huang, J.S.; He, J.; Wang, H.F.; Zhang, S.Y.; Chen, S.X. Assessing air-quality in Beijing-Tianjin-Hebei region: The method and mixed tales of $PM_{2.5}$ and O_3. *Atmos. Environ.* **2018**, *193*, 290–301. [CrossRef]

40. Lang, J.L.; Cheng, S.Y.; Ying, W.W.; Zhou, Y.; Wei, X.; Chen, D.S. A study on the trends of vehicular emissions in the Beijing-Tianjin-Hebei (BTH) regin, China. *Atmos. Environ.* **2012**, *62*, 605–614. [CrossRef]

41. Lang, J.L.; Cheng, S.Y.; Han, L.H.; Zhou, Y.; Liu, Y.T. Vehicular emission characteristics in Beijing-Tianjin-Hebei (BTH) region. *J. Beijing Univ. Technol.* **2012**, *38*, 1716–1723. (In Chinese)

42. Lang, J.L.; Zhou, Y.; Cheng, S.Y.; Zhang, Y.Y.; Dong, M.; Li, S.Y.; Wang, G.; Zhang, Y.L. Unregulated pollutant emissions from on-road vehicles in China, 1999–2014. *Sci. Total Environ.* **2016**, *573*, 974–984. [CrossRef] [PubMed]

43. Ministry of Ecology and Environmental of the People's Republic of China. *Limits and Measurement Methods for Emissions from Diesel Fuelled Heavy-Duty Vehicles (CHINA V)*; Ministry of Ecology and Environmental of the People's Republic of China Press: Beijing, China, 2013. (In Chinese)

44. Ministry of Ecology and Environmental of the People's Republic of China. *Limits and Measurement Methods for Emissions from Light-Duty Vehicles (CHINA V)*; Ministry of Ecology and Environmental of the People's Republic of China Press: Beijing, China, 2013. (In Chinese)

45. Ministry of Ecology and Environmental of the People's Republic of China. *Limits and Measurement Methods for Emissions from Diesel Fuelled Heavy-Duty Vehicles (CHINA VI)*; Ministry of Ecology and Environmental of the People's Republic of China Press: Beijing, China, 2018. (In Chinese)

46. Ministry of Ecology and Environmental of the People's Republic of China. *Limits and Measurement Methods for Emissions from Light-Duty Vehicles (CHINA VI)*; Ministry of Ecology and Environmental of the People's Republic of China Press: Beijing, China, 2018. (In Chinese)

47. Cai, H.; Xie, S.D. Determination of emission factors from motor vehicles under different emission standards in China. *Acta Scientiarum Naturalium Universitatis Pekinensis* **2010**, *46*, 319–326. (In Chinese)

48. Li, Y.Y. *Research on Total Amount of Vehicular Emissions in Beijing-Tianjin-Hebei (BTH) Region and Its Abatement Control Stratety*; Tianjin University of Technology: Tianjin, China, 2015. (In Chinese)

49. *Editorial Committee of China Electric Power Yearbook*; China Electric Power Press: Beijing, China, 2017. (In Chinese)

50. Hebei Statistics Bureau. *Hebei Economic Year book*; China Statistics Press: Beijing, China, 2016. (In Chinese)

51. Tianjin Statistics Bureau. *Tianjin Statistic Yearbook*; China Statistics Press: Beijing, China, 2016. (In Chinese)

52. Beijing Statistics Bureau. *Beijing Statistic Yearbook*; China Statistics Press: Beijing, China, 2016. (In Chinese)

53. IPCC. *Coefficient of Greenhouse Gas Emissions*; IPCC: Geneva, Switzerland, 2006.

54. Li, G.C.; Sun, W.; Huang, G.H.; Lv, Y.; Liu, Z.F. An CJ. Planning of integrated energy-environment systems under dual interval uncertainties. *Electr. Power Energy Syst.* **2018**, *100*, 287–298. [CrossRef]

55. Chen, W.T.; Li, Y.P.; Huang, G.H.; Chen, W.T.; Chen, X.; Li, Y.P. A two-stage inexact-stochastic programming model for planning carbon dioxide emission trading under uncertainty. *Appl. Energy* **2010**, *87*, 1033–1047. [CrossRef]

56. Dai, C.; Cai, Y.P.; Ren, W.; Xie, Y.F.; Guo, H.C. Identification of optimal placements of best management practices through an interval-fuzzy possibilistic programming model. *Agric. Water Manag.* **2016**, *165*, 108–121. [CrossRef]

57. Zhang, X.D.; Huang, G.H.; Nie, X.H. Robust stochastic fuzzy possibilistic programming for environmental decision making under uncertainty. *Sci. Total Environ.* **2009**, *408*, 192–201. [CrossRef] [PubMed]

MDPI

St. Alban-Anlage 66

4052 Basel

Switzerland

Tel. +41 61 683 77 34

Fax +41 61 302 89 18

www.mdpi.com

Applied Sciences Editorial Office

E-mail: applsci@mdpi.com

www.mdpi.com/journal/applsci

www.ingramcontent.com/pod-product-compliance
Lightning Source LLC
Chambersburg PA
CBHW051721210326
41597CB00032B/5553